Enhanced Quantization

Particles, Fields & Gravity

Enhanced
Quantization

Particles, Fields & Gravity

John R Klauder
University of Florida, USA

 **World Scientific**

NEW JERSEY • LONDON • SINGAPORE • BEIJING • SHANGHAI • HONG KONG • TAIPEI • CHENNAI

Published by

World Scientific Publishing Co. Pte. Ltd.

5 Toh Tuck Link, Singapore 596224

USA office: 27 Warren Street, Suite 401-402, Hackensack, NJ 07601

UK office: 57 Shelton Street, Covent Garden, London WC2H 9HE

Library of Congress Cataloging-in-Publication Data
Klauder, John R., author.
    Enhanced quantization : particles, fields and gravity / by John R. Klauder (University of Florida, USA).
        pages cm
    Includes bibliographical references and index.
    ISBN 978-9814644624 (hardcover : alk. paper)
    1. Quantum theory. I. Title.
    QC174.12.K584 2015
    530.12--dc23
                        2015000747

British Library Cataloguing-in-Publication Data
A catalogue record for this book is available from the British Library.

Printed in Singapore

To my family

Preface

In a certain sense, every study of *Quantum Mechanics* consists of two parts: the first part, which we (and others) refer to as "quantization", takes a classical theory and assigns a quantum theory to it; the second part is an effort to solve the quantum dynamical equations of motion—exactly if possible, approximately otherwise—to obtain some of the consequences and properties of the quantum theory. Normally, in most texts, the quantization part is treated quickly because the canonical quantization procedures have (literally) been *canonized* because they work so well; thus most attention is devoted to ways to solve the quantum equations. Instead, the present text reanalyzes the conventional process of quantization itself and finds, in some cases, that it can lead to serious problems. After a careful examination, we introduce different quantization procedures—*ones that still yield all the good results of conventional quantization procedures*—yet ones that we believe are superior to the conventional procedures because the new procedures resolve those serious problems which arise from conventional procedures. Regarding the second topic, how one goes about solving a given set of quantum dynamical equations is essentially unchanged, and since that aspect is well treated elsewhere, we will have relatively little to say about that issue.

As emphasized above, the principal purpose of this monograph is to describe and apply a new way to quantize classical systems. As already noted, this new way retains *all* the good results that conventional canonical quantization procedures produce, and in addition, it offers *new* procedures when conventional canonical quantization procedures yield unsatisfactory results. While there are many systems that can be helped by the new viewpoint, one important set of examples refers to the "triviality" of conventionally quantized, covariant, scalar field models φ_n^4, with spacetime dimensions $n \geq 5$

(and possibly for $n = 4$ as well). When conventionally quantized, these models all become *free* quantum models—i.e., with a *vanishing* nonlinear interaction—and thus they clearly pass to *free* classical models in the classical limit, thereby contradicting their original nonlinear classical behavior. We believe this is unnatural behavior and calls for a partial revision of the very process of quantization. In fact, the new approach for quantization of such models leads to *non*trivial quantum results, which pass to the original nonlinear classical model in the classical limit. Positive implications of the new procedures apply also to suitable fermion interactions and to the "granddaddy" of all problems: quantum gravity.

These important results are presented in several later chapters, but before that we need to develop and refine the new procedures on simpler examples. Indeed, for maximum clarity, the new procedures should be developed slowly and carefully to show how the new viewpoint is entirely plausible and even has certain advantages, and, in addition, to show how the new procedures are able to yield conventional results when needed, as well as *non*-conventional results when *they* are needed. Many of these properties can already be made clear on very simple systems, and therefore we shall start with such examples.

However, before we begin anything, let us recall that, traditionally, all conventional quantization procedures involve one unusual property, which was clearly identified in [LaL77] (page 3), namely:

> "Thus quantum mechanics occupies a very unusual place among physical theories: it contains classical mechanics as a limiting case, yet at the same time it requires this limiting case for its own formulation."

We emphasize that this unusual property is completely eliminated in the new procedures advocated in this text, and instead classical mechanics is shown to be a natural subset of quantum mechanics in which Planck's constant remains positive throughout—*just as it is in the real world!*

This last remark deserves further elaboration. In the conventional view, and for a single degree of freedom, classical mechanics resides within a two-dimensional phase space, while the associated quantum mechanics requires an infinite-dimensional Hilbert space. In the former situation, Planck's constant $h = 0$, while in the latter situation, $h > 0$. But a world where $h = 0$ is fictitious because Nature has chosen the reduced Planck's constant $\hbar \equiv h/2\pi \simeq 10^{-27} erg\,sec$, and thus we require an enhanced version of classical mechanics for which $\hbar > 0$ and which necessarily lives in the quantum mechanical Hilbert space; in addition, our new procedure should—*and will*—recreate all the successes of conventional quantization. Seeking an appropriate, alternative version of the classical/quantum connection is a principal motivation, and finding an acceptable, alternative version of the classical/quantum connection is a principal result of the program of *Enhanced Quantization*, the properties of which are developed in this monograph. As already remarked, this viewpoint also resolves several long-standing issues with conventional quantization procedures as highlighted later in this text.

The basic ideas underlying the new approach to the classical/quantum connection arose in 1962–1964 [Kla62; Kl63a; Kl63b; Kla64] when the author regarded them as curiosities. However, the use of these ideas in resolving conventionally insoluble quantization problems in several idealized models [Kla65; Kla70] suggested the utility of the new procedures alongside canonical quantization. This view was central to the discussion in the author's earlier book, *Beyond Conventional Quantization* [Kla00], and the present monograph can be considered as a sequel—and, in some sense, a prequel as well—to that earlier work. However, only in 2012 was it finally recognized that the new procedures deserve to be considered more broadly as a worthy alternative to traditional quantization procedures [Kl12c]. At that point preparation of the present monograph was deemed worthwhile.

Following a brief chapter entitled "Introduction and Overview", this monograph is divided into five Parts. In Part 1 we briefly discuss conventional classical mechanics and conventional quantum mechanics with an emphasis on highly singular potentials and the issues they invoke. In Part 2 we discuss the new procedures of enhanced quantization for simple systems, and, in fact, we have chosen to present the essential elements *twice:* first, in a functional notation, and second, in an abstract notation, to ensure that every reader will understand the new story. A few simple model problems that are aided by enhanced quantization procedures are also included. In Part 3 we study the quantization of specialized systems, such as fermions and anyons, as well as several, relatively simple, models involving infinitely-many degrees of freedom each of which is a clear example of an unnatural quantization when treated conventionally but which lead to acceptable results when they are quantized by the new procedures. Part 4 is devoted to the study of ultralocal models for bosons and fermions. These models completely fail to lead to proper solutions when viewed conventionally, but they lead to acceptable results when quantized by the new procedures. In addition, they are genuine field theories and their high degree of symmetry leads to solutions that are complete, up to quadrature, and which clearly show the advantages of enhanced quantization. Their mode of solution also provides inspiration about how to treat other models, such as covariant scalar models. Indeed, Part 5 is devoted to the analysis of covariant scalar fields as well as quantum gravity using the new methods. In our study of covariant scalar field theories we examine—as remarked earlier—the so-called "triviality of φ_n^4" in spacetime dimensions $n \geq 5$ (and possibly for $n = 4$ as well) that arises from a conventional approach toward their quantization. A proposal for a *non*trivial quartic-interaction, covariant, scalar field theory for all spacetime dimensions is offered as well, part of which has received preliminary Monte Carlo support. In Part 5 we shall also find that the gravitational field is, in some sense, an ideal candidate for enhanced quantization. Unlike earlier parts of the text, the topics in Part 5 are sufficiently complicated technically that we are unable to carry out our analysis to completion as was the case in the earlier parts of the text; nevertheless, it would seem that enhanced quantization has much to say about the quantization of such difficult problems.

Some of the examples we discuss rely on published analysis done several decades ago, while the discussion for gravity, for fermions, for matrices, etc., is more recent. An appreciation that the principles of *Enhanced Quantization* have far-reaching consequences as well as acceptable results for various problems, particularly for those kinds of problems that lead to unnatural results when quantized in a conventional manner, has provided a principal motivation for the preparation of this monograph.

Over the years the author has profited from various forms of interaction with a number of people, and it is fitting to thank some of them on this occasion. They include: T. Adorno, E. Aslaksen, J. Ben Geloun, C.M. Bender, W.R. Bomstad, I. Daubechies, E. Deumens, M. Fanuel, J.-P. Gazeau, L. Hartmann, G.C. Hegerfeldt, M. Hynek, M. Karasev, K.V. Kuchař, C.B. Lang, H. Leutwyler, J.S. Little, D. Maslov, H. Narnhofer, E. Onofri, T. Paul, J.J. Stankowicz, D. Tanner, C.B. Thorn, T.T. Truong, G. Watson, R. Woodard, and S. Zonetti.

Special thanks are extended to Erik Deumens and James Stankowicz for their important contributions to preliminary Monte Carlo studies of potentially nontrivial formulation of φ_4^4 models studied in Chap. 9. Glenn Watson is also thanked for a careful review of the final text.

Gainesville, FL   *John R. Klauder*
                                                       November 2014

Contents

Part 2. Enhanced Quantization of Simple Systems   57

Chapter 1

Introduction and Overview

1.1 Motivation for this Text

Classical physics has helped our understanding of the world in which we live in very many ways. However, classical physics does not properly describe certain aspects of the world, especially those dealing with atoms. This fact may lead one to believe that quantum physics is only necessary at the microscopic level, but that is incorrect since the macroscopic world of trees, houses, people, and the like, is all composed of atoms. That means that the quantum laws have an implicit impact on our everyday world, even if it may be unseen. Although it is an evident observation, it is undoubtedly true that Planck's constant h and the equations that contain it are of fundamental importance for the quantum theory and, as already noted above, therefore—at least implicitly—for the classical world as well. However, the normal practice when discussing classical physics is to ignore quantum effects and thereby set Planck's constant to zero, i.e., choosing equations for which $h = 0$. While this leads to descriptions of classical phenomena that are often wonderfully accurate, these equations nevertheless describe a fictitious world, one in which quantum effects do not exist. This approach does not provide an accurate description of the world in which we find ourselves. In short, we live in a classical world in which Planck's constant is *not* zero but has a measurable value, and a proper theory of the classical world must account for the fact that Planck's constant has a role to play. Planck's original constant h is very commonly replaced by the reduced Planck's constant, written as $\hbar \equiv h/2\pi$, with the measured value of $\hbar \simeq 10^{-27}$ *erg sec*; hereafter, we will exclusively use $\hbar$ and, following the usual custom, we may refer to $\hbar$ simply as Planck's constant.

The modification needed to incorporate the influence of Planck's constant would seem to require some sort of "semi-classical" story. Indeed, there is already an established kind of discipline that deals with semi-classical matters, but this domain often deals with approximate solutions of Schrödinger's equation, such as, for example, the WKB (Wentzel, Kramers, and Brillouin) approximation. This kind of modification introduced by semi-classical concepts is typically quite different from a genuine classical theory that incorporates Planck's constant. Indeed, we do not need an approximate form of quantum theory but instead an enhanced version of classical theory that includes Planck's constant in a rational way. Consequently, we are driven to find a version of classical mechanics that is consistent with quantum theory and is derived as part of the quantum formalism. We seek a legitimate restriction of the quantum expressions that fits exactly within a formalism that describes classical mechanics. Quantum theory unfolds within a Hilbert space, typically having infinitely many dimensions, and yet somehow it can be limited to a domain of two dimensions for each degree of freedom, as is customary in a phase-space formulation of classical mechanics. This means that we are asking for a *projection* onto a 'two-dimensional subset' and the question immediately arises: which two-dimensional subset do we use? It would seem natural that any choice of a subset will require disturbing the microscopic system, but that is not our intention. Is there *any* way to extract information from the microscopic quantum world without actually disturbing the system? The answer is *yes*, and indeed there are several ways to do so. It is these special, non-disturbing ways that we seek to exploit. In a sense it is not unlike trying to find how to describe what is the dynamics at the middle of the Sun. No human is likely to ever visit the center of the Sun, and so there, too, we must rely on information that we can obtain without disturbing the system. How do we proceed? We imagine some inertial coordinates in which Newtonian mechanics is most simply formulated, and adopt the universal equation that force equals mass times acceleration, i.e., $F = ma$. Reasonable assumptions about the material that resides at the center of the Sun are the rest of the input. We need to find an analogous set of inertial coordinates that allows us to partially describe the dynamics inside a microscopic system, such as an atom. Yes, it will be that simple!

1.2   Scope of this Text

Traditional procedures for classical and quantum mechanics are well known
and well developed. There is a certain realm in which classical mechanics
proves to be adequate and another realm in which quantum mechanics
proves to be adequate. While the set of mathematical problems suitable
for classical mechanics is clear, there is, apparently, not a corresponding
set of mathematical problems that lead to a natural quantization procedure. In this text we focus on choosing a new way to bridge the classical
- quantum divide that has the virtue of making clear the set of mathematical problems that can be successfully quantized. It is important to
understand that for many problems *the new procedures exactly recover the
results of conventional canonical quantization.* However, for some models
conventional procedures fail to produce acceptable results, and it is these
examples we find most interesting. To show the benefit of introducing a
new procedure we introduce a variety of examples. These examples include
several small-scale problems with a few degrees of freedom, which already
show the superiority of the new methods when compared with the usual
procedures. One of these examples deals with a simplified model of the
cosmological singularity that enjoys an enhanced classical theory in which
quantum corrections eliminate the singularity when the model universe is
tiny and may also show an enlarged acceleration beyond that generated
by the cosmological constant alone. Other examples, dealing with many
degrees of freedom include completely soluble models that are trivial (=
free) when quantized conventionally, but enjoy acceptable behavior when
quantized by the new procedures. These models offer clear proof of the
value of the new quantization procedures. We also include idealized field
theory models that exhibit similar good results for enhanced quantization.
In order to make the case for the new procedures, it is clearly useful to
have a number of models that can be solved essentially to completion so
the value of the new techniques can be more easily appreciated.

**It is important to appreciate that many of the model problems
listed below are representative of infinitely many other models
that are implicitly also resolved but are largely not discussed.
In particular, model types (4) − (11) can also be solved to the
same extent as the example shown for general potentials that are
lower-bounded polynomials of the illustrated kind.**

To see at a glance the array of models we consider in this text, we offer
the classical (c) Hamiltonian for each one:

(1) $H_c(p, q) = q\, p^2$, $q > 0$

(2) $H_c(p, q) = \frac{1}{2}p^2 - e^2/q$, $q > 0$

(3) $H_c(p, q) = -\frac{1}{2}p^2/q - \frac{1}{2}kq + \frac{1}{6}\Lambda q^3$, $q > 0$

(4) $H_c(\vec{p}, \vec{q}) = \frac{1}{2}[\vec{p}^2 + m_0^2\,\vec{q}^2] + \lambda_0(\vec{q}^2)^2$,

         $(\vec{p}; \vec{q}) = (p_1, \cdots, p_N; q_1, \ldots, q_N)$, $1 \leq N \leq \infty$

(5) $H_c(p, q) = \frac{1}{2}[\mathrm{Tr}(p^2) + m_0^2\mathrm{Tr}(q^2)] + \lambda_0\,\mathrm{Tr}(q^4)$,

         $p = \{p_{ab}\}$, $p_{ba} = p_{ab}$, $q = \{q_{ab}\}$, $q_{ba} = q_{ab}$, $1 \leq a, b \leq N \leq \infty$

(6) $H_c(\pi, \phi) = \int \{\frac{1}{2}[\pi(x)^2 + m_0^2\phi(x)^2] + g_0\phi(x)^4\}\, d^s x$

(7) $H_c(\pi, \phi) = \int \{\frac{1}{2}[\pi(x)^2 + m_0^2\phi(x)^2] + g_0\phi(x)^4\}\, d^s x + g_0'\{\int \phi(x)^2\, d^s x\}^2$

(8) $H_c(\vec{\pi}, \vec{\phi}) = \int \{\frac{1}{2}[\vec{\pi}(x)^2 + m_0^2\vec{\phi}(x)^2] + g_0[\vec{\phi}(x)^2]^2\}\, d^s x$,

         $(\vec{\pi}; \vec{\phi}) = \{\pi_1, \cdots, \pi_N; \phi_1, \cdots, \phi_N\}$, $1 \leq N \leq \infty$

(9) $H_c(\psi^*, \psi) = \int \{\mu_0\psi^*(x)\beta\psi(x) + g_0[\psi^*(x)\mathcal{M}\psi(x)]^2\}\, d^s x$

(10) $H_c(\psi^*, \psi, \pi, \phi) = \int \{\frac{1}{2}[\pi(x)^2 + m_0^2\phi(x)^2] + \mu_0\psi^*(x)\beta\psi(x)$

              $+ g_0\psi^*(x)\phi(x)\psi(x) + g_0'\phi(x)^4\}\, d^s x$

(11) $H_c(\pi, \phi) = \int \{\frac{1}{2}[\pi(x)^2 + (\nabla\phi)(x)^2 + m_0^2\phi(x)^2] + g_0\phi(x)^4\}\, d^s x$

(12) $H_c(\pi, g) = \int \{g(x)^{-1/2}[\pi_b^a(x)\pi_a^b(x) - \frac{1}{2}\pi_a^a(x)\pi_b^b(x)]$

            $+ g(x)^{1/2}R(x)\}\, d^3 x$

Comments:

(1) The classical solutions with positive energy all have singularities that disappear when $\hbar$ corrections are included (Chap. 4).

(2) Singular classical solutions are tamed when a quantum correction is introduced with a potential minimum approximately at the Bohr radius (Chap. 4).

(3) The Hamiltonian is constrained to vanish. The parameter $k = 0, \pm 1$ describes different geometries, and Λ denotes the cosmological constant (Chap. 5).

(4) This model is invariant under orthogonal rotations, which helps in finding the solution when $N = \infty$ (Chap. 7).

(5) This matrix model—Tr means trace—is also rotationally invariant, which is helpful in finding the solution when $N = \infty$ (Chap. 7).

(6) This ultralocal scalar model can be solved up to quadrature (Chap. 8).

(7) This model is a combination of model (4) and model (6); it can be

solved up to quadrature (Chap. 8).

(8) An extension of model (6) to include $N \leq \infty$ components at each point of space; it can be solved up to quadrature (Chap. 8).

(9) An ultralocal spinor model that can be solved up to quadrature (Chap. 8). (Strictly speaking, this model should be regarded as 'first' quantized.)

(10) Another ultralocal model with fermions and a scalar with a Yukawa-like interaction that can be solved up to quadrature (Chap. 8). (Strictly speaking, this model should be regarded as 'first' quantized.)

(11) A covariant scalar field theory with a quartic interaction in n space-time dimensions and, therefore, $s = n - 1$ space dimensions. It is studied for all $n \geq 2$ ($s \geq 1$) (Chap. 9).

(12) This gravitational Hamiltonian is constrained to vanish, and enhanced quantization is actually quite natural for this problem (Chap. 10).

IMPORTANT REMARKS: Conventional quantization of the classical models (4) – (10) ALL lead to unnatural, trivial (= free) quantum theories. Enhanced quantization procedures, on the other hand, resolve ALL these difficulties, and offer a clear benefit of reexamining the exclusive use of canonical quantization procedures. Indeed, what better evidence can be offered for an alternative quantization procedure than 'solving insoluble problems'!

For the scalar field models in (11), conventional methods lead to triviality for $n \geq 5$ (and possibly for $n = 4$). New solutions are proposed for all $n \geq 2$. For (12), kinematical issues for gravity are dealt with, and some remarks are made about dynamics.

1.3   Approach of this Text

Our goal is to derive classical physics from quantum physics without invoking the limit $\hbar \to 0$ and to obtain this result within the quantum formalism as it stands. As usual, the quantum (Q) formalism is embodied within the quantum action functional

$$A_Q = \int_0^T \langle \psi(t)| \left[i\hbar(\partial/\partial t) - \mathcal{H}\right] |\psi(t)\rangle \, dt \, , \tag{1.1}$$

and a stationary variation of A_Q leads to the usual quantum equation of motion. Correspondingly, the classical (C) formalism, for a single degree of freedom, is embodied within the classical action functional

$$A_C = \int_0^T [p(t)\dot{q}(t) - H_c(p(t), q(t))] \, dt , \qquad (1.2)$$

and a stationary variation of A_C leads to the usual classical equations of motion. How do we bridge the clear gap that separates these two, widely different modes of description? That topic is developed in Chap. 4. However, before we get there, some dead wood needs to be cleared from the path.

To achieve our long term goal, it will be necessary to challenge a piece of conventional and well-accepted wisdom. The sacred cow we need to demolish is the assumption that a classical action functional $A_C = A_0 + g A_I$, a sum of a free term (A_0) and an interaction term (A_I), always has the natural and evident property that, as the coupling constant $g \to 0$, the action functional for the interacting model passes continuously to the classical action functional for the free system A_0. This assumption is 'evident' because, as we all learned in school, the equation $y = ax + b$ intersects the y-axis at $y = b$. The failure of the interacting classical action functional to be continuously connected to its own free classical action arises when the *domain* of the action functional, i.e., the set of paths or field histories allowed in the action functional, prohibits the naive outcome. Specifically, if $D(A_0)$ denotes the set of elements that constitutes the domain of the free action and $D(A_C) = D(A_0) \cap D(A_I) \equiv D(A_0') \subset D(A_0)$, it follows that $\lim_{g \to 0}[A_0 + g A_I] = A_0' \neq A_0$. We call A_0', the classical free action functional endowed with the domain $D(A_C) \subset D(A_0)$, a *pseudofree theory*. Concrete examples of such systems are discussed in Chap. 2. In brief, for a sufficiently singular interaction, the interacting classical action functional is *not* continuously connected to its own free classical action functional. It is plausible, then, that a similar situation holds in the quantum theory. Namely, that the quantum propagator for an interacting system does not reduce to the quantum propagator for the free system as the coupling constant vanishes, but instead converges to the propagator for a pseudofree quantum system. Examples of this behavior will be illustrated in Chap. 3. Indeed, many examples of classical and quantum discontinuous perturbations will appear in subsequent chapters as well. If the interacting models are not continuously connected to their own free models, it stands to reason that a perturbation analysis of the interacting theory about the free theory is incorrect and will lead to divergences. Instead, one must perform a perturbations analysis about the appropriate pseudofree model and not about the free model.

The essence of the previous paragraph is the following: Models with singular potentials that involve discontinuous perturbations exhibit the property that the limiting behavior of the interacting model, as the coupling constant is reduced to zero, is a pseudofree model and not the free model. Moreover, if the interaction is reapplied to the pseudofree theory, it continuously reduces to the same pseudofree theory as the coupling constant is again reduced to zero. Thus a sufficiently singular interaction is a discontinuous perturbation of the free theory but it is a continuous perturbation of the proper pseudofree theory. This situation can apply to both classical and quantum systems.

PART 1
Selected Topics in
Classical and Quantum Mechanics

Chapter 2

Selected Topics in Classical Mechanics

WHAT TO LOOK FOR

The principal goal of this monograph is to present a *new way to quantize canonical systems* that we call *enhanced quantization*. This new way to quantize will *maintain and preserve* all the "good" examples for which conventional canonical quantization has proven to be correct. But, importantly, it will also offer a new way to quantize the "bad" examples for which conventional canonical quantization has "failed"; the several meanings of the word failed that we have in mind will be explained when appropriate. However, even before undertaking the story of enhanced quantization, it is important that we start with a review of the conventional procedures used to quantize classical systems so that the reader can appreciate the similarities as well as the differences between the old and the new ways, in order to gain an appreciation for why such a new prescription for quantization is useful. As already noted, when the old way of quantization works well, then the new way of quantization must replicate the traditional results. However, when the old way fails (in ways we will discuss), the new way will offer a better way to quantize. This chapter is devoted to a rapid survey of several historical procedures which have led to a classical mechanics formalism that is traditionally used to connect a classical theory to an associated quantum theory. We also stress some uncommon features that normally receive little attention.

2.1   Newtonian Classical Mechanics

The principal problem in classical mechanics is to find the motion of a particle in space (or any other object idealized as a point particle) under the influence of external forces. Initially specializing to one-dimensional motion for simplicity at this point, we seek the trajectory, namely the function $x(t)$, where x is the position and t is the time, assuming some initial starting information at $t = 0$. In the absence of an external force, and in suitable coordinates, called *inertial coordinates*, Newton's First Law of Motion asserts that a particle at rest will stay at rest while a particle in uniform motion will remain in uniform motion. In equation form, this statement means that in the absence of any force, the trajectory is $x(t) = x_0 + v_0 t$, where x_0 is the initial position and v_0 is the initial velocity at $t = 0$. This trajectory satisfies the differential equation, $d^2 x/dt^2 = 0$, where $dx(t)/dt = v(t)$ is the velocity and $d^2 x(t)/dt^2 = dv(t)/dt = a(t)$ is the acceleration. We can also say that $x(t) = x_0 + v_0 t$ is a solution to the equation of motion $d^2 x(t)/dt^2 = 0$, and indeed it is the most general solution to that equation of motion.

If we deal with an N-dimensional configuration space, with inertial position coordinates given by $x \equiv \{x_1, x_2, \ldots, x_N\}$, then the force-free equation of motion is given by $d^2 x_n(t)/dt^2 = 0$, for $1 \le n \le N$, with the most general solution given by $x_n(t) = x_{0\,n} + v_{0\,n} t$.

However, when an external force is present, the basic equation of motion must change to take that force into account. In the general case, the fundamental equation of mechanics is Newton's Second Law of Motion, which, for a simple, one-dimensional motion takes the form $dp/dt = F(x)$, where again t is the time, $p = mv$ is the momentum, with m the mass and v the velocity. In this equation, $F(x)$ is the externally applied force at the position x. For a conservative system, which, in particular, leads to no loss of energy to the surrounding environment, the force $F(x) = -\partial V(x)/\partial x$, where $V(x)$ is called the potential. The property of no loss of energy comes about from the fact that, for a time-independent mass m, it follows that Newton's equation, i.e., $m(d^2 x/dt^2) + \partial V(x)/\partial x = 0$, leads to

$$0 = (dx/dt) \cdot \{m(d^2 x/dt^2) + \partial V(x)/\partial x\}$$
$$= (d/dt)\{\tfrac{1}{2}m(dx/dt)^2 + V(x)\}, \tag{2.1}$$

which implies that $\tfrac{1}{2}mv^2 + V(x) = E$ is a constant of the motion called the energy; in particular, $\tfrac{1}{2}mv^2$ is the kinetic energy, and $V(x)$ is now called the potential energy.

For an N-dimensional space, Newton's Second Law of Motion is given by $dp_n(t)/dt = F_n(x)$, for $1 \leq n \leq N$. A conservative set of forces fulfills the property that $F_n(x) = -\partial V(x)/\partial x_n$, from which it follows that $\partial F_n(x)/\partial x_l = \partial F_l(x)/\partial x_n$, etc., which leads, for a time-independent mass m, to $(d/dt)\{\frac{1}{2}m\Sigma_n v_n(t)^2 + V(x)\} = 0$, and thus to conservation of the energy $E = \frac{1}{2}m\Sigma_n v_n(t)^2 + V(x)$, where $v_n(t) = dx_n(t)/dt$.

The important set of equations of motion given by

$$m\, d^2 x_n(t)/dt^2 = F_n(x(t))\,, \tag{2.2}$$

which applies for both conservative and non-conservative forces, are a set of differential equations with solutions depending on initial data, such as $x_n(0) = x_{0\,n}$ and $v_n(0) = v_{0\,n}$, which determine the several trajectories $x_n(t)$ and describe the motion of the particle in the configuration space $\mathbb{R}^N$, i.e., where $-\infty < x_n < \infty$, for all n, $1 \leq n \leq N$. Observe, however, that a trajectory is not determined just by giving an initial point x_0 in $\mathbb{R}^N$ but by giving an initial point x_0 and an initial velocity $v_0 = dx_0/dt$.

2.2   Lagrangian Classical Mechanics

The foregoing description of motion is tied to special coordinates, the inertial coordinates, but there are decided advantages to be gained if we can incorporate more general coordinates. To that end, and again starting with a single coordinate, we introduce the Lagrangian, initially defined by $L(\dot{x}, x) = K(\dot{x}) - V(x)$, i.e., the kinetic energy minus the potential energy, where $\dot{x} \equiv dx/dt$ is a common notational shorthand for the velocity. We also introduce the momentum p by the rule

$$p \equiv \partial L(\dot{x}, x)/\partial \dot{x} = m\dot{x}\,, \tag{2.3}$$

and then Newton's equation of motion becomes

$$dp/dt + \partial V(x)/\partial x = (d/dt)\,[\partial L(\dot{x}, x)/\partial \dot{x}] - \partial L(\dot{x}, x)/\partial x = 0\,. \tag{2.4}$$

Although it may seem that we have only complicated the original approach, this last form of the equation of motion based on the Lagrangian offers some useful advantages.

Consider the temporal integral, called the *action*, or more precisely the *action functional*, given by

$$A(x) \equiv \int_0^T L(\dot{x}, x)\, dt\,. \tag{2.5}$$

Here, the notation $A(x)$ is understood to mean the result of this integral depends on the whole function $x(t)$, $0 \leq t \leq T$, $T > 0$, and a function of a function is generally called a functional. We want to make a *stationary variation* of this functional, which means that we take the difference of the action for two infinitesimally close functions, e.g., $A(x+\delta x) - A(x) \equiv \delta A(x)$, which amounts to the variation part. For the stationary part we insist that $\delta A(x) = 0$ for arbitrarily small δx, which means that we retain only linear terms in δx. Let us compute the required variation by treating the arguments $\dot{x}$ and x as independent, which leads to

$$\delta A(x) = \int_0^T \{\partial L(\dot{x}, x)/\partial \dot{x} \cdot \delta \dot{x} + \partial L(\dot{x}, x)/\partial x \cdot \delta x\} \, dt$$
$$= \int_0^T \{-(d/dt)\,\partial L(\dot{x}, x)/\partial \dot{x} + \partial L(\dot{x}, x)/\partial x\} \cdot \delta x \, dt$$
$$+ \partial L(\dot{x}, x)/\partial \dot{x} \cdot \delta x \, \Big|_0^T \, , \tag{2.6}$$

where the latter equation has been obtained by an integration by parts; note that we have used the notation $\cdot$ to separate the derivative from the differential. If we now insist that the variation of the paths (i.e., the functions), denoted by $\delta x(t)$, must vanish at the end points, i.e.,

$$\delta x(0) = \delta x(T) = 0 \, , \tag{2.7}$$

then it follows that requiring that $\delta A(x) = 0$ for small, but arbitrary, variations $\delta x(t)$, with the end points held fixed, the result is that the integrand must be zero, namely that

$$(d/dt)\,[\partial L(\dot{x}, x)/\partial \dot{x}] - \partial L(\dot{x}, x)/\partial x = 0 \, , \tag{2.8}$$

which is just the equation of motion that we had obtained in (2.4).

In addition, we note that if we modify the action so that

$$A' = \int_0^T [L(\dot{x}, x) + \dot{f}(x)] \, dt \, , \tag{2.9}$$

the new action A' serves to derive the same equations of motion since the added term is really just $f(x(T)) - f(x(0))$, an expression that vanishes under the variation since $\delta x(T) = \delta x(0) = 0$. We can derive the same equation of motion even if we put stronger conditions on the variational paths. In particular, besides holding $\delta x(0) = \delta x(T) = 0$, we could also impose $\delta \dot{x}(0) = \delta \dot{x}(T) = 0$ as well. This extension has the virtue that we could extend the class of action functionals that lead to the same equation of motion, namely

$$A'' = \int_0^T [L(\dot{x}, x) + \dot{f}(\dot{x}, x)] \, dt \, . \tag{2.10}$$

This extension holds because the added term remains unvaried as is clear from the relation

$$\int_0^T \dot{f}(\dot{x}, x) \, dt = f(\dot{x}(T), x(T)) - f(\dot{x}(0), x(0)) \, , \tag{2.11}$$

which is fixed by design. Of course, it would be a rare problem that has a valid solution in which the values of $\dot{x}(T), x(T), \dot{x}(0)$, and $x(0)$ had all been specified beforehand; indeed, for most sets of such data there is *no* solution to the equations of motion at all. That fact serves to demonstrate the clear difference between (i) *deriving the equations of motion* and (ii) *solving the equations of motion!* In fact, holding just $x(T)$ and $x(0)$ fixed, which is the customary procedure, does *not* guarantee a unique solution to the equations of motion in the general situation. One need only consider the example of an harmonic oscillator that leads to the equation of motion $\ddot{x}(t) = -x(t)$. If we choose $x(0) = 0$ and $x(\pi) = 1$ we find *no solution*, but if we choose $x(0) = 0$ and $x(\pi) = 0$, we find *infinitely many solutions*, all because the general solution to the problem with $x(0) = 0$ is given by $x(t) = A\sin(t)$ for any $A \in \mathbb{R}$. To ensure a unique solution to such problems, it suffices to choose $x(0)$ and $\dot{x}(0)$ as initial conditions. In that sense, it is noteworthy that a *subset* of the data $\dot{x}(T), x(T), \dot{x}(0)$, and $x(0)$ held fixed above, such as the last two entries, *always* suffices to determine a unique solution.

Of course, nothing can guarantee that such a solution will be satisfactory for all time because some problems will instead insure that divergences appear within a finite time; such is the case for the Lagrangian $L = \frac{1}{2}\dot{x}^2 + x^4$ which gives rise to the equation of motion $\ddot{x}(t) = 4x^3(t)$. The solution to such an equation has a fixed energy $E = \frac{1}{2}\dot{x}(t)^2 - x(t)^4$ and, for example, for $x(0) > 0$, $\dot{x}(0) > 0$, and $E > 0$, it follows that $\dot{x}(t) = \{2[E + x(t)^4]\}^{1/2}$, which leads to

$$\tau \equiv \int_0^\tau dt = \int_{x(0)}^\infty \{2[E + x^4]\}^{-1/2}\, dx < \infty \,, \tag{2.12}$$

a relation that ensures the trajectory reaches $x = \infty$ in a finite time τ.

2.2.1 *Multiple degrees of freedom*

The extension of the foregoing discussion to multiple degrees of freedom is straightforward. In this case, we let $x(t) \equiv \{x_1(t), x_2(t), \ldots, x_N(t)\}$ and likewise $\dot{x}(t) \equiv \{\dot{x}_1(t), \dot{x}_2(t), \ldots, \dot{x}_N(t)\}$. The Lagrangian then becomes a function of these two vectors, $L(\dot{x}, x)$, and the action functional becomes

$$A = \int_0^T L(\dot{x}(t), x(t))\, dt \,. \tag{2.13}$$

After a stationary variation of the action holding the endpoints fixed, i.e., $\delta x(T) = \delta x(0) = 0$ for each component of the vectors, the equations of motion become

$$(d/dt)\,[\partial L(\dot{x}, x)/\partial \dot{x}_n(t)] - \partial L(\dot{x}, x)/\partial x_n(t) = 0 \,. \tag{2.14}$$

A similar story applies to higher dimensions for a single particle.

As was previously the case, a trajectory is still not determined just by an initial point x in $\mathbb{R}^N$ but by an initial point x and an initial velocity $\dot{x}$.

2.2.2 *Virtues of the Lagrangian formulation*

The real benefit of introducing the action functional involving the Lagrangian function is that we can take advantage of the well-known fact that changing integration variables in evaluating an integral does not change the value of the integral. For example, consider the simple integral $I = \int_0^1 u^2 \, du = 1/3$. Now let us change the variable of integration to v where $u = v^3$ so that I becomes $\int_0^1 v^6 \, 3v^2 \, dv = 3 \int_0^1 v^8 \, dv = 3/9 = 1/3$, which is just the previous answer. (Of course, our choice of new coordinates kept the original limits of integration the same, and that was done deliberately since those limits will not change in the analogous cases of interest.) The implication of our simple exercise is that the integral of the Lagrangian (2.5) will have the same evaluation after a change of integration variables, and thus the equation of motion derived by a stationary variation will be equivalent to the original equation of motion even though the variables have been changed.

2.2.3 *Example*

Let us present a useful two-dimensional example involving (inertial) Cartesian coordinates and (non-inertial) polar coordinates. We start with the Lagrangian (note the special form of the potential)

$$L = \tfrac{1}{2} m (\dot{x}^2 + \dot{y}^2) - V(\sqrt{x^2 + y^2}) \,, \tag{2.15}$$

which leads to the Lagrangian equations of motion given by

$$m\ddot{x} = -\partial V/\partial x \,, \qquad m\ddot{y} = -\partial V/\partial y \,. \tag{2.16}$$

Next, we reexpress the Lagrangian in polar coordinates r and θ, where $x(t) = r(t) \cos(\theta(t))$ and $y(t) = r(t) \sin(\theta(t))$; observe, in the integral that defines the action, no change of the integration limits, 0 and T, occurs. In these variables, it follows that

$$L = \tfrac{1}{2} m (\dot{r}^2 + r^2 \dot{\theta}^2) - V(r) \,, \tag{2.17}$$

and the Lagrangian equations of motion become

$$m[\ddot{r} - r\dot{\theta}^2] = -\partial V/\partial r \,, \qquad m(d/dt)[r^2 \dot{\theta}] = -\partial V/\partial \theta \,. \tag{2.18}$$

Since the potential does not depend on θ, it follows that $\partial V(r)/\partial\theta = 0$, and thus from that equation we learn that $mr^2\dot\theta \equiv \ell$ is a constant of the motion representing the angular momentum about the center. This fact allows us to eliminate $\dot\theta^2$ from the first equation, which leads to

$$m\ddot r = \ell^2/(mr^3) - \partial V(r)/\partial r\ , \tag{2.19}$$

where the new term on the right-hand-side arises as the gradient of the centrifugal potential $-\frac12\ell^2/(mr^2)$ which arises from the Lagrangian itself when we substitute $\frac12\ell^2/(mr^2)$ in place of $\frac12 mr^2\dot\theta^2$.

2.3   Hamiltonian Classical Mechanics

The fact that a point x in the configuration space does not determine a trajectory all by itself provides a motivation for the next stage of development. To emphasize that we are focusing on a new approach, we also change the notation for position from x to q. We also make use of the momentum $p \equiv \partial L(\dot q, q)/\partial\dot q$, which under most conditions can be solved for $\dot q = \dot q(p, q)$. Next we consider the combination

$$U(p, \dot q, q) \equiv p\dot q - L(\dot q, q)\ , \tag{2.20}$$

where, at first glance, U seems to depend on all three variables. Our initial task is to show that in fact U does *not* depend on $\dot q$. To that end, we note that

$$\partial U(p, \dot q, q)/\partial\dot q = p - \partial L(\dot q, q)/\partial\dot q \equiv 0\ , \tag{2.21}$$

and thus we assert that

$$p\,\dot q(p, q) - L(\dot q(p, q), q) \equiv H(p, q)\ , \tag{2.22}$$

which is called the *Hamiltonian*. It follows that we can reexpress the action integral as

$$A = \int_0^T [p(t)\dot q(t) - H(p(t), q(t))]\, dt\ . \tag{2.23}$$

A stationary variation of this form of the action involves treating p and q as separate and independent variables, and in particular, an infinitesimal variation leads to

$$\delta A = \int_0^T \{[\dot q - \partial H(p, q)/\partial p] \cdot \delta p - [\dot p + \partial H(p, q)/\partial q] \cdot \delta q\}\, dt + p\,\delta q\big|_0^T\ .$$
$$\tag{2.24}$$

Insisting that $\delta A = 0$, when $\delta q(0) = \delta q(T) = 0$, then guarantees that we are led to Hamilton's equations of motion, namely

$$\dot{q}(t) = \partial H(p,q)/\partial p(t) , \qquad \dot{p}(t) = -\partial H(p,q)/\partial q(t) . \qquad (2.25)$$

Moreover, further generalizations follow if besides the requirement that $\delta q(0) = \delta q(T) = 0$, we also require that $\delta p(0) = \delta p(T) = 0$ as well; this step is somewhat analogous to adding $\delta \dot{x}(T) = \delta \dot{x}(0) = 0$ in the Lagrangian story. In the present case, it follows that the very same equations of motion arise from a stationary variation of the modified action functional given by

$$A = \int_0^T [p(t)\dot{q}(t) + \dot{G}(p(t),q(t)) - H(p(t),q(t))] \, dt , \qquad (2.26)$$

for a general function $G(p,q)$ since the added term is effectively unvaried because

$$\int_0^T \dot{G}(p(t),q(t)) \, dt = G(p(T),q(T)) - G(p(0),q(0)) . \qquad (2.27)$$

As a simple example, consider $G(p,q) = -p\,q$, which then leads to

$$A' = \int_0^T [-q(t)\dot{p}(t) - H(p(t),q(t))] \, dt , \qquad (2.28)$$

a variation that requires that $\delta p(0) = \delta p(T) = 0$, but makes no demands on $\delta q(0)$ and $\delta q(T)$. Another simple example arises if $G(p,q) = -\frac{1}{2}p\,q$ for which

$$A'' = \int_0^T \{ \tfrac{1}{2}[p(t)\dot{q}(t) - q(t)\dot{p}(t)] - H(p(t),q(t)) \} \, dt , \qquad (2.29)$$

for which both $\delta p(0) = \delta p(T) = 0$ and $\delta q(0) = \delta q(T) = 0$ are needed. Thus the three action functionals, A, A', and A'', while strictly different from each other, all lead, based on a stationary variation, to one and the same set of Hamilton's equations of motion thanks to the fact that we have agreed to require that variations of both p and q vanish at the end points of the integral.

2.3.1   *Canonical coordinate transformations*

Unexpectedly, we have just opened Pandora's Box! Let us consider a change of variables from (p,q) to $(\widetilde{p},\widetilde{q})$ where $\widetilde{p} = \widetilde{p}(p,q)$ and $\widetilde{q} = \widetilde{q}(p,q)$ represents a very special class of coordinate transformations, the so-called *canonical coordinate transformations*. These coordinate transformations are not arbitrary but instead have to fulfill a relation of the form

$$p \, dq = \widetilde{p} \, d\widetilde{q} + d\widetilde{G}(\widetilde{p},\widetilde{q}) , \qquad (2.30)$$

where $\widetilde{G}(\widetilde{p},\widetilde{q})$ is referred to as the *generator of the coordinate transformation*. Indeed, the simple example above where $G(p,q) = -pq$ in effect amounts to the coordinate change $q \to \widetilde{p}$ and $p \to -\widetilde{q}$. The coordinate changes that take place fulfill the relation that

$$d\widetilde{p}\, d\widetilde{q} = dp\, dq \,, \qquad (2.31)$$

asserting that the elemental surface element on phase space is *invariant under a canonical coordinate transformation*. This fact is most easily seen by taking the exterior differential (e.g., see [Wiki-a]) of the relation (2.30). In simpler terms, it means that the Jacobian of the coordinate transformation $(p, q) \to (\widetilde{p}, \widetilde{q})$ is unity, namely

$$(\partial\widetilde{p}/\partial p)(\partial\widetilde{q}/\partial q) - (\partial\widetilde{p}/\partial q)(\partial\widetilde{q}/\partial p) = 1 \,. \qquad (2.32)$$

The chosen restrictions on canonical coordinate transformations have the virtue that the equations of motion derived from a stationary variation of the action functional in any set of canonical coordinates is form invariant. This fact follows from the form invariance of the action functional itself along with the vanishing of the variations of both coordinate and momentum, i.e., both p and q, or equally both $\widetilde{p}$ and $\widetilde{q}$, at the beginning and end of the temporal integration, i.e., at $t = 0$ and $t = T$. Specifically, the action functional expressed in arbitrary canonical coordinates is given by

$$A = \int_0^T \{\widetilde{p}\,\dot{\widetilde{q}} + \dot{\widetilde{G}}(\widetilde{p}, \widetilde{q}) - \widetilde{H}(\widetilde{p}, \widetilde{q})\}\, dt \,, \qquad (2.33)$$

where $\widetilde{H}(\widetilde{p}, \widetilde{q}) \equiv H(p, q)$, meaning that the Hamiltonian transforms as a scalar under a canonical coordinate transformation. Under a stationary variation, the general action functional (2.33) yields the equations of motion

$$\dot{\widetilde{q}}(t) = \partial\widetilde{H}(\widetilde{p}, \widetilde{q})/\partial\widetilde{p} \,, \qquad \dot{\widetilde{p}}(t) = -\partial\widetilde{H}(\widetilde{p}, \widetilde{q})/\partial\widetilde{q} \,, \qquad (2.34)$$

which illustrates the form equivalence of the resulting equations of motion in any choice of canonical coordinates. Note well that $\widetilde{G}(\widetilde{p}, \widetilde{q})$ plays no role in determining the equations of motion thanks to our choice to hold both $\widetilde{p}(t)$ and $\widetilde{q}(t)$ fixed at $t = 0$ and $t = T$. While this convention serves to derive the equations of motion, almost no solution to those equations will satisfy all of the boundary conditions. But, once again, there always is a *subset* of that data, e.g., $\widetilde{p}(0)$ and $\widetilde{q}(0)$, that will guarantee a unique solution. Of course, that solution may not exist for all time, as was the case when we discussed a similar situation in the Lagrangian story.

2.3.2 *Poisson brackets*

Poisson brackets, denoted by $\{\cdot, \cdot\}$, offer another way to rewrite several properties of classical mechanics including the equations of motion. They are considered to be a 'Royal Route' to canonical quantization as well, and they are generally studied as a helpful tool for the purpose of quantization. For a single degree-of-freedom and its phase-space variables p and q, let us consider functions of these variables, such as $A = A(p, q)$ and $B = B(p, q)$. In that case, the Poisson bracket of A and B is defined by

$$\{A, B\} \equiv \frac{\partial A}{\partial q} \frac{\partial B}{\partial p} - \frac{\partial A}{\partial p} \frac{\partial B}{\partial q} . \tag{2.35}$$

Clearly, the Poisson bracket satisfies the property that $\{B, A\} = -\{A, B\}$ and, less obviously, it fulfills the relation (known as the Jacobi identity)

$$\{A, \{B, C\}\} + \{C, \{A, B\}\} + \{B, \{C, A\}\} \equiv 0 , \tag{2.36}$$

as a detailed calculation readily confirms.

Obviously the canonical variables satisfy $\{q, p\} = 1$, and any two quantities $\widetilde{q} = \widetilde{q}(p, q)$ and $\widetilde{p} = \widetilde{p}(p, q)$ that satisfy $\{\widetilde{q}, \widetilde{p}\} = 1$ are called *canonically conjugate variables*. The new canonical coordinates $\widetilde{p}$ and $\widetilde{q}$ are just as valid as the original canonical coordinate pair p and q. Such variables also fulfill the one form relation (2.30,) which also leads to the invariance of the two form, the element of surface area, $d\widetilde{p}\, d\widetilde{q} = dp\, dq$, namely, the change of variables involves a Jacobian that is unity, which also may be seen as just the quantity evaluated by the Poisson bracket. The example $\widetilde{q} = q^3$ and $\widetilde{p} = p/3q^2$ belongs to a special class of canonically conjugate variables, and for this case it is seen that $\widetilde{G}(\widetilde{p}, \widetilde{q}) = 0$. Indeed, canonical coordinate transformations of the form $\widetilde{q} = f(q)$ and $\widetilde{p} = p/f'(q)$ form a special subset of transformations, and they all have a vanishing generator, $\widetilde{G}(\widetilde{p}, \widetilde{q}) \equiv 0$.

The equations of motion, as expressed by a Poisson bracket relation, are given by

$$\dot{q}(t) = \{q(t), H(p(t), q(t))\} , \qquad \dot{p}(t) = \{p(t), H(p(t), q(t))\} . \tag{2.37}$$

Indeed, the equation of motion for the general quantity $A(p(t), q(t))$ may be given as

$$\dot{A}(p(t), q(t)) = \{A(p(t), q(t)), H(p(t), q(t))\} , \tag{2.38}$$

and, as a special case, if the Hamiltonian does not explicitly depend on t, then

$$\dot{H}(p(t), q(t)) = \{H(p(t), q(t)), H(p(t), q(t))\} = 0 , \tag{2.39}$$

implying that $H(p(t), q(t)) = E$ is a constant of the motion, which we have already identified as the energy.

Multiple degrees of freedom

For multiple degrees of freedom, where we set $p \equiv \{p_1, p_2, \ldots, p_N\}$ and $q \equiv \{q_1, q_2, \ldots, q_N\}$, the action functional is given by

$$A = \int_0^T \{\Sigma_{n=1}^N p_n \dot{q}_n - H(p,q)\} \, dt \, . \tag{2.40}$$

A new set of canonical coordinates $\widetilde{p}$ and $\widetilde{q}$ must satisfy

$$\Sigma_{n=1}^N p_n \, dq_n = \Sigma_{n=1}^N \widetilde{p}_n \, d\widetilde{q}_n + d\widetilde{G}(\widetilde{p}, \widetilde{q}) \, . \tag{2.41}$$

In turn, an exterior differential (e.g., see [Wiki-b]) of this last relation leads to the fundamental symplectic identity

$$\sum_{n=1}^N d\widetilde{p}_n \, d\widetilde{q}_n = \sum_{n=1}^N dp_n \, dq_n \, . \tag{2.42}$$

An exterior square (e.g., see [Wiki-c]) of this equality leads to

$$\sum_{m=1,n=m+1}^{N-1,N} d\widetilde{p}_m \, d\widetilde{q}_m \, d\widetilde{p}_n \, d\widetilde{q}_n = \sum_{m=1,n=m+1}^{N-1,N} dp_m \, dq_m \, dp_n \, dq_n \, , \tag{2.43}$$

and this kind of exterior multiplicative relation extends all the way up to

$$d\widetilde{p}_1 \, d\widetilde{q}_1 \, d\widetilde{p}_2 \, d\widetilde{q}_2 \cdots d\widetilde{p}_N \, d\widetilde{q}_N = dp_1 \, dq_1 \, dp_2 \, dq_2 \cdots dp_N \, dq_N \, . \tag{2.44}$$

Additionally, the Poisson bracket is given by

$$\{A(p,q), B(p,q)\} \equiv \sum_{n=1}^N \left[\frac{\partial A(p,q)}{\partial q_n} \frac{\partial B(p,q)}{\partial p_n} - \frac{\partial A(p,q)}{\partial p_n} \frac{\partial B(p,q)}{\partial q_n} \right] . \tag{2.45}$$

2.4 Free & Pseudofree Models

This section is devoted to exploring some issues that arise when singular interactions are involved. For the sake of clarity we will primarily focus on harmonic oscillators as well as suitable anharmonic oscillators. Our discussions will involve single degree-of-freedom action functionals associated with the Lagrangian formulation, but that by no means is meant to exclude similar discussions that hold for other examples.

A simple free model

We start with the action functional

$$A_0 = \tfrac{1}{2} \int_0^T [\dot{x}(t)^2 - \omega^2 x(t)^2] \, dt \, , \tag{2.46}$$

where $0 < \omega < \infty$, and which leads to a linear equation of motion, $\ddot{x}(t) = -\omega^2 x(t)$, that is readily solved in terms of trigonometric functions, e.g.,

$x(t) = x_0 \cos(\omega t) + (\dot{x}_0/\omega) \sin(\omega t)$, where x_0 denotes the initial position at $t = 0$ and and $\dot{x}_0$ denotes the initial velocity at $t = 0$, and both x_0 and $\dot{x}_0$ are arbitrary.

In deriving the equations of motion from this action we implicitly restricted our variations to functions $x(t)$ and their time derivatives $\dot{x}(t)$ such that the separate integrals, $\int_0^T \dot{x}(t)^2 \, dt$ and $\int_0^T x(t)^2 \, dt$, were both finite. For example, the function $x(t) = [t - (T/2)]^{3/5}$ is included in the set of possible variations, but the function $x(t) = [t - (T/2)]^{2/5}$ is excluded from it. The general criteria for the suitable domain of functions is simply stated as the set $D(A_0) = \{x : \int_0^T [\dot{x}^2 + x^2] dt < \infty\}$, or equivalently, $D(A_0) = \{x : x \in L^2([0,T])\} \cap \{\dot{x} : \dot{x} \in L^2([0,T])\}$. It is readily shown (see next paragraph) that any function $x(t)$ that satisfies the given requirements is uniformly bounded as well as continuous.

To prove continuity and the existence of a uniform bound, we first assume that the functions of interest are actually defined for all t, $-\infty < t < \infty$. The desired inequality holds because $\tilde{x}(\omega)$, the Fourier transform of $x(t)$, defined by $\tilde{x}(\omega) \equiv (2\pi)^{-1/2} \int e^{-i\omega t} x(t) \, dt$ as well as $x(t) = (2\pi)^{-1/2} \int e^{i\omega t} \tilde{x}(\omega) \, d\omega$, satisfies

$$\begin{aligned}
|x(t)| &\leq (2\pi)^{-1/2} \int |\tilde{x}(\omega)| \, d\omega \\
&= (2\pi)^{-1/2} \int \sqrt{\omega^2 + 1} \, |\tilde{x}(\omega)| \, d\omega / \sqrt{\omega^2 + 1} \\
&\leq (2\pi)^{-1/2} \{ \int (\omega^2 + 1) |\tilde{x}(\omega)|^2 \, d\omega \cdot \int (\omega^2 + 1)^{-1} \, d\omega \}^{1/2} \\
&= \{ 2^{-1} \int [\dot{x}(t)^2 + x(t)^2] \, dt \}^{1/2} ,
\end{aligned} \tag{2.47}$$

which is finite by assumption. Continuity of $x(t)$ follows from the fact, for any $\epsilon > 0$, that

$$\begin{aligned}
|x(t) - x(s)| &\leq (2\pi)^{-1/2} \int |1 - e^{-i\omega(t-s)}| \, |\tilde{x}(\omega)| \, d\omega \\
&= (2\pi)^{-1/2} \{ \int_{-\infty}^{-B} + \int_{-B}^{A} + \int_A^\infty \} |1 - e^{-i\omega(t-s)}| \, |\tilde{x}(\omega)| \, d\omega \\
&\leq 2\epsilon/3 + (2\pi)^{-1/2} \int_{-B}^{A} |1 - e^{-i\omega(t-s)}| \, \tilde{x}(\omega)| \, d\omega \leq \epsilon , \quad (2.48)
\end{aligned}$$

where $A > 0$ and $B > 0$ are chosen so that the two end integrals are each bounded by $\epsilon/3$, while the middle integral can also be bounded by $\epsilon/3$ by choosing s close to t. Under the same conditions, a closely related calculation establishes that $x(t) \to 0$ when $|t| \to \infty$, a property known as the Riemann Lebesgue lemma.

The given action functional (2.46) applies to the free harmonic oscillator, which is simply called the *free model* for present purposes.

Continuous perturbations: non-singular interacting models

As a first example of an interacting model we consider the action functional for an anharmonic oscillator given by

$$A_g = \int_0^T \{ \tfrac{1}{2}[\dot{x}(t)^2 - \omega^2 x(t)^2] - g x(t)^4 \}\, dt \,, \tag{2.49}$$

which leads to a non-linear equation of motion. We shall not focus on the solutions of the equation of motion, but rather on the domain of functions that is allowed by the interacting action functional (= free + interaction). Since the domain of the free model action functional consists of uniformly bounded functions $x(t)$, it follows that *all* functions in the domain of the interacting model are in the domain of the free model. This conclusion also applies to any interaction term $g \int_0^T x(t)^p\, dt$ for any integer $p > 0$. Of course, if p is an *even* integer and $g > 0$, full-time solutions of the equations of motion are the norm; while if p is an *odd* integer larger that 2, full-time solutions are not possible. However, our concern here is centered on the domain of the action functional and not on any possible solutions to the resultant equations of motion.

Whenever the domain of the allowed functions in an interacting model is exactly the same as the domain of functions in the free model, namely $D(A_g) = D(A_0)$, an important (and obvious) property holds. In particular, it follows that

$$\lim_{g \to 0} A_g \equiv \lim_{g \to 0} [A_0 + g A_I] = A_0 \,, \tag{2.50}$$

namely, the interacting model action (A_g) is continuously connected to the free model action (A_0). This property may seem self evident, but the next section will be devoted to cases where (2.50) is false!

Discontinuous perturbations: singular interacting models

We now consider alternative interactions that will change the domain of the interacting action functional. As our initial example let us consider

$$A_g \equiv \int_0^T \{ \tfrac{1}{2}[\dot{x}(t)^2 - \omega^2 x(t)^2] - g x(t)^{-4} \}\, dt \,, \tag{2.51}$$

where the interaction involves the potential $x(t)^{-4}$, in contrast to the principal example of the previous section where the potential was $x(t)^4$. For the present case, we see that the domain of the interacting model is a fundamentally different domain then that of the free model. In particular, for any $g > 0$, the domain of the interacting model is $D_{-4}(A_g) = \{x : \int [\dot{x}^2 + x^2 + x^{-4}]\, dt < \infty\}$, which clearly is markedly different from the domain $D(A_0)$ of the free model. In particular, in the domain $D_{-4}(A_g)$ *no* path that reaches or crosses the axis $x = 0$ is permitted; *all* allowed paths in $D_{-4}(A_g)$ lie in the interval $x(t) > 0$ or $x(t) < 0$. A corresponding change of the interaction to $|x(t)|^{-\alpha}$, $\alpha > 2$, leads to a corresponding, different domain $D_{-\alpha}(A_g) \equiv \{x : \int [\dot{x}^2 + x^2 + |x|^{-\alpha}] \, dt < \infty\}$. While solutions to the relevant equations of motion are difficult to determine in terms of known functions, such solutions, would, in every case where $\alpha > 2$, lead to expressions with $x(t) > 0$ or $x(t) < 0$.

Interacting models may be defined in many different ways . In particular, we may consider models in which the interacting potential is given, for $\alpha > 2$, by $g|x(t) - c|^{-\alpha}$, or $g_- |x(t) - c|^{-\alpha} + g_+ |x(t) + c|^{-\alpha}$, or $g_1 |x(t)^2 - c_1^2|^{-\alpha} + g_2 |x(t)^2 - c_2^2|^{-\alpha}$, etc. In each case, observe that for different values of c the appropriate domains are distinct from one another. It is easy to imagine many additional examples of such distinct potentials and associated domains!

Pseudofree models

The general form of the several action functionals given above is $A_g = A_0 + g A_I$, where A_0 denotes a free model with an action made up of quadratic elements. The interaction action A_I involved a variety of examples that can also be added together to make many types of interactions. For those action functionals for which the domain of allowed functions or paths is identical to the domain of the associated free model (for $g \equiv 0$), i.e., $D(A_g) = D(A_0)$, it follows that

$$\lim_{g \to 0} A_g \equiv \lim_{g \to 0} [A_0 + g A_I] = A_0 . \tag{2.52}$$

In such cases we say that the interacting models are continuously connected to their own free models, or equivalently, that we are dealing with a *continuous perturbation* as one would intuitively expect and is implicitly assumed whenever we study an interacting model as a perturbation of a free model. In the present case, we also say that the free model and the pseudofree model are the same, a remark that will be understood better when we define the meaning of "pseudofree" in the next paragraph.

On the other hand, there are a great many interacting models for which the interacting model is *not* continuously connected to its own free model; stated otherwise, in such situations, it means that

$$\lim_{g \to 0} A_g \equiv \lim_{g \to 0} [A_0 + g A_I] = A_0' \neq A_0 , \tag{2.53}$$

where A_0' is called the pseudofree model and it is *not* identical to the free model. In such cases, it is important to understand that the functional form of the action functional A_0' is *exactly* that of the free action functional A_0; the difference between A_0' and A_0, however, is the *domain of functions that are allowed*. When this situation arises, we say we are dealing with a *discontinuous perturbation* of A_0. This domain situation is changed if we consider $A_g' \equiv A_0' + gA_I$, and in this case, it follows that

$$\lim_{g \to 0} A_g' = \lim_{g \to 0}[A_0' + gA_I] = A_0' , \qquad (2.54)$$

namely, A_I is a *continuous perturbation* of A_0'.

As a basic free model, let us again choose an harmonic oscillator with an action functional $A_0 = \frac{1}{2}\int_0^T [\dot{x}^2 - \omega^2 x^2] \, dt$, and a domain given by $D(A_0) = \{x : \int [\dot{x}^2 + x^2] \, dt < \infty\}$, which applies for any choice of ω, $0 < \omega < \infty$. For an interaction, let us initially choose an inverse quartic potential $x(t)^{-4}$ so that

$$A_g = \int_0^T \{\tfrac{1}{2}[\dot{x}(t)^2 - \omega^2 x(t)^2] - gx(t)^{-4}\} \, dt , \qquad (2.55)$$

with a domain $D(A_g) = \{x : \int[\dot{x}^2 + x^2 + x^{-4}] dt < \infty\}$. When $g \to 0$, it follows that

$$\lim_{g \to 0} A_g = \lim_{g \to 0}[A_0 + gA_I] \, dt = A_0' \equiv \tfrac{1}{2}\int_0^T [\dot{x}(t)^2 - \omega^2 x(t)^2] \, dt , \quad (2.56)$$

which "appears like" A_0 but is *fundamentally different from* A_0 because $D(A_0') \subset D(A_0)$, i.e., specifically

$$D(A_0') \equiv \{x : \int \dot{x}^2 + x^2 + x^{-4} \, dt < \infty\}$$
$$\subset D(A_0) \equiv \{x : \int[\dot{x}^2 + x^2] \, dt < \infty\} . \qquad (2.57)$$

The alternative domain of the pseudofree model affects solutions of the equations of motion. In particular, the equations of motion that follow from a stationary variation of the pseudofree action functional are *exactly* the same as the equations of motion of the free action functional, namely in the present case, $\ddot{x}(t) = -\omega^2 x(t)$. While the solutions to this equation of motion for the free action are

$$x(t) = x_{cs}(t) \equiv x_0 \cos(\omega t) + (\dot{x}_0/\omega) \sin(\omega t) , \qquad (2.58)$$

the solutions of the pseudofree action involving the restriction $\int x(t)^{-4} dt < \infty$ are

$$x(t) = \pm |x_0 \cos(\omega t) + (\dot{x}_0/\omega) \sin(\omega t| = \pm |x_{cs}(t)| \, (\neq 0) , \qquad (2.59)$$

which guarantees that the solution does not cross the axis $x = 0$.

To find solutions of models with a nonlinear interaction, it is natural to rely on a perturbation analysis developed around solutions for the free model, especially when the coupling constant is small. For nonsingular interactions, for which the pseudofree model and the free model are the same, this procedure is generally successful. However, for singular interactions for which the pseudofree and free models are different, a perturbation analysis around the free theory is *doomed*. Instead, any perturbation analysis of such a situation must be taken around solutions of the pseudofree model and *not* around solutions of the free model since it is the pseudofree model that is continuously connected to the interacting models and not the free model.

As another example, suppose the interacting model was

$$A_g = \int_0^T \{ \tfrac{1}{2} [\dot{x}(t)^2 - \omega^2 x(t)^2] - g(x(t)^2 - c^2)^{-4} \} \, dt \,, \qquad (2.60)$$

$c > 0$, which leads to a pseudofree model given by $A_0'' = \tfrac{1}{2} \int_0^T [\dot{x}(t)^2 - \omega^2 x(t)^2] \, dt$ with a domain $D(A_0'') = \{ x : \int [\dot{x}^2 + x^2 + (x^2 - c^2)^{-4}] \, dt < \infty \}$. In this case the solutions to the equations of motion for the pseudofree model have three varieties: $x(t) = x_{cs}(t) < -c$, $-c < x(t) = x_{cs}(t) < c$, and $c < x(t) = x_{cs}(t)$. While we have focussed on such models based on the introduction and removal of a singular perturbation, it is also true that for a large value of c, e.g., the size of the Solar System, then for all physical applications such a pseudofree model exhibits no difference from that of the free model since no realistic harmonic oscillator of that size could ever exist. Clearly, other singular interactions could have similar behavior like that of the chosen example.

The domain of the pseudofree theory is *smaller* than the domain of the free theory when a discontinuous perturbation is involved. Figurately speaking, it may be helpful to say that the pseudofree theory is an "amputated" version of the free theory, as if the the free theory has lost a "finger", or a "hand", or an "arm", etc.—and that loss can never be reversed!

A similar story applies to additional, but different, examples of discontinuous perturbations for which the pseudofree action functional A_0' "appears like" the free action functional A_0 but differs from it because the associated domains are *not* equal, and specifically $D(A_0') \subset D(A_0)$. Moreover, the interaction action A_I, which is a *discontinuous perturbation* of the free action, is a *continuous perturbation* of the pseudofree action functional. This story can be repeated with a *new* discontinuous perturbation (e.g., $[x(t) - \pi]^{-4}$), in which case a *new* pseudofree model arises with an action functional given by $A_0'' = \tfrac{1}{2} \int_0^T [\dot{x}(t)^2 - \omega^2 x(t)^2] \, dt$ now with the domain

$D(A_0'') = \{x : \int [\dot{x}^2 + x^2 + x^{-4} + (x - \pi)^{-4}] dt < \infty\}$. Here the pseudofree theory is a free theory that is missing a "toe" and a "finger"; and so on.

In summary, we have emphasized that continuous perturbations of a free model—such as $x(t)^4$ for an harmonic oscillator—have the virtue of fully restoring the harmonic oscillator action when the coupling constant vanishes. However, discontinuous perturbations—such as $x(t)^{-4}$ for an harmonic oscillator—leave an *indelible imprint* on the system after the coupling constant vanishes; such perturbations can also be considered to be *hard core interactions* since they forbid some histories that are allowed in the free model. Specifically, we have seen that singular potentials inevitably change action functionals, changing free actions into pseudofree actions and thereby also changing discontinuous perturbations of the free model into continuous perturbations of the pseudofree model. We emphasize that while there is only *one* free action for a given physical system, there are *arbitrarily many*, distinct, pseudofree actions associated with each and every free system. A pseudofree theory can also lead to a "compound pseudofree theory"—also covered just by the name "pseudofree theory"—after introducing a new discontinuous perturbation to a pseudofree theory, etc.

We have purposely emphasized the free/pseudofree story here because it is very important to understanding how classical theories are connected to each other, and particularly, how sending a nonlinear interaction coupling constant to zero may *not* remove all the influence of that interaction! We emphasize once again: The usual free theory with its simple domain is the right interpretation of the "free" model for continuous perturbations. However, for discontinuous perturbations, the right interpretation of the "free" model is as a "pseudofree model", namely the free model with an amputated domain such that the perturbation becomes a continuous perturbation. Clearly, the domain of an action functional for a free model is fundamentally important, and it should not always be taken as the naive, natural domain; instead, one should keep an open mind about the domain until the kind of interactions that may be introduced to the model are established. Of course, if the interaction is of the form $[x(t)^2 - c^2]^{-4}$ and $c > 0$ is enormous, e.g., Solar System in size, then for all intents and purposes, solutions of the free and pseudofree models are essentially the same.

The general story laid our here regarding free and pseudofree classical models has a parallel story in the quantum world as well, as we will illustrate in the next chapter.

2.5   Classical Field Theories

The previous discussion started with a single degree of freedom, either $x(t)$ when discussing Newton's or Lagrange's contributions, or a single canonical pair $p(t)$ and $q(t)$ when discussing Hamilton's formulation. In each case we briefly introduced N degrees of freedom, where implicitly $N < \infty$. In discussing a field, however, we need to consider situations where $N = \infty$. In many cases, the infinitely many degrees of freedom are encoded within a field language, e.g., the field $\phi(t, x)$, where $x \in \mathbb{R}^s$, $s \geq 1$, which is suitable for a Newtonian or a Lagrangian approach, or in the case of Hamilton's approach, two fields, namely $\pi(t, x)$ and $\phi(t, x)$. This brief section is devoted to an introductory account of some field theoretic examples. While relativistic scalar fields are the most important examples physically—and they will be discussed in Chap. 9—it is useful to begin with a kinematically simpler set of examples. These simplified models will be particularly important when it comes time to study their quantization in Chap. 8.

2.5.1   *Ultralocal scalar field models*

The simplest examples of classical field theory models would seem to be what are called *ultralocal* scalar field models, and a suitable action functional of a quartic interacting ultralocal scalar field model is given by

$$A_g = \int\int_0^T \{\tfrac{1}{2}[\dot{\phi}(t, x)^2 - m_0^2\phi(t, x)^2] - g\phi(t, x)^4\} \, dt \, d^s x \,, \qquad (2.61)$$

where the integration over x is all of $\mathbb{R}^s$. The equation of motion that follows from this action functional is given by

$$\ddot{\phi}(t, x) - m_0^2\phi(t, x) - 4g\phi(t, x)^3 = 0 \qquad (2.62)$$

or, more explicitly, recognizing that x is only a "spectator" variable

$$\ddot{\phi}_x(t) - m_0^2\phi_x(t) - 4g\phi_x(t)^3 = 0 \,. \qquad (2.63)$$

This equation appears to just be an independent anharmonic oscillator for each point $x \in \mathbb{R}^s$, and, as time evolves, there is no connection between what happens at x and any other point $x' \neq x$. If we introduce the momentum field $\pi(t, x) \equiv \dot{\phi}(t, x)$, or $\pi_x(t) = \dot{\phi}_x(t)$, for this problem, the energy density is given, for any t, by

$$E_x = \tfrac{1}{2}[\pi_x(t)^2 + m_0^2\phi_x(t)^2] + g\phi_x(t)^4 \,, \qquad (2.64)$$

which is subject only to the requirement that $E = \int E_x \, d^s x < \infty$. This means that for x in a suitable set of measure zero, it is possible that $E_x = \infty$

on that set such that E_x remains an integrable function for which $E < \infty$; this possibility shows that an ultralocal scalar field model is more than just a family of independent anharmonic oscillators.

In deriving this equation of motion, we gave only a cursory glance to the domain of the action functional. For the free ultralocal model, where $g \equiv 0$, it follows that the domain of the free action functional $D(A_0) = \{\phi(t,x) : \iint [\dot{\phi}_x(t)^2 + \phi_x(t)^2] \, dt \, d^s x < \infty\}$. On the other hand, for the quartic interacting model with the classical action functional (2.61), the domain is given by

$$D(A_g) = \{\phi(t,x) : \iint [\dot{\phi}_x(t) + \phi_x(t)^2 + \phi_x(t)^4] \, dt \, d^s x < \infty\} \,, \quad (2.65)$$

and it follows that $D(A_g) \subset D(A_0)$ since there are many function $\phi(t,x)$ such that $\iint \int \phi(t,x)^4 \, dt \, d^s x = \infty$ and $\iint [\dot{\phi}(t,x)^2 + \phi(t,x)^2] \, dt \, d^s x < \infty$, e.g., $\phi_{sing}(t,x) = |x|^{-s/3} \exp(-t^2 - x^2)$, etc. As a consequence, it follows that the quartic interaction is a discontinuous perturbation of the free ultralocal model and is, instead, a continuous perturbation of the pseudofree model for which the action functional is that of the free ultralocal model restricted to the domain $D(A_g)$. Stated otherwise, and despite appearances, the quartic interacting ultralocal model is *not* continuously connected to the free ultralocal model! Just as was the case for the discussion of a single degree of freedom, it follows that a perturbation analysis of solutions of the equations of motion with a quartic interaction should be made around solutions of the pseudofree model and not around solutions of the free model. However, that need not be necessary for all solutions since there are still many solutions to the equations of motion that are common to both the free model and the pseudofree model, just not all solutions. It is noteworthy in the field case that we can find discontinuous perturbations as positive powers, such as $\phi(t,x)^4$, without the recourse to negative powers, such as $x(t)^{-4}$, that were necessary in the particle problems.

Focussing only on positive powers of the field leads us to continuous perturbations in certain specialized cases. If the perturbation under consideration is, for example, $\phi(t,x)^6$, it follows that the pseudofree domain is given by

$$D(A_g'') = \{\phi(t,x) : \iint [\dot{\phi}(t,x)^2 + \phi(t,x)^2 + \phi(t,x)^6] \, dt \, d^s x\} \,. \quad (2.66)$$

Suppose now that we wish to introduce another perturbation, namely, a quartic one $\phi(t,x)^4$. In this case we would find that the quartic perturbation was a continuous one of the pseudofree theory that had already seen the interaction $\phi(t,x)^6$. Specifically, the quartic perturbation would be a continuous perturbation since

$$[\int\int\phi(t,x)^4\,dt\,d^sx]^2 = [\int\int\phi(t,x)\cdot\phi(t,x)^3\,dt\,d^sx]^2$$
$$\leq [\int\int\phi(t,x)^2\,dt\,d^sx]\cdot[\int\int\phi(t,x)^6\,dt\,d^sx]\;. \quad (2.67)$$

2.5.2   *Relativistic scalar field models*

A typical example of a field of interest is given by the covariant φ_n^4 models. The Lagrangian action functional for such models is given by

$$A = \int\{\tfrac{1}{2}[(\partial_\mu\phi(x))^2 - m_0^2\phi(x)^2] - \lambda\phi(x)^4\}\,d^nx\;, \quad (2.68)$$

where $\partial_\mu\phi(x) \equiv \partial\phi(x)/\partial x^\mu$, and $x = \{x^0, x^1, x^2, \ldots, x^s\}$, $s = n - 1$. Here, we use the convention that x^0 refers to the time coordinate while the s others, $s \geq 1$, refer to space coordinates. In addition, the implicit sum on the index μ involves the (indefinite) Lorentz metric, and in particular

$$(\partial_\mu\phi(x))^2 \equiv (\partial\phi(x)/\partial x^0)^2 - \Sigma_{k=1}^s(\partial\phi(x)/dx^k)^2\;, \quad (2.69)$$

where $x_0 = t$ in units where the speed of light is unity. In a more transparent notation,

$$A = \int\{\tfrac{1}{2}[\dot\phi(x)^2 - (\vec\nabla\phi(x))^2 - m_0^2\phi(x)^2] - \lambda\phi(x)^4\}\,d^nx\;, \quad (2.70)$$

in which the spatial gradient is made explicit. So long as $\lambda \geq 0$, global solutions for such equations may be expected. Indeed, such equations have been studied classically and it has been shown [Ree76] that given sufficiently smooth initial conditions, these models have global solutions for all spacetime dimensions $n \geq 2$. Variations on these models include changing the power of the nonlinear interaction term from 4 to p, where p is even. Also local polynomials in the field $\phi(x)$ may be considered so long as the largest power is even with a positive coupling constant.

The formulation given above refers to Lagrange, but there is an alternative formulation that follows the Hamiltonian prescription. In this case we introduce a canonically conjugate field $\pi(x)$ known as the momentum and the equal-time Poisson brackets such that $\{\phi(x), \pi(y)\} = \delta(x - y)$, where the s-dimensional delta function involves only the s spatial coordinates. In terms of such fields, Hamilton's form of the action functional is given by

$$A = \int\{\pi(x)\dot\phi(x) - \tfrac{1}{2}[\pi(x)^2 + (\vec\nabla\phi(x))^2 + m_0^2\phi(x)^2] - \lambda\phi(x)^4\}\,d^nx\;, (2.71)$$

and apart from the term $\pi(x)\dot\phi(x)$, the rest of the integrand is (minus) the Hamiltonian density and the Hamiltonian itself involves only a complete spatial integral all at a fixed value of time $t = x^0$. As was the case for finitely many degrees of freedom, the equations of motion ensure that the value of the Hamiltonian is a constant, the energy, for the solution at hand, based on the initial data, e.g., $\pi(x)$ and $\phi(x)$ both evaluated at $x^0 = 0$.

The domain of functions that are allowed for the action functional for covariant scalar fields is given for the free scalar field, and for $x \in \mathbb{R}^n$, $n = s + 1 \geq 2$, by

$$D(A_0) = \{\phi(x) : \int [(\nabla \phi)(x)^2 + \phi(x)^2] \, d^n x < \infty\} , \qquad (2.72)$$

and for the quartic interacting covariant scalar field by

$$D(A_\lambda) = \{\phi(x) : \int [(\nabla \phi)(x)^2 + \phi(x)^2 + \phi(x)^4] \, d^n x < \infty\} . \qquad (2.73)$$

It is noteworthy that $D(A_\lambda) = D(A_0)$ provided $n \leq 4$, while $D(A_\lambda) \subset D(A_0)$ when $n \geq 5$; in particular, it follows that

$$\{\int \phi(x)^4 \, d^n x\}^{1/2} \leq C \int [(\nabla \phi)(x)^2 + m^2 \phi(x)^2] \, d^n x , \qquad (2.74)$$

where $C = (4/3)[m^{(n-4)/2}]$ provided $n \leq 4$ while $C = \infty$ for $n \geq 5$. This inequality is proved in [LSU68; Kla00]. This fact also means that the quartic interaction is a *continuous perturbation* of the classical free theory for $n \leq 4$, while the quartic interaction is a *discontinuous perturbation* of the classical free theory for $n \geq 5$. An example of functions that are in $D(A_0)$ but not in $D(A_\lambda)$ is given by

$$\phi(x) = |x|^{-\alpha} e^{-x^2} , \qquad (n/4) \leq \alpha < (n-2)/2 , \qquad (2.75)$$

which has solutions for $n > 4$. It also follows that a perturbation analysis of solutions to the equations of motion with a quartic interaction should take into account the domain limitation that applies to the interacting theory.

Other classical field theories are of interest as well. Examples of models with an infinite number of degrees of freedom are considered in Chaps. 7, 8, and 9, and, specifically, Chap. 10, which deals with an enhanced version of the quantum theory of gravity. In each case a brief discussion of the classical behavior of the model is considered as well.

Chapter 3

Selected Topics in Quantum Mechanics

WHAT TO LOOK FOR

The rules of conventional canonical quantization, outlined in this chapter, are quite simple and have been fully developed many years ago. Among the procedures is the choice of "Cartesian phase-space coordinates" singled out to promote to Hermitian operators. This process is complicated because phase space does not have a metric in order to distinguish such coordinates. Even when suitable coordinates are chosen, the quantization may fail in the sense that the classical limit in which Planck's constant $\hbar \to 0$ does not recover the original classical theory. Attention is also paid to singular perturbations for which an interacting quantum theory does not reduce to the free quantum theory as the coupling constant vanishes.

3.1 Basic Canonical Quantization

The basic rules of conventional canonical quantization are simple and quite direct; let us describe them for a system with a single degree of freedom. They are: (i) choose a system described by canonical phase-space coordinates p and q which are deemed to be Cartesian coordinates (more on this topic later), and choose $H_c(p, q)$ as the classical (c) Hamiltonian as well as Hamilton's equations of motion, $\dot{q}(t) = \partial H_c(p, q)/\partial p(t)$ and $\dot{p}(t) = -\partial H_c(p, q)/\partial q(t)$, to study the classical dynamics for the system under study; (ii) promote these classical variables to Hermitian operators, i.e., $p \to P$ and $q \to Q$, subject to the Heisenberg commutation relation $[Q, P] \equiv QP - PQ = i\hbar \mathbb{1}$, where $\mathbb{1}$ is the unit operator in the set

33

of operators, and choose a Hermitian Hamiltonian operator—actually, if possible, choose a self-adjoint operator (the difference is reviewed below)—$\mathcal{H} = \mathcal{H}(P,Q) = H_c(P,Q)$ apart from terms of order $\hbar$ representing the usual ambiguities of conventional canonical quantization; (iii) let the operators act on general vectors $|\psi\rangle \in \mathfrak{H}$, an infinite-dimensional Hilbert space, and for the dynamics, adopt an abstract version of Schrödinger's equation of motion given by $i\hbar\partial|\psi(t)\rangle/\partial t = \mathcal{H}(P,Q)|\psi(t)\rangle$. Any uncertainty about which Hamiltonian correction terms of order $\hbar$ to use is to be decided by experiment. The choice of which canonical coordinates to promote to operators is discussed below.

In the previous paragraph we have introduced both Hermitian and self-adjoint operators. It is important for the reader to appreciate the meaning of these terms since Hermitian operators do *not* generally have real spectra and only self-adjoint operators have real spectra and can serve as generators of unitary transformations, e.g., if Y is self adjoint then $U(b) \equiv e^{ibY}$, $-\infty < b < \infty$, is a unitary operator for which $U(b)^\dagger U(b) = U(0) = \mathbb{1}$, the unit operator.

We interrupt our story about canonical quantization for a brief review.

3.1.1   *A brief review of selected operator properties*

Let us pause briefly to recall some properties of operators as they appear in quantum studies. A general operator Y is called Hermitian provided $\langle\phi|Y|\psi\rangle = \langle\psi|Y|\phi\rangle^*$ when both $|\phi\rangle$ and $|\psi\rangle$ are in the domain of Y, $\mathcal{D}(Y)$. This criterion implies that the Hermitian adjoint of Y, $Y^\dagger$, acts exactly like Y when confined to the domain $\mathcal{D}(Y)$, but that information does not address what is the domain of $Y^\dagger$ itself. Specifically, we can only conclude that $\mathcal{D}(Y^\dagger) \supseteq \mathcal{D}(Y)$. If in fact $Y^\dagger = Y$ *and* $\mathcal{D}(Y^\dagger) = \mathcal{D}(Y)$, then we say that the operator Y is *self adjoint*. Sometimes the domain of an Hermitian operator can be enlarged, with a corresponding reduction of the domain of its adjoint, so that the resultant operator is self adjoint; the result may be unique, or there may be a variety of different self-adjoint extensions. Also there are cases where no extension is possible that would result in a self-adjoint operator.

We can illustrate these different properties of operators as follows. Let $P = -i\partial/\partial x$ and $Q = x$—known as the Schrödinger representation (here suitable units are chosen so that $\hbar = 1$)—which then act on various functions of x. We consider three different examples:

(1) First, we consider the space $L^2(\mathbb{R}) = \{\psi : \int_{-\infty}^\infty |\psi(x)|^2\,dx < \infty\}$. For

P we choose the domain $\mathcal{D}(P) = \{\psi : \int_{-\infty}^{\infty}[|\psi'|^2 + |\psi|^2]dx < \infty, \psi(-\infty) = \psi(\infty) = 0\}$. Integration by parts leads to

$$\int_{-\infty}^{\infty}\phi(x)^*[(-i)\psi'(x)]\,dx = -i\phi(x)^*\psi(x)|_{-\infty}^{\infty} + \int_{-\infty}^{\infty}[(-i)\phi'(x)]^*\psi(x)\,dx. \tag{3.1}$$

Since $\psi(-\infty) = \psi(\infty) = 0$ is assumed, the first term on the right vanishes, and it follows that $P^\dagger = P$ on the domain $\mathcal{D}(P)$, but then the domain of $P^\dagger$ is $\mathcal{D}(P^\dagger) = \{\phi : \int_{-\infty}^{\infty}[|\phi'|^2 + |\phi|^2]dx < \infty\}$ which is larger that $\mathcal{D}(P)$. However, this apparent distinction is false since if ψ satisfies $\int_{-\infty}^{\infty}[|\psi'|^2 + |\psi|^2]dx < \infty$ then it *already* satisfies $\psi(-\infty) = \psi(\infty) = 0$ (thanks to the Riemann Lebesgue lemma—see Chap. 2). Hence, $P^\dagger = P$ and $\mathcal{D}(P^\dagger) = \mathcal{D}(P)$, and the operator P is self adjoint and unique.

For the same interval, suppose we had started with $D_S(P) = \{\psi : \psi \in \mathbb{S}, i.e., \psi(x) \in C^\infty$ and $|\psi(x)| \cdot |x|^p \to 0$ as $|x| \to \infty$ *for all* $p \in \mathbb{N} = \{0, 1, 2, ...\}\}$. This domain is far smaller than our original $\mathcal{D}(P)$, but it suffices to determine that $\mathcal{D}(P^\dagger)$ is unchanged. We can enlarge $D_S(P)$ up to $\mathcal{D}(P^\dagger)$ and arrive at the former self-adjoint operator. In this case we say that P and $D_S(P)$ is *essentially self adjoint* since it can be extended to a unique self-adjoint operator.

(2) For our second example, our space is $L^2([0,1]) = \{\psi : \int_0^1 |\psi(x)|^2\,dx < \infty\}$. For the operator P we choose the domain $\mathcal{D}(P) = \{\psi : \int_0^1[|\psi'|^2 + |\psi|^2]dx < \infty, \psi(0) = \psi(1) = 0\}$. Once again we have

$$\int_0^1 \phi(x)^*[(-i)\psi'(x)]\,dx = -i\phi(x)^*\psi(x)|_0^1 + \int_0^1[(-i)\phi'(x)]^*\psi(x)\,dx. \tag{3.2}$$

Therefore, $P^\dagger = P$ on $\mathcal{D}(P)$, but $\mathcal{D}(P^\dagger) = \{\psi : \int_0^1[|\psi'|^2 + |\psi|^2]dx < \infty\}$, and now $\mathcal{D}(P^\dagger) \supset \mathcal{D}(P)$. Indeed, at present, $P^\dagger$ has normalizable eigenfunctions $\psi_{\pm\beta}(x) = N_{\pm\beta}e^{\mp\beta x}$, $0 < \beta < \infty$, with imaginary eigenvalues, where $N_{\pm\beta}$ is a normalizing factor. Thus P is Hermitian but it is not presently self adjoint. To make it self adjoint we need to enlarge $\mathcal{D}(P)$ and shrink $\mathcal{D}(P^\dagger)$ until they are equal; and the result must be such that the first term on the right side of (3.2) vanishes. This leads to $D_\alpha(P^\dagger) = D_\alpha(P) = \{\psi : \int_0^1[|\psi'|^2 + |\psi|^2]dx < \infty, \psi(1) = e^{i\alpha}\psi(0)\}$. Thus we find a *family* of self-adjoint operators P_α, $0 \le \alpha < 2\pi$, each with the spectrum $\mathsf{spec}(P_\alpha) = \alpha + 2\pi n$, $-\infty < n < \infty$, which means that each value of α labels an *inequivalent* self-adjoint extension. The physics behind this distinction lies in a charged particle circumnavigating an infinitely long magnetic solenoid with a total flux value proportional to α [AhB59].

(3) In the final case we consider the space $L^2([0,\infty)) = \{\psi : \int_0^\infty |\psi(x)|^2\,dx < \infty\}$. We choose $\mathcal{D}(P) = \{\psi : \int_0^\infty[|\psi'|^2 + |\psi|^2]dx <$

$\infty, \psi(0) = \psi(\infty) = 0\}$, and we consider the equation

$$\int_0^\infty \phi(x)^* \left[(-i)\psi'(x)\right] dx = -i\phi(x)^*\psi(x)\big|_0^\infty + \int_0^\infty \left[(-i)\phi'(x)\right]^*\psi(x)\,dx. \quad (3.3)$$

In fact the requirement that $\psi(\infty) = 0$ is not necessary because it is automatically true, but the requirement that $\psi(0) = 0$ must be maintained to make $P^\dagger = P$ on $\mathcal{D}(P)$. Moreover, the operator $P^\dagger$ has imaginary eigenvalues for normalizable eigenfunctions given by $\psi_\beta(x) = N_\beta e^{-\beta x}$, for all real $\beta > 0$, where N_β is a normalization factor. As it turns out, there is *no* way to make the domains equal and maintain that $P^\dagger = P$. Thus, we are faced with the fact that in this case the operator P remains Hermitian and it *cannot* be made self adjoint. The failure of any self-adjoint extension is clear since if P could be made self adjoint it would serve as the generator of unitary transformations that shift functions toward larger or smaller x values; but if they are shifted to smaller values they will eventually lose normalization since the norm of all functions still involves an integral from zero to infinity.

[Remark: The reader may wish to read Sec. 38 in Dirac [Dir58] where he seeks a proper canonically conjugate operator for the radial variable $r \geq 0$ in three dimensions, where $r^2 = x^2 + y^2 + z^2$.]

Lesson concluded. We now return to canonical quantization.

3.1.2   Canonical quantization, continued

The bare outline above of basic details describes the canonical quantization of a classical canonical system. Some additional and standard features are: (i) the inner product of the Hilbert space is denoted by $\langle \phi | \psi \rangle$, which is chosen to be linear in the second element and anti-linear in the first element; (ii) the state vector $|\psi(t)\rangle$ is a unit vector, i.e., $\| \, |\psi(t)\rangle \|^2 \equiv \langle \psi(t) | \psi(t) \rangle = 1$; (iii) expectations of P and Q, respectively, denote the *mean momentum*, $\overline{p}(t) \equiv \langle \psi(t) | P | \psi(t) \rangle$, and the *mean position*, $\overline{q}(t) \equiv \langle \psi(t) | Q | \psi(t) \rangle$, of the system under study. With the Hamiltonian operator $\mathcal{H}$ chosen to be self adjoint, it serves to generate a unitary transformation such that $U(t) \equiv e^{-i\mathcal{H}t/\hbar}$, $-\infty < t < \infty$, and therefore $|\psi(t)\rangle \equiv U(t)|\psi\rangle$, where $|\psi\rangle$ is chosen as the initial state at time $t = 0$. This means that the temporal development of the mean position is given by

$$\overline{q}(t) = \langle \psi | U(t)^\dagger Q U(t) | \psi \rangle \equiv \langle \psi | Q(t) | \psi \rangle , \quad (3.4)$$

where we have introduced

$$Q(t) \equiv e^{i\mathcal{H}t/\hbar} Q e^{-i\mathcal{H}t/\hbar} . \quad (3.5)$$

A time derivative of this equation leads to

$$i\hbar \dot{Q}(t) \equiv [Q(t), \mathcal{H}] , \tag{3.6}$$

which is part of the system of equations that describes the Heisenberg formulation of quantum dynamics, leading to an equivalent description of the quantum system; observe the similarity of this equation with the classical equation of motion written in terms of a Poisson bracket. In the Heisenberg picture, the operators are time dependent, and the wave functions are time independent; in the Schrödinger picture, the wave functions are time dependent, and the operators are time independent. Indeed, there are multiple variations regarding how to describe the quantum theory itself [Sty02], and as formulated, they all yield the same result. For the most part, we shall adopt the Schrödinger picture as well as its most frequently used formalism to describe the quantum theory.

However, there is an important problem we have overlooked.

3.1.3 *A gap in the process of canonical quantization*

We start with a short quote from Dirac [Dir58] (page 114) that addresses the choice of classical canonical coordinates and momenta to be promoted to Hermitian operators in a canonical quantization procedure:

> *(text)* "... if the system does have a classical analogue, its connexion with classical mechanics is especially close and one can usually assume that the Hamiltonian is the same function of the canonical coordinates and momenta in the quantum theory as in the classical theory.†"
> *go to footnote*

> *(footnote)* "† This assumption is found in practice to be successful only when applied with the dynamical coordinates and momenta referring to a Cartesian system of axes and not to more general curvilinear coordinates."

It is important to recognize that we have ignored a very basic question at this point, namely: How was the particular quantum Hamiltonian, $\mathcal{H}(P, Q)$, chosen in the first place? The answer of course is that we had originally chosen a particular classical system described by its classical Hamiltonian $H_c(p, q)$, which we assumed describes the dynamics of the system under study from a classical point of view. Therefore the answer to how we decided on the quantum Hamiltonian is that we 'quantized' the chosen classical system and thus the classical Hamiltonian led to the quantum

Hamiltonian, i.e., $H_c(p,q) \to \mathcal{H}(P,Q) = H_c(P,Q)$, modulo any $\mathcal{O}(\hbar)$ ambiguities that are to be decided from experiments made on the quantum system itself. But that leads to the fundamental question about which particular set of phase-space coordinates were chosen since that determines the functional form of the Hamiltonian itself. Did we use the phase-space coordinates (p,q) with the classical Hamiltonian $H_c(p,q)$ or did we use the phase-space coordinates $(\widetilde{p}, \widetilde{q})$ and thus the classical Hamiltonian $\widetilde{H}_c(\widetilde{p}, \widetilde{q})$. Why does this matter? We do believe that $P^2 + Q^2$ and $P^2 + Q^4$ denote two different operators with very different spectra. Since the spectrum of an operator is observable, in principle, it means that the functional form of the Hamiltonian operator carries *physics*. On the other hand, the classical Hamiltonian carries *no physics* because the system described by the classical Hamiltonian $H_c(p,q)$ can also be described by the classical Hamiltonian $\widetilde{H}_c(\widetilde{p}, \widetilde{q}) \left[= H_c(p,q) \right]$ whenever the phase-space coordinates (p,q) and $(\widetilde{p}, \widetilde{q})$ are related by a canonical transformation. For example, we could choose $\widetilde{p} = \frac{1}{2}[\omega^{-1}p^2 + \omega q^2]$ and $\widetilde{q} = \arctan(p/\omega q)$, which, for an harmonic oscillator, leads to the classical Hamiltonian given by $\widetilde{H}(\widetilde{p}, \widetilde{q}) = \omega \widetilde{p}$ with dynamical solutions given by $\widetilde{p}(t) = \widetilde{p}_0$ and $\widetilde{q}(t) = \widetilde{q}_0 + \omega t$. If we were to choose $\widetilde{P}$ and $\widetilde{Q}$ as the right choice, the Hamiltonian operator would be $\widetilde{\mathcal{H}} = \omega \widetilde{P}$ which has a continuous spectrum rather than the well known—and correct—discrete spectrum for an harmonic oscillator. Thus choosing the "wrong" canonical set of coordinates to promote can make a huge difference in the proposed quantum story. Summarizing, we can not just choose *any* phase-space coordinates to promote to operators if we seek to achieve a proper physical solution to our problem; instead, we must choose phase-space coordinates that will lead to the *correct* Hamiltonian operator for the system under consideration. The favored phase-space coordinates have been identified by Dirac as "Cartesian coordinates" (see the quote above). However, Cartesian coordinates require a *metric*, e.g., such as $dx^2 + dy^2$, when dealing with a flat two-dimensional surface; indeed, Cartesian coordinates can *only* exist on a flat plane. Does Dirac mean that q is Cartesian with a metric dq^2 and p is Cartesian with a separate metric dp^2? That cannot be because *one* dimension, by itself, cannot be called Cartesian (think of a garden hose all curled up). He must mean something like $\omega^{-1} dp^2 + \omega dq^2$, but that cannot arise just from a phase space. Of course, in higher dimensions, e.g., two dimensions, and influenced by the usual form of a kinetic energy proportional to $(p_1^2 + p_2^2)$, some authors have seized on that form and assumed that what Dirac meant was that Cartesian coordinates involves two separate two-dimensional flat spaces such as $dq_1^2 + dq_2^2$ and $dp_1^2 + dp_2^2$. But that view does not explain how to treat the case of a *single* degree of freedom with one q and one p. Thus we are asked to consider that the phase-space coordinates belong to a flat metric space, but, strictly speaking, that is impossible since the classical phase-space coordinates belong to a symplectic space, a space without a metric, that has a "symplectic geometry" described by a symplectic form, which is a two-form $\widetilde{\omega}$. In canonical coordinates, $\widetilde{\omega} = dp\, dq = d\widetilde{p}\, d\widetilde{q}$, a differential surface area, or more transparently, using an exterior product, by $\widetilde{\omega} = dp \wedge dq = d\widetilde{p} \wedge d\widetilde{q}$, emphasizing the antisymmetric nature of the two form. A rubber sheet offers one example of such a surface: a rubber sheet can be flat when lying on a flat floor; however, a rubber sheet can also be stretched around an automobile fender in which case it is no longer flat. Thus we are faced with a dilemma: suitable phase-space coordinates are needed when promoting them to operators; however, the rule for choosing these phase-space coordinates is not available in the realm of a classical phase space. Faced with this gap, we are forced to simply "designate" which are the proper phase-space coordinates that are to be promoted to operators. In this regard, it is noteworthy that Dirac (see quote above) made his remark stressing the preference for "Cartesian coordinates" in a *footnote* apparently signifying his inability to establish this fundamental requirement by a logical argument.

Fortunately, the gap represented by how to choose the proper phase-space coordinates will be satisfactorily filled when we discuss enhanced quantization in the following chapter.

3.1.4 *Successes of conventional quantization*

Let us employ Schrödinger's representation, i.e., $P \to -i\hbar\partial/\partial x$ and $Q \to x$, and use the fact that the self-adjoint position operator Q has eigenvectors $|x\rangle$, such that $Q|x\rangle = x|x\rangle$, and which are normalized according to $\langle x|x'\rangle = \delta(x - x')$, in terms of the Dirac delta function having the usual properties: $\delta(x) = 0$ if $x \neq 0$, and $\int \delta(x)\,dx = 1$ when the range of integration includes $x = 0$. The abstract Schrödinger's equation is recast as a differential equation given by

$$i\hbar\partial\langle x|\psi(t)\rangle/\partial t = \langle x|\mathcal{H}(P,Q)|\psi(t)\rangle$$
$$= \mathcal{H}(-i\hbar\partial/\partial x, x)\langle x|\psi(t)\rangle \,, \tag{3.7}$$

or, as most commonly written, using $\psi(x,t) \equiv \langle x|\psi(t)\rangle$,

$$i\hbar\partial\psi(x,t)/\partial t = \mathcal{H}(-i\hbar\partial/\partial x, x)\psi(x,t) \,, \tag{3.8}$$

where $\psi(x,t)$ is the usual Schrödinger wave function. Assuming that P enters $\mathcal{H}$ as a polynomial, then we are led to a partial differential equation. A great many physical systems have quantum Hamiltonians of the form $\mathcal{H}(P,Q) = P^2/2m + V(Q)$, and for such problems we are led to study the partial differential equation

$$i\hbar \partial \psi(x,t)/\partial t = -(\hbar^2/2m)\partial^2 \psi(x,t)/\partial x^2 + V(x)\psi(x,t) , \qquad (3.9)$$

which is the commonly encountered form of Schödinger's equation for a single degree of freedom. Relevant solutions of this equation must be normalizable, i.e., $\int |\psi(x,t)|^2\, dx < \infty$ for all t, $-\infty < t < \infty$. Moreover, if the operator $\mathcal{H}(P,Q)$ is self adjoint—and *only* if $\mathcal{H}(P,Q)$ is self adjoint—it follows that the norm of the wave function is independent of the time and, in fact, if the wave function is normalized at $t = 0$, it follows that $\int |\psi(x,t)|^2\, dx = 1$ for all t.

The generally accepted understanding of the meaning of the wave function is as a *probability amplitude*. This means that the mean position at time t is given by $\langle Q(t)\rangle = \int x|\psi(x,t)|^2\, dx$, in which $|\psi(x,t)|^2$ is interpreted as the probability density of finding the particle at position x at time t. Likewise $\langle Q^r\rangle = \int x^r|\psi(x,t)|^2\, dx$ has a similar meaning for the mean (position)^r, etc. If we introduce the Fourier transform of the wave function given by

$$\widetilde{\psi}(p,t) = (1/\sqrt{2\pi\hbar})\int_{-\infty}^{\infty} e^{-ipx/\hbar}\,\psi(x,t)\, dx , \qquad (3.10)$$

it follows that $|\widetilde{\psi}(p,t)|^2$ is interpreted as the probability density to find the particle's momentum value to be p at time t, and, additionally, $\int |\widetilde{\psi}(p,t)|^2\, dp = 1$ for all t. It also follows that $\widetilde{\psi}(p,t) = \langle p|\psi(t)\rangle$, where $|p\rangle$ are eigenvectors of the operator P, i.e., $P|p\rangle = p|p\rangle$, $-\infty < p < \infty$, and for which $\langle p|p'\rangle = \delta(p - p')$, where $\delta(p) = 0$ for $p \neq 0$ and $\int \delta(p)\, dp = 1$ provided that the range of integration includes $p = 0$. In that case, we have the mean (momentum)^r at time t given by $\langle P^r(t)\rangle = \int p^r|\widetilde{\psi}(p,t)|^2\, dp$. This evaluation of the mean $\langle P^r\rangle$ can also be written as

$$\langle P^r\rangle = (-i\hbar)^r \int \psi(x,t)^* (\partial/\partial x)^r\, \psi(x,t)\, dx , \qquad (3.11)$$

which follows from properties of a Fourier transformation.

In addition to the time-dependent Schrödinger equation (3.8), it is important to recall the time-independent version of that equation as well. Let us assume that $\psi(x,t) = \phi_E(x)e^{-iEt/\hbar}$, which then leads to the equation

$$E\phi_E(x) = -(\hbar^2/2m)\,\partial^2 \phi_E(x)/\partial x^2 + V(x)\phi_E(x) , \qquad (3.12)$$

which becomes an acceptable differential equation of motion when the potential is not time dependent. The new equation is an eigenvalue equation for which there are suitable square-integrable solutions $\phi_E(x)$ for certain values of E, which are then the associated eigenvalues of the Hamiltonian operator. On the other hand, for example, if $V(x) = 0$, then the solutions of (3.12) are given by $\phi_E(x) = (1/\sqrt{2\pi\hbar}) \exp(\pm ipx/\hbar)$, where $E = p^2/2m$, $-\infty < p < \infty$, with a delta-function normalization suitable to an operator with a continuous spectrum. If we assume that the potential $V(x)$ only leads to normalizable eigenfunctions, then the solutions of (3.12) take the form $\xi_n(x)$, $0 \leq n < \infty$ (or $1 \leq n < \infty$, etc.), where the eigenfunctions are orthonormal, i.e., $\int_{-\infty}^{\infty} \xi_m(x)^* \xi_n(x) \, dx = \delta_{m,n}$, and $\{\xi_n(x)\}_{n=0}^{\infty}$ form a complete set of functions for which every square-integrable function $\phi(x) = \Sigma_{n=0}^{\infty} a_n \xi_n(x)$, where $\Sigma_{n=0}^{\infty} |a_n|^2 < \infty$; and, for example, $E_0 \leq E_1 \leq E_2 \leq \cdots$. Given the representation of a general wave function $\phi(x)$ as a series involving the eigenfunctions of the problem, it is a simple matter to offer the time-dependent solution of the Schrödinger equation, namely,

$$\phi(x,t) = \sum_{n=0}^{\infty} a_n \xi_n(x) e^{-iE_n t/\hbar} , \tag{3.13}$$

based on the initial wave function $\phi(x) = \phi(x,0)$ at time $t = 0$.

It is noteworthy that the (normalizable) ground state of a given Schrödinger equation implicitly defines all the other solutions of a given model up to a single constant (E_0). Let $\xi_0(x)$ denote the ground state, which has the property that it never vanishes, as is typically the case. Thus, denoting derivatives by primes, it follows that

$$-(\hbar^2/2m)\xi_0''(x) + V(x)\xi_0(x) = E_0 \xi_0(x) . \tag{3.14}$$

This relation can be used to eliminate the potential leading to

$$-(\hbar^2/2m)\psi_E''(x) + [E_0 + (\hbar^2/2m)\xi_0''(x)/\xi_0(x)]\psi_E(x) = E\psi_E(x) , \tag{3.15}$$

an equation that implicitly determines all solutions to the given Schrödinger equation.

3.1.5 *Propagators and path integrals*

As an equation that involves a single time derivative, the time-dependent Schrödinger equation needs only an initial wave function to uniquely determine its temporal behavior for all time. In particular, if the wave function at time t' is given by $\psi(x', t')$, it follows that there is some integral kernel

$K(x, t; x', t')$, called the *propagator*, which determines the wave function $\psi(x, t)$ for any time t, by means of the equation

$$\psi(x, t) = \int K(x, t; x', t') \psi(x', t') \, dx' \,. \tag{3.16}$$

For a system with only discrete energy eigenfunctions, one expression for the propagator is given by

$$K(x, t; x', t') = \sum_{n=0}^{\infty} \xi_n(x) \xi_n(x')^* \, e^{-i(t-t')E_n/\hbar} \,. \tag{3.17}$$

More generally, it is also clear that the propagator is a solution of Schrödingr's equation, i.e.,

$$i\hbar \, \partial K(x, t; x', t')/\partial t \tag{3.18}$$

$$= -(\hbar^2/2m)\partial^2 K(x, t; x', t')/\partial x^2 + V(x) K(x, t; x', t') \,,$$

subject to the initial condition that $K(x, t'; x', t') = \delta(x - x')$.

Many textbooks exist that discuss how to find the solution, or even an approximate solution, of Schrödinger's equations with nonlinear potentials. In line with the focus of this monograph, the reader is directed to such sources for the purpose of finding solutions—analytic as well as numerical—to those equations.

However, the harmonic oscillator is so basic to quantum mechanics we will present the solution for that one model. Beyond that, we will also derive additional solutions in later chapters to certain models that are regarded as "insoluble" when viewed conventionally!

Harmonic oscillator

It is no exaggeration to say that the harmonic oscillator is one of the most important examples, if not *the* most important example, to study. As discussed earlier, the classical Hamiltonian for an harmonic oscillator is given by $H_c(p, q) = \frac{1}{2}(p^2 + \omega^2 q^2)$, in units where the mass $m = 1$. The solution to the equation of motion is $q(t) = q_0 \cos(\omega t) + (\dot{q}_0/\omega) \sin(\omega t)$, where q_0 and $\dot{q}_0$ are the initial position and velocity of the particle at $t = 0$; also, the solution for $p(t) = \dot{q}(t)$. Of course, from a classical point of view, these phase-space coordinates are just one example of the family of canonical coordinates that could be used to describe the classical Hamiltonian and the classical solutions. For example, as discussed already, we could choose $\widetilde{p} = \frac{1}{2}[\omega^{-1}p^2 + \omega q^2]$ and $\widetilde{q} = \arctan(p/\omega q)$, which leads to the classical Hamiltonian given by $\widetilde{H}(\widetilde{p}, \widetilde{q}) = \omega \widetilde{p}$ with dynamical solutions given by $\widetilde{p}(t) = \widetilde{p}_0$ and $\widetilde{q}(t) = \widetilde{q}_0 + \omega t$. However, it is well established that the original phase-space coordinates (p, q) are already the favored "Cartesian coordinates" suitable to promote to operators in the process of quantization. This means that the quantum Hamiltonian is given by $\mathcal{H}(P,Q) = \frac{1}{2}(P^2 + \omega^2 Q^2)$ plus a possible extra term of $\mathcal{O}(\hbar)$. For example, if this oscillator refers to a conduction band of a material, then the extra term vanishes; however, if the oscillator refers to one of the degrees of freedom of a free field, then the extra term should be $-\frac{1}{2}\omega\hbar$. In principle, these correction terms are determined by "experiment", either a real laboratory experiment, as in the case of the conduction band, or a mathematical consistency, as in the case of the free field.

The time-independent Schrödinger equation of motion for an harmonic oscillator is given (with $m = 1$) by

$$-(\hbar^2/2)\partial^2 \phi_E(x)/\partial x^2 + (\omega^2/2)x^2 \phi_E(x) = E\phi_E(x) . \tag{3.19}$$

For large values of $|x|$, the value of E is not of primary importance, and the solution is roughly proportional to $\exp(\pm \omega x^2/2\hbar)$ since $(\hbar^2/2)$ times the second derivative leads to $[\frac{1}{2}\omega^2 x^2 \pm \frac{1}{2}\omega\hbar]\exp(\pm\omega x^2/2\hbar)$. If we choose the wave function to be proportional to $\exp(-\omega x^2/2\hbar)$ for $x \ll -1$, then we need to choose E so that the solution is also proportional to $\exp(-\omega x^2/2\hbar)$ when $x \gg 1$, i.e., the solution for the proper E value must have *no* component of the alternative behavior, $\exp(+\omega x^2/2\hbar)$, at large $x \gg 1$ in order for the solution to be square integrable. This is the underlying requirement that selects the proper values of E. It is well known that the eigensolutions for the harmonic oscillator are Hermite functions $\{h_n(x)\}_{n=0}^{\infty}$, and it is useful to introduce the generating function for Hermite functions given (for $\hbar = \omega = 1$) by

$$\exp[-s^2 + 2sx - \tfrac{1}{2}x^2] = \pi^{1/2}\sum_{n=0}^{\infty}(n!)^{-1/2}(s\sqrt{2})^n h_n(x) , \tag{3.20}$$

and it follows that

$$\begin{aligned}
[-\tfrac{1}{2}\partial^2/\partial x^2 &+ \tfrac{1}{2}x^2]\,e^{-s^2 + 2sx - \frac{1}{2}x^2} \\
&= [2sx - 2s^2 + \tfrac{1}{2}]e^{-s^2 + 2sx - \frac{1}{2}x^2} \\
&= [s(\partial/\partial s) + \tfrac{1}{2}]e^{-s^2 + 2sx - \frac{1}{2}x^2} \\
&= \pi^{1/2}\sum_{n=0}^{\infty}(n!)^{-1/2}[n + \tfrac{1}{2}](s\sqrt{2})^n h_n(x) .
\end{aligned} \tag{3.21}$$

After restoring ω and $\hbar$, we learn that

$$\begin{aligned}
-(\hbar^2/2)\partial^2 h_n((\omega/\hbar)^{1/2}x)/\partial x^2 &+ (\omega^2/2)x^2 h_n((\omega/\hbar)^{1/2}x) \\
&= \omega\hbar(n + \tfrac{1}{2})h_n((\omega/\hbar)^{1/2}x) ,
\end{aligned} \tag{3.22}$$

and thus $E_n = \omega\hbar(n + \frac{1}{2})$, $0 \le n < \infty$, for the harmonic oscillator.

The Hermite functions form an orthonormal set of functions. Specifically, this means that

$$\int h_m(x)\, h_n(x)\, dx = \delta_{m,n} \,, \tag{3.23}$$

and

$$\sum_{n=0}^{\infty} h_n(x)\, h_n(x') = \delta(x - x') \,, \tag{3.24}$$

As a delta function is involved, it formerly follows that

$$\psi(x) = \sum_{n=0}^{\infty} \int h_n(x)\, h_n(x')\, \psi(x')\, dx' \,. \tag{3.25}$$

Finally, we introduce the propagator $K(x,T;x',0)$ which takes a general wave function $\psi(x')$ at time 0 and generates the wave function $\psi(x,T)$ at a different time T. In particular, again with $\omega = \hbar = 1$,

$$K(x,T;x',0) \equiv \sum_{n=0}^{\infty} h_n(x)\, h_n(x')\, e^{-i(n+1/2)T} \,, \tag{3.26}$$

and therefore

$$\psi(x,T) = \int K(x,T;x',0)\, \psi(x',0)\, dx' \,. \tag{3.27}$$

It is also clear that the propagator satisfies the *folding identity* given by

$$K(x'',T+S;x',0) = \int K(x'',T+S;x,T)\, K(x,T;x',0)\, dx \,. \tag{3.28}$$

Path integrals

A phase-space path integral representation for the propagator was given by Feynman in 1952 [Fey52] that has proved to be an important way of studying properties of the propagator. In particular, the propagator is formally given by

$$K(x'',t'';x',t') = \mathcal{M} \int e^{(i/\hbar)\int_{t'}^{t''} [p\dot{q} - H(p,q)]\, dt}\, \mathcal{D}p\, \mathcal{D}q \,, \tag{3.29}$$

integrated over all paths $q(t)$ that satisfy $q(t'') = x''$ and $q(t') = x'$ and all paths $p(t)$ without any boundary restriction; thus this procedure is called a *path integral*. As a special case, if $H(p,q) = p^2/2m + V(q)$, the integral over p can be readily carried out leading to the formal expression, proposed even earlier in 1948 by Feynman [Fey48],

$$K(x'',t'';x',t') = \mathcal{N} \int e^{(i/\hbar)\int_{t'}^{t''} [m\dot{q}^2/2 - V(q)]\, dt}\, \mathcal{D}q \,. \tag{3.30}$$

In these two expressions it should be noted that the integrand involves the classical action, first in the form for Hamiltonian mechanics and second in the form for Lagrangian mechanics. The factors $\mathcal{M}$ and $\mathcal{N}$ are formal normalization factors chosen so that in each case

$$\lim_{t'' \to t'} K(x'', t''; x', t') = \delta(x'' - x') . \tag{3.31}$$

Indeed, besides the normalization factors, the whole expression for the path integral, while symbolically appealing, is only formal in nature and further clarification of exactly how such formal integrals are defined is required. As one may imagine, there are several different ways to give meaning to these formal expressions.

One of the more popular procedures is to regularize each path integral and remove the regularization as the final step. This procedure is quite common in the study of integrals. Two elementary examples follow that apply to suitable functions: First, the integral $\int_0^\infty f(x)\,dx \equiv \lim_{X \to \infty} \int_0^X f(x)\,dx$; second, the integral $\int_0^1 g(x)\,dx = \lim_{N \to \infty} \Sigma_{n=1}^N g(n/N)\,N^{-1}$—provided the limits exist. The second example is an example of Riemann's definition of the integral and we will use it in a path integral regularization. For the second version of the path integral it follows (for $t'' = T$ and $t' = 0$) that

$$K(x'', T; x', 0) = \lim_{\epsilon \to 0} M_\epsilon \int \cdot \int \exp\{(i/\hbar)\Sigma_{n=0}^N [m(x_{n+1} - x_n)^2/2\epsilon$$
$$-\epsilon V(x_n)]\}\textstyle\prod_{n=1}^N dx_n , \tag{3.32}$$

where $\epsilon = T/N$, M_ϵ is a normalization factor, and $x_{N+1} = x''$ and $x_0 = x'$. A similar expression for the phase-space path integral is somewhat different. Although the formal measure $\mathcal{D}p\,\mathcal{D}q$ appears to be invariant under canonical transformations, analogous to $dp\,dq$, that feature is not true since the formal measure contains one more integral over p than q, a fact that follows directly from the folding identity (3.28). In fact, the physical meaning of each q is as an eigenvalue of the operator Q, and likewise the physical meaning of each p is as an eigenvalue for P. Thus it is imprecise to say p_n and q_n as that would signify that they both held at the same time, which can not be true. To solve that dilemma, as well as respect the presence of one more p integral, we put the q variables on one temporal lattice, represented by q_n, and the p values on a shifted temporal lattice, represented by $p_{n+1/2}$. With this notation, the regularized form of the phase-space path integral is given (again for $t'' = T$ and $t = 0$) by

$$K(x'', T; x', 0) = \lim_{\epsilon \to 0} \int \cdot \int \exp\{(i/\hbar)\Sigma_{n=0}^N [p_{n+1/2}(q_{n+1} - q_n) \tag{3.33}$$
$$-\epsilon H(p_{n+1/2}, (q_{n+1} + q_n)/2)]\} \textstyle\prod_{n=0}^N dp_{n1/2}/(2\pi\hbar) \prod_{n=1}^N dq_n ,$$

where again $q_{N+1} = x''$ and $q_0 = x'$. We will give yet another regularization for the phase-space path integral below when we discuss coherent states and their properties.

There is also some benefit in discussing a 'Euclidean', or 'imaginary time', path integral in which the time variable $t \to -it$ leading to a strictly real set of equations. In particular, the time-dependent Schrödinger equation assumes (for $m = 1$) the form

$$\hbar \partial \rho(x,t)/\partial t = (\hbar^2/2)\partial^2 \rho(x,t)/\partial x^2 - V(x)\rho(x,t) \ . \qquad (3.34)$$

In turn, for such a problem, this equation leads to real, time-dependent solutions, and for a discrete spectrum, the eigenfunctions may be called $\sigma_n(x)$, $0 \le n < \infty$. The propagator for the Euclidean version, $L(x'', T; x', 0)$, is given, for $T \ge 0$. by

$$L(x'', T; x', 0) = \sum_{n=0}^{\infty} \sigma_n(x'') \sigma_n(x') e^{-E_n T/\hbar} \ , \qquad (3.35)$$

and it is also given by a real (formal) path integral,

$$L(x'', T; x', 0) = \mathcal{N} \int e^{-(1/\hbar)\int_0^T [\frac{1}{2}\dot{x}^2 + V(x)] \, dt} \mathcal{D}x \qquad (3.36)$$

integrated over all paths such that $x(T) = x''$ and $x(0) = x'$. A regularized version of this formal equation follows along the lines of (3.32) with $T \to -iT$ and $\epsilon \to -i\epsilon$, leading to

$$L(x'', T; x', 0) = \lim_{\epsilon \to 0} M'_\epsilon \int \cdot \int \exp\{-(1/\hbar)\sum_{n=0}^{N}[(x_{n+1} - x_n)^2/2\epsilon$$
$$+\epsilon V(x_n)]\}\prod_{n=1}^{N} dx_n \ . \qquad (3.37)$$

As an integral with a real, positive integrand, this expression is ideal for approximate evaluation by means of Monte Carlo studies. Extended to many degrees of freedom, such integrals—apart from the limit that $\epsilon \to 0$— are used to analyze Euclidean quantum field theories, and we will draw on that fact later in Chap. 9.

We emphasize that the formal expression (3.36) can also be regularized with the help of an appropriately pinned Wiener measure. In particular, such a Wiener measure $dW_{x'', x'}^T(x)$ is implicitly defined by the fact that

$$\int e^{i\int_0^T f(t)x(t) \, dt} \, dW_{x'', x'}^T(x) \qquad (3.38)$$

$$= e^{i\int_0^T f(t) [x' + (x'' - x')(t/T)] \, dt - \frac{1}{2}\int_0^T \int_0^T f(s) C(s,t) f(t) \, ds \, dt} \ ,$$

where $C(s,t) = \min(s,t) - st/T$. The Euclidean propagator then becomes

$$L(x'', T; x', 0) = \int e^{-\int_0^T V(x) \, dt} \, dW_{x'', x'}^T(x) \ , \qquad (3.39)$$

which is a genuine—and *not* formal—integral, commonly known as a Feynman-Kac path integral.

For the interested reader, there are a number of excellent books that are devoted to path integrals. Among these, we mention [Sch05; Roe96; Kl10a].

3.1.6 *Perturbation theory: successes and failures*

Successes

There are many problems of the type in which $\mathcal{H} = \mathcal{H}_0 + g\mathcal{V}$, where the eigenvalues and eigenfunctions of $\mathcal{H}_0$ are known and one wishes to find the eigenvalues and eigenfunctions of $\mathcal{H}$. In view of the fact that $\lim_{g\to 0} \mathcal{H} = \mathcal{H}_0$ for these problems, it is common to develop a *power-series expansion* of the eigen-elements of $\mathcal{H}$ about those of $\mathcal{H}_0$. For this purpose, there are standard expressions that involve the eigenvalues and eigenfunctions of $\mathcal{H}_0$ as well as the matrix elements of the potential $\mathcal{V}$ in the basis of the eigenfunctions of $\mathcal{H}_0$. As one example, such expressions may be found on Wikipedia [Wiki-d]; we do not pursue successes further.

Failures

Along with the many successes of traditional perturbation expansions involving continuous perturbations (say $\mathcal{V}$), there also are many perturbations (say $\mathcal{V}'$) in which a traditional perturbation analysis *fails* for the simple reason that the $\lim_{g\to 0}[\mathcal{H}_0 + g\mathcal{V}'] \equiv \mathcal{H}'_0 \neq \mathcal{H}_0$ thus involving a *discontinuous perturbation $\mathcal{V}'$*. Even as the perturbation $\mathcal{V}'$ is a discontinuous perturbation for $\mathcal{H}_0$, it is a *continuous perturbation* for $\mathcal{H}'_0$; this statement means that $\lim_{g\to 0} \mathcal{H}'_0 + g\mathcal{V}' = \mathcal{H}'_0$, and thus there may well be a successful, valid perturbation analysis of the interacting system with Hamiltonian $\mathcal{H}' = \mathcal{H}'_0 + g\mathcal{V}'$ expressed in terms of the eigen-elements of $\mathcal{H}'_0$. Let us focus on the discontinuous nature of the perturbation $\mathcal{V}'$ embodied in the formula $\lim_{g\to 0} \mathcal{H}_0 + g\mathcal{V}' = \mathcal{H}'_0$ and discuss the general properties of the pseudofree system $\mathcal{H}'_0$.

We have encountered examples of this kind already in Chap. 2, not for quantum systems, but already for classical systems. In particular, we saw that the classical action functional

$$A_g = \int_0^T \{ \tfrac{1}{2}[\dot{x}(t)^2 - \omega^2 x(t)^2] - g x(t)^{-4} \} \, dt \tag{3.40}$$

had the property that $\lim_{g\to 0} A_g \equiv \lim_{g\to 0}[A_0 + gA_I] \equiv A'_0 \neq A_0$. Here we imply that while the action functionals A_0' and A_0 have the same 'algebraic form'—specifically, $\frac{1}{2}\int_0^T [\dot{x}(t)^2 - \omega^2 x(t)^2]\, dt$—with the domain $D(A_0) \equiv \{x : \int [\dot{x}^2 + x^2]\, dt < \infty\}$ while the domain $D(A_0') \equiv \{x : \int [\dot{x}^2 + x^2 + x^{-4}]\, dt < \infty\}$, and thus, $D(A_0') \subset D(A_0)$, leading to the assertion that $A_0' \neq A_0$. It is important to appreciate that this classical example of a discontinuous perturbation also leads to a discontinuous perturbation in the quantum story as well.

The simplest description of the quantum differences between the free quantum system and the pseudofree quantum system is the following [Sim73]: As noted earlier when discussing the harmonic oscillator, the time-independent Schrödinger equation was given (for $m = 1$) by

$$\mathcal{H}_0 \psi_E(x) = -(\hbar^2/2)\partial^2 \psi_E(x)/\partial x^2 + (\omega^2/2)x^2 \psi_E(x) = E\psi_E(x) .$$

(3.41)

On the other hand, the time-independent Schrödinger equation for the pseudofree system is given by

$$\mathcal{H}_0' \phi_E(x) = -(\hbar^2/2)\partial^2 \phi_E(x)/\partial x^2_{DBC\,(0)} + \omega^2 x^2 \phi_E(x) = E\phi_E(x) ,$$

(3.42)

which is the same differential equation but subject to different boundary conditions. In particular, the notation $DBC\,(0)$ means vanishing 'Dirichlet boundary conditions' [Wiki-e] at $x = 0$; i.e., all eigen-solutions $\phi_E(x)$ to this differential equation satisfy $\phi_E(0) = 0$. These eigenfunctions must still span the same space of square integrable functions. Note that the Hermite functions with an *odd* index, i.e., $h_{2l+1}(x)$, $l \in \mathbb{N}$, have the property that they vanish at $x = 0$. On the other hand the Hermite functions with an *even* index, i.e., $h_{2l}(x)$, do not vanish at $x = 0$. Thus the odd functions are allowed but the even functions are disallowed in the new set of eigenfunctions for $\mathcal{H}_0'$. Nevertheless, we need to replace those even functions with other even functions that still satisfy the harmonic oscillator differential equation away from the origin. The obvious solution is to take the *odd* eigenfunctions and change their behavior to *even* functions simply by introducing $\tilde{h}_{2l}(x) \equiv sign(x)h_{2l+1}(x)$. Thus, using the fact that $h_n(-x) = (-1)^n h_n(x)$, we find the propagator for the pseudofree Hamiltonian $\mathcal{H}_0'$ is given (for $\hbar = \omega = 1$) by

$$K'(x'',T;x',0) = \theta(x''x')\sum_{n=0}^{\infty}[1 - (-1)^n]\,h_n(x'')h_n(x')e^{-i(n+1/2)T} ,$$

(3.43)

where $\theta(y) = 1$ when $y > 0$ and $\theta(y) = 0$ when $y < 0$. It follows that the eigenvalues are doubly degenerate having the values (properly all multiplied by $\hbar\omega$) $3/2, 7/2, 11/2, \ldots$. As promised, the pseudofree propagator, $K'(x'', T; x', 0)$, for $\mathcal{H}_0'$, clearly differs from the free propagator, $K(x'', T; x', 0)$, for $\mathcal{H}_0$.

As another example of a discontinuous perturbation, we consider

$$A_g'' = \int_0^T \{\tfrac{1}{2}[\dot{x}(t)^2 - \omega^2 x(t)^2] - g\,(x(t)^2 - c^2)^{-4}\}\, dt \,. \tag{3.44}$$

Suppose that $c = \pi$. It so happens that *no* Hermite function vanishes at $x = \pi$, and therefore *no* eigenfunction or eigenvalue for the usual harmonic oscillator will apply to the eigen-solutions of the new pseudofree Hamiltonian operator for this example $\mathcal{H}_0''$. Instead, the proper solutions are given as solutions to the differential equation

$$-(\hbar^2/2)\partial^2 \phi_E(x)/\partial x_{DBC\,(-c,c)}^2 + (\omega^2/2)x^2 \phi_E(x) = E\phi_E(x) \,, \tag{3.45}$$

where the boundary conditions require that the eigenfunctions $\phi_E(x)$ all vanish (for $c = \pi$) at $x = -\pi$ and $x = \pi$. Such eigenfunctions and eigenvalues exist, but the former are not given by known functions.

One can imagine other discontinuous interactions that lead to still other pseudofree Hamiltonian operators with any number of vanishing points required for the eigenfunctions. Each and every one of these pseudofree models illustrate examples of situations in which the interacting models are *not* continuously connected to their own free model but instead are connected to a suitable pseudofree model. Nevertheless, just as was the case for the classical models, if one perturbs the right pseudofree Hamiltonian operator with the right perturbation, the result, as the coupling constant vanishes, is that the interacting model is indeed continuously connected to the right pseudofree model in the quantum case as well. **This property of pseudofree models is a fundamentally important concept to embrace.**

3.2 Coherent States

Formal eigenvectors of the position operator Q, namely $|x\rangle$, where $Q|x\rangle = x|x\rangle$ and $\langle x|x'\rangle = \delta(x - x')$ have an important role to play in transforming the abstract expressions of quantum theory into familiar functional equations as well as differential equations. A similar role can be played by eigenvectors of the operator P, namely $|\kappa\rangle$, where—*changing previous notation*—$P|\kappa\rangle = \kappa|\kappa\rangle$ and $\langle\kappa|\kappa'\rangle = \delta(\kappa - \kappa')$, although they are so used only in special cases. These two sets of vectors are especially useful because they both generate natural resolutions of unity, specifically, when integrated over the whole real line, $\int|x\rangle\langle x|\,dx = \int|\kappa\rangle\langle\kappa|\,d\kappa = \mathbb{1}$, when the operators Q and P are jointly irreducible; secondly, despite not being genuine elements of Hilbert space, they are important because they are 'orthonormal' in a formal sense. Some other bases are discrete, such as the eigenvectors and eigenvalues of an harmonic oscillator $|n\rangle$, defined by $N|n\rangle = n|n\rangle$, $0 \leq n < \infty$, for the usual number operator N. These vectors satisfy $\Sigma_{n=0}^{\infty}|n\rangle\langle n| = \mathbb{1}$ and $\langle n|n'\rangle = \delta_{n,n'}$. Still other bases involve partly discrete and partly continuum eigenvalues and eigenvectors.

Coherent states, on the other hand, offer a different perspective on the issues as compared to the usual sets of eigenvectors and eigenvalues of self-adjoint operators such as Q, P, and N. In particular, coherent states, such as (the yet-to-be-defined vectors) $|p, q\rangle$ for all $(p, q) \in \mathbb{R}^2$, are genuine vectors in Hilbert space, and are typically—but not universally—normalized to unity. These vectors also exhibit a natural resolution of unity, e.g., as given by $\int |p, q\rangle\langle p, q| \, dp \, dq / 2\pi\hbar = \mathbb{1}$ under suitable conditions. Moreover, such vectors are *continuously parameterized*, i.e., if $(p, q) \to (p_0, q_0)$ as real numbers, then $|p, q\rangle \to |p_0, q_0\rangle$ in the strong sense, i.e., $\| \, |p, q\rangle - |p_0, q_0\rangle \, \| \to 0$. What is invariably lacking is the orthogonality of such vectors; while $\langle p, q|p', q'\rangle$ may sometimes vanish, that normally occurs on a set of measure zero, i.e., it almost never happens as examples will confirm. The lack of orthogonality proves to be less important than might be imagined at first glance. For general studies of coherent states, see, e.g., [KlS85; Per86].

Our first set of coherent states are known as *canonical coherent states* and are chosen as

$$|p, q\rangle \equiv e^{-iqP/\hbar} e^{ipQ/\hbar} |0\rangle \, , \tag{3.46}$$

where we select the unit vector $|0\rangle$—generally known as the *fiducial vector*—as the solution of the equation $(\omega Q + iP)|0\rangle = 0$. In the x-representation, it follows that $\langle x|0\rangle = (\omega/\pi\hbar)^{1/4} \exp(-\omega x^2/2\hbar)$; shortly, we will choose different vectors in place of $|0\rangle$, but for the moment, we stick with $|0\rangle$. The vector $|0\rangle$ is normalized, and we require that each of the coherent states $|p, q\rangle$ be normalized as well. This condition requires that the transformations of the fiducial vector be unitary transformations, and, in turn, the generators P and Q must be self-adjoint operators (and *not* simply Hermitian operators).

The physical meaning of the variables p and q follows from the fact that

$$\langle p, q| \, P \, |p, q\rangle = p \, , \qquad \langle p, q| \, Q \, |p, q\rangle = q \, . \tag{3.47}$$

The x-representation of the coherent states is given by

$$\langle x|p, q\rangle = (\omega/\pi\hbar)^{1/4} \, e^{ip(x - q)/\hbar - \omega(x - q)^2/2\hbar} \, , \tag{3.48}$$

which shows a translation of the vector by q and a phase change by p. Indeed, the meaning of these transformations is clarified if we consider the Fourier transform of $\langle x | p, q \rangle$ given by

$$\langle \kappa | p, q \rangle = (1/2\pi\hbar)^{1/2} \int e^{-i\kappa x/\hbar} \langle x | p, q \rangle \, dx$$

$$= (1/\pi\omega\hbar)^{1/4} e^{-iq\kappa/\hbar - (\kappa - p)^2/2\omega\hbar} . \tag{3.49}$$

In the κ-representation we see that p plays the role of a translation of the Fourier transform while q enters via a phase factor. That we still retain a Gaussian expression after the Fourier transform is a special property of the chosen fiducial vector; however, the validity of q serving to translate in the x-representation and for p serving to translate in the κ-representation is a general feature and does not depend on the fact that we have chosen a Gaussian function for the fiducial vector.

The overlap function of two of these coherent states is given by

$$\langle p, q | p', q' \rangle \tag{3.50}$$

$$= \exp\{i(p + p')(q - q')/2\hbar - (1/4\hbar)[\omega^{-1}(p - p')^2 + \omega(q - q')^2]\} ,$$

while the resolution of unity

$$\int |p, q\rangle \langle p, q| \, dp \, dq/(2\pi\hbar) = \mathbb{1} , \tag{3.51}$$

when integrated over the entire phase space, is easily verified (just evaluate $\int \langle x | p, q \rangle \langle p, q | x' \rangle \, dp \, dq/2\pi\hbar$).

Another interesting feature of coherent states is their interpolation between the x-representation and the κ-representation. To see this let us consider the *coherent-state representation* of a general state vector $|\psi\rangle$ given by

$$\psi(p, q) \equiv \langle p, q | \psi \rangle = (\omega/\pi\hbar)^{1/4} \int e^{-(\omega/2\hbar)(x - q)^2 - ip(x - q)/\hbar} \psi(x) \, dx$$

$$= (\omega/\pi\hbar)^{1/4} \int e^{-(\omega/2\hbar)x^2 - ipx/\hbar} \psi(x + q) \, dx , \tag{3.52}$$

where, as before, $\psi(x) = \langle x | \psi \rangle$. Observe that as ω becomes very large, the weighting function tends toward a delta function, while, on the other hand, as ω becomes very small, the Gaussian weighting tends to disappear. Introducing the proper rescaling, it follows that

$$\lim_{\omega \to \infty} (\omega/4\pi\hbar)^{1/4} \psi(p, q) = \psi(q) , \tag{3.53}$$

while in the other limit,

$$\lim_{\omega \to 0} (1/4\pi\hbar\omega)^{1/4} \psi(p, q) = e^{ipq/\hbar} \tilde{\psi}(p) , \tag{3.54}$$

where $\tilde{\psi}(p)$ denotes the Fourier transform of $\psi(x)$. Thus, in this sense, the coherent-state representation interpolates between the 'x' and the 'κ' representations (up to a phase factor).

3.2.1   *Other fiducial vectors*

Here we discuss the use of different fiducial vectors than $|0\rangle$ as used above. In particular, we rename the fiducial vector as $|\eta\rangle$, which is taken as a general unit vector, and the coherent states are now defined as

$$|p,q\rangle \equiv |p,q;\eta\rangle \equiv e^{-iqP/\hbar}\,e^{ipQ/\hbar}|\eta\rangle\,, \qquad (3.55)$$

where—letting $|\eta\rangle$ be implicit—the x-representation is given by

$$\langle x|p,q\rangle = e^{ip(x-q)/\hbar}\,\eta(x-q)\,, \qquad (3.56)$$

which illustrates the translation effect of q. In like manner, the coherent-state representation of a general vector $|\psi\rangle$ is given by

$$\langle p,q|\psi\rangle = \int \langle p,q|x\rangle \langle x|\psi\rangle\,dx = \int \eta(x)^*\,e^{-ipx/\hbar}\,\psi(x+q)\,dx\,. \qquad (3.57)$$

In turn, this equation implies the interesting relation

$$(2\pi\hbar)^{-1}\eta^*(0)^{-1}\int\psi(p,q)\,dp = \psi(q)\,, \qquad (3.58)$$

meaning that the Schrödinger representation is readily recovered from the coherent-state representation by just 'integrating out the unwanted variable p' (provided that $\eta^*(0) \neq 0$; otherwise add a phase factor).

Although $|\eta\rangle$ is a general vector and changes the coherent states markedly, it it commonly the case that $|\eta\rangle$ is left implicit in the notation, as is clear in $|p,q\rangle$, instead of the more informative notation $|p,q;\eta\rangle$. However, that is both a 'good point' and a 'bad point' of the Dirac bra-ket notation; for example, the notation for the vector $|17\rangle$ is good because it presumably only deals with essentials; however, '17' tells you nothing about what it refers to because the meaning is assumed to be clear from the context.

The physical meaning of the variables p and q is much as before since

$$\langle p,q|P|p,q\rangle = p + \langle P\rangle\,, \qquad \langle p,q|Q|p,q\rangle = q + \langle Q\rangle\,, \qquad (3.59)$$

where $\langle(\cdot)\rangle \equiv \langle\eta|(\cdot)|\eta\rangle$. It is often useful to choose $|\eta\rangle$ so that $\langle P\rangle = \langle Q\rangle = 0$, a process called *physical centering*.

The resolution of unity for the coherent states is exactly as before for any choice of the fiducial vector, namely

$$\int |p,q\rangle\langle p,q|\,dp\,dq/(2\pi\hbar) = \mathbb{1}\,, \qquad (3.60)$$

and the coherent-state vectors are continuously parameterized as before. The coherent-state representation $\psi(p,q) \equiv \langle p,q|\psi\rangle$ implicitly depends on the choice of $|\eta\rangle$, and thus matrix elements such as $\langle p,q|\mathcal{H}|p',q'\rangle$ clearly also depend on the fiducial vector. However, there is a pleasant surprise in this story. From the definition of the coherent states, it follows that $\psi(p,q) = \langle \eta | e^{-ipQ/\hbar} e^{iqP/\hbar} | \psi \rangle$. In this case we see that

$$-i\hbar(\partial/\partial q)\,\psi(p,q) = \langle \eta | e^{-ipQ/\hbar} e^{iqP/\hbar} P | \psi \rangle \,, \qquad (3.61)$$

$$(q + i\hbar(\partial/\partial p))\,\psi(p,q) = \langle \eta | e^{-ipQ/\hbar} e^{iqP/\hbar} Q | \psi \rangle \,, \qquad (3.62)$$

which holds for a general fiducial vector. Thus, for example, Schrödinger's time-dependent equation in the coherent-state representation becomes

$$i\hbar\partial\psi(p,q,t)/\partial t = \mathcal{H}(-i\hbar(\partial/\partial q), q + i\hbar(\partial/\partial p))\,\psi(p,q,t) \,, \qquad (3.63)$$

which apparently does not involve the fiducial vector. However, when one seeks to impose the initial condition, say at $t = 0$, it is necessary to assert that $\psi(p,q,0) \equiv \langle p,q;\eta|\psi\rangle$, which is how the fiducial vector enters the story; note that we have added the fiducial vector explicitly this time to ensure that it enters the story properly.

In fact, the equation (3.63) actually implies the usual Schrödinger equation. From (3.58) it follows that we can recover the Schrödinger equation by integrating out the unwanted variable p. Assuming that the potential term is a polynomial for clarity, we can perform multiple integrations by parts to eliminate the derivatives with respect to p leaving behind the precise form of the traditional time-dependent Schrödinger equation!

The coherent-state representation of the propagator follows the form of the configuration-space representation of the propagator. In particular,

$$\psi(p',q',T) = \int K(p',q',T;p,q,0)\,\psi(p,q,0)\,dp\,dq/(2\pi\hbar) \,. \qquad (3.64)$$

For a system with a discrete spectrum, we may assume that the x-representation of the eigenfunctions is $\xi_n(x) = \langle x|\xi_n\rangle$, $0 \le n < \infty$, and the coherent-state representation of these vectors is given by $\xi_n(p,q) = \int \langle p,q|x\rangle \xi_n(x)\,dx$. In particular, the coherent-state propagator in this case is given by

$$K(p,q,T;p',q',0) = \sum_{n=0}^{\infty} \xi_n(p,q)\,\xi_n(p',q')^* \, e^{-iE_nT/\hbar} \,, \qquad (3.65)$$

and this relation holds for any choice of $|\eta\rangle$.

There are also coherent-state path-integral formulations as well. In a coherent-state representation, the variables p and q are *mean* values in the coherent states and *not* eigenvalues of any operator, Thus they can be specified together at the same time, which on a temporal lattice implies they could be indexed together as (p_n, q_n). Applying a different lattice regularization to the formal expression (3.29) provides the key to giving a regularized expression for the coherent-state propagator. In one version, it follows that

$$K(p'', q'', T; p', q', 0) = \lim_{\epsilon \to 0} \int \cdot \int \exp\{(i/\hbar)\textstyle\sum_{n=0}^{N}[(p_{n+1} + p_n)(q_{n+1} - q_n)/2$$

$$-\epsilon H((p_{n+1} + p_n)/2, (q_{n+1} + q_n)/2)]\} \textstyle\prod_{n=1}^{N} dp_n \, dq_n/(2\pi\hbar) \,, \tag{3.66}$$

where $\epsilon = T/N$, $(p_{N+1}, q_{N+1}) = (p'', q'')$, and $(p_0, q_0) = (p', q')$.

There is another way to generate the coherent-state propagator for the fiducial vector $|0\rangle$ defined by $(Q + iP)|0\rangle = 0$ using the coherent-state vectors $|p, q\rangle = e^{-iqP/\hbar} e^{ipQ/\hbar} |0\rangle$. For the Hilbert-space functional representation defined by $\psi(p, q) \equiv \langle p, q|\psi\rangle$ it follows that

$$[i\hbar(\partial/\partial p) - \hbar(\partial/\partial q) + ip]\,\psi(p, q)$$

$$= [i\hbar(\partial/\partial p) - \hbar(\partial/\partial q) + ip]\langle 0|e^{-ipQ/\hbar} e^{iqP/\hbar}|\psi\rangle$$

$$= \langle 0|(Q - iP)\,e^{-ipQ/\hbar} e^{iqP/\hbar}|\psi\rangle$$

$$= 0 \tag{3.67}$$

for all elements $\psi(p, q)$! Such a condition is called a *complex polarization*. Observe that the operator $B \equiv [i\hbar(\partial/\partial p) - \hbar(\partial/\partial q) + ip]$ is first order in derivatives, and thus the operator

$$A \equiv B^\dagger B = (-i\hbar\partial/\partial p)^2 + (-i\hbar\partial/\partial q - p)^2 - \hbar \,, \tag{3.68}$$

is a nonnegative, second order, self-adjoint operator on $L^2(\mathbb{R}^2)$, which has the appearance of a Hamiltonian operator for *two coordinates*. Moreover, the zero eigenspace for the operator A is the Hilbert space composed of the elements $\psi(p, q) \equiv \langle p, q|\psi\rangle$. We can create a projection onto the zero subspace by considering $\lim_{\nu \to \infty} \exp[-\frac{1}{2}\nu A T/\hbar]$ for $0 < T < \infty$. It follows that

$$(2\pi\hbar) \lim_{\nu \to \infty} e^{-\frac{1}{2}\nu A T/\hbar} \delta(p - p')\delta(q - q') = \langle p, q|p', q'\rangle \,, \tag{3.69}$$

i.e., the coherent-state overlap function which is a projection operator onto the space of functions $\{\psi(p, q)\}$. A path integral for this expression can be given as

$$\langle p'', q''|p', q'\rangle$$

$$= \lim_{\nu \to \infty} \mathcal{M} \int \cdot \int e^{i\int_0^T (x\dot{p} + k\dot{q})\,dt/\hbar - \frac{1}{2}\nu\int_0^T [(k-p)^2 + x^2]\,dt/\hbar}$$

$$\times \mathcal{D}k\,\mathcal{D}x\,\mathcal{D}p\,\mathcal{D}q \tag{3.70}$$

$$= \lim_{\nu \to \infty} \mathcal{N} \int \cdot \int e^{i\int_0^T p\dot{q}\,dt/\hbar - \frac{1}{2}\nu^{-1}\int_0^T [\dot{p}^2 + \dot{q}^2]\,dt/\hbar}\,\mathcal{D}p\,\mathcal{D}q \,,$$

with boundary conditions that $(p(T), q(T)) = (p'', q'')$ and $(p(0), q(0)) = (p', q')$. This result resembles a Feynman-Kac integral for a vanishing potential. We can include a potential, which we choose to call $h(p, q)$, still regarding both p and q as 'configuration space coordinates'. In particular, a potential term leads to

$$K(p'', q'', T; p', q', 0) = \langle p'', q''| e^{-i\mathcal{H}(P,Q)T/\hbar} |p', q'\rangle \qquad (3.71)$$

$$= \lim_{\nu \to \infty} \mathcal{M} \int \cdot \int e^{i\int_0^T [p\dot{q} - h(p,q)] \, dt/\hbar} e^{-\frac{1}{2}\nu^{-1}\int_0^T [\dot{p}^2 + \dot{q}^2] \, dt/\hbar} \, \mathcal{D}p\,\mathcal{D}q \,,$$

subject to the previous boundary conditions. In this case the function $h(p, q)$ is related to the quantum operator $\mathcal{H}(P, Q)$ according to the relation

$$\mathcal{H}(P, Q) = \int h(p, q) |p, q\rangle\langle p, q| \, dp\,dq/(2\pi\hbar) \,. \qquad (3.72)$$

A precise description of the coherent-state propagator (3.71) involves a pinned Wiener measure $dW_{p'',q'';p',q'}^{\nu,T}(p, q)$ concentrated on continuous phase-space paths $p(t)$ and $q(t)$, $0 \le t \le T$, pinned so that $(p(T), q(T)) = (p'', q'')$ and $(p(0), q(0)) = (p', q')$, and which involves the diffusion constant ν for which, for example,

$$\int [p(s'') - p(s')]^2 dW_{p'',q'';p',q'}^{\nu,T}(p, q) = \nu |s'' - s'| \hbar \,. \qquad (3.73)$$

Based on this measure, the coherent-state propagator is given, for $|\eta\rangle = |0\rangle$ and where $(Q + iP)|0\rangle = 0$, by

$$K(p'', q'', T; p', q', 0) = \langle p'', q''; 0| e^{-i\mathcal{H}T/\hbar} |p', q'; 0\rangle \qquad (3.74)$$

$$= \lim_{\nu \to \infty} (2\pi\hbar) \, e^{\nu T/2} \int e^{(i/\hbar)\int_0^T [p\,dq - h(p,q)\,dt]} \, dW_{p'',q'';p',q'}^{\nu,T}(p, q) \,,$$

where $\int p\,dq$ denotes a (midpoint) Stratonovich stochastic integral [Hid70]. In this case $h(p, q)$ is again related to the Hamiltonian operator $\mathcal{H}(P, Q)$ by

$$\mathcal{H}(P, Q) = \int h(p, q) |p, q\rangle\langle p, q| \, dp\,dq/(2\pi\hbar) \,, \qquad (3.75)$$

and for all $\nu < \infty$, this expression represents a genuine path integral as signalled by a finite (non-formal) normalization factor for all $\nu < \infty$. For purposes of canonical coordinate transformations, the Stratonovich rule for the stochastic integral $\int p(t) \, dq(t)$ is convenient since, in that case, the rules of ordinary calculus hold. Consequently, the expression (3.74) is covariant under general canonical coordinate transformations. Such a representation is said to involve a *continuous-time regularization*. For additional discussion of this expression, see [DaK85; Kl10a].

We will revisit coherent states in the following chapter, and we take that opportunity to introduce several other varieties of coherent states.

PART 2
Enhanced Quantization of Simple Systems

Chapter 4

Essentials of Enhanced Quantization

WHAT TO LOOK FOR

Although classical mechanics and quantum mechanics are separate disciplines, we live in a world where Planck's constant $\hbar > 0$, meaning that the classical and quantum world views must both come from the same principles, i.e., they must *coexist*. Traditionally, canonical quantization procedures postulate a direct linking of various c-number and q-number quantities that lie in disjoint realms, along with the quite distinct interpretations given to each realm. In this chapter we propose a different association of classical and quantum quantities that renders classical theory a natural subset of quantum theory letting them coexist as required. This proposal also shines light on alternative linking assignments of classical and quantum quantities that offer different perspectives on the very meaning of quantization. In this chapter we focus on elaborating the general principles, while in later chapters we offer examples of what this alternative viewpoint can achieve; these examples include removal of singularities in classical solutions to certain models, and an alternative quantization of several field theory models that are trivial when quantized by traditional methods but become well defined and nontrivial when viewed from the new viewpoint.

4.1 Scope of the Problem

The most common approach to a quantum theory is through Schrödinger's equation

$$i\hbar\, \partial\psi(x,t)/\partial t = \mathcal{H}(-i\hbar\, \partial/\partial x, x)\, \psi(x,t) \,, \tag{4.1}$$

illustrated for a single degree of freedom. Here the function $\mathcal{H}(p, q)$ generally differs from the classical (c) Hamiltonian $H_c(p, q)$ by terms of order $\hbar$, the coordinates $p \to -i\hbar\partial/\partial x$ and $q \to x$, and $\int_{-\infty}^{\infty} |\psi(x, t)|^2 \, dx = 1$. A similar prescription applies to classical systems with N degrees of freedom, $N \leq \infty$. Although this scheme is widely successful, there are certain questionable aspects. Generally, the given procedure works well only for certain canonical coordinate systems, namely, for "Cartesian coordinates" [Dir58] (page 114), which is reproduced in Sec. 3.1.1, despite the fact that the classical phase space has a symplectic structure (e.g., $dp \, dq$, interpreted as an element of surface area) but *no* metric structure (e.g., $\omega^{-1}dp^2 + \omega \, dq^2$) [Wiki-f] with which to identify Cartesian coordinates. Moreover, for certain classical systems, and even when using the correct coordinates to provide a canonical quantization, a subsequent classical limit in which $\hbar \to 0$ leads to a manifestly *different* classical system from the original one, thus violating the eminently natural rule [LaL77] (page 3), which is quoted in our preface, that de-quantization should lead back to the original classical system. Finally, the classical framework for which $\hbar = 0$ is fundamentally different from the quantum framework for which $\hbar > 0$, and it is the latter realm that characterizes the real world.

There are several arguments that support a new approach. First, by analogy, note that the real world is relativistic in character, but a nonrelativistic approximation to classical mechanics can be made within relativistic classical mechanics *without changing the formalism* and keeping the speed of light c fixed and finite. Likewise, since the real world is also governed by quantum mechanics, it is necessary that classical mechanics somehow be contained within quantum mechanics in such a way that *it involves the same formalism* and keeps the reduced Planck's constant $\hbar = h/2\pi$ fixed and nonzero. Second, the canonical prescription on how a classical system should be quantized leads, for some problems—e.g., self-interacting scalar fields φ_n^4, for spacetime dimensions $n \geq 5$ (and possibly $n = 4$ as well)—to unnatural behavior in that the classical limit of a conventional quantization does *not* reduce to the original classical model when $\hbar \to 0$! As we shall illustrate, this serious discrepancy can be overcome with procedures that are discussed in the present monograph. However, we do not need to invoke complicated examples to learn some of the advantages of a new way to link classical and quantum systems together.

The next two sections are, in a certain sense, *the heart of this work*. We present (essentially) the same material **twice**: first, in the commonly used functional form of quantum mechanics with wave functions and differential equations, and second, in the abstract formulation of quantum mechanics with bras, kets, and operators. This duplication is done to ensure that at least one of the presentations fits the readers preferred taste. Also, there is always the advantage of receiving an important message twice.

4.2 Enhanced Quantization: Take One

4.2.1 *Enhanced canonical quantization*

As discussed above, we shall propose a different manner of quantization—called *Enhanced Quantization*—that *keeps all the good results of conventional canonical quantization, but offers better solutions when needed.* To establish enhanced quantization, we start with the quantum action functional A_Q given, for a single degree of freedom and normalized wave functions, by

$$A_Q = \int_0^T \int_{-\infty}^{\infty} \psi(x,t)^* \left[i\hbar(\partial/\partial t) - \mathcal{H}(-i\hbar\partial/\partial x, x) \right] \psi(x,t)\, dx\, dt \quad (4.2)$$

from which Schrödinger's equation (4.1) may be derived by a general stationary variation, $\delta A_Q = 0$, provided that $\delta\psi(x,0) = 0 = \delta\psi(x,T)$. Such variations correspond to variations that can actually be realized in practice, but, in some situations, not all variations are possible. As an example, consider a *microscopic system*: then while *microscopic observations* can make sufficiently general variations to deduce Schrödinger's equation, *macroscopic observations* are confined to a much smaller subset of possible variations. For example, we include only those variations that can be realized *without disturbing the observed system*, such as changes made in accordance with Galilean invariance, namely, a change in position by q and a change in momentum by p (as realized by a change in velocity); note that no disturbance of the observed system need occur since we can instead translate the reference system of the observer. Choosing a foundation on which to build, we select a basic normalized function—call it $\eta(x)$—which, when transported by p and q as noted above, gives rise to a family of functions $\eta_{p,q}(x) \equiv e^{ip(x-q)/\hbar}\eta(x-q)$, where $-\infty < p, q < \infty$ [note: $\eta_{0,0}(x) \equiv \eta(x)$]. Within the context of quantum mechanics, and as discussed in the previous chapter, these functions are also well known as *canonical coherent states* [KlS85], and the basic function $\eta(x)$ is generally referred to as the *fiducial vector*, all expressed here in the Schrödinger representation. While not required, it is useful to impose $\int_{-\infty}^{\infty} x|\eta(x)|^2 dx = 0$ as well as $\int_{-\infty}^{\infty} \eta(x)^* \eta'(x)\, dx = 0$, called "physical centering", which then leads to

$$\int_{-\infty}^{\infty} x \, |\eta_{p,q}(x)|^2 \, dx = q \,, \qquad -i\hbar \int_{-\infty}^{\infty} \eta_{p,q}(x)^* \, \eta'_{p,q}(x) \, dx = p \,, \qquad (4.3)$$

two relations that fix the physical meaning of p and q, independently of the basic function $\eta(x)$. Finally, being unable as macroscopic observers to vary the functions $\psi(x,t)$ in (4.2) arbitrarily, we restrict (R) the set of allowed variational states so that $\psi(x,t) \rightarrow \eta_{p(t),q(t)}(x)$, which leads to

$$A_{Q(R)} = \int_0^T \int_{-\infty}^{\infty} \eta_{p(t),q(t)}(x)^* \, [i\hbar(\partial/\partial t) - \mathcal{H}(-i\hbar\partial/\partial x, x)] \, \eta_{p(t),q(t)}(x) \, dx \, dt$$

$$= \int_0^T [p(t)\dot{q}(t) - H(p(t), q(t))] \, dt \,, \qquad (4.4)$$

where

$$\int_{-\infty}^{\infty} \eta_{p(t),q(t)}(x)^* \, [i\hbar(\partial/\partial t)] \eta_{p(t),q(t)}(x) \, dx$$

$$= \int_{-\infty}^{\infty} \eta(x)^* \{ [p(t) - i\hbar\partial/\partial x] \dot{q}(t) - \dot{p}(t) x \} \eta(x) \, dx$$

$$= p(t)\dot{q}(t) \qquad (4.5)$$

and

$$H(p,q) \equiv \int_{-\infty}^{\infty} \eta_{p,q}(x)^* \, \mathcal{H}(-i\hbar\partial/\partial x, x) \eta_{p,q}(x) \, dx$$

$$= \int_{-\infty}^{\infty} \eta(x)^* \mathcal{H}(p - i\hbar\partial/\partial x, q + x) \, \eta(x) \, dx \,. \qquad (4.6)$$

Because $\hbar > 0$ still, we call (4.4) [including (4.5) and (4.6)] the *enhanced classical action functional*. Assuming, for clarity, that the Hamiltonian operator is a polynomial, it readily follows that

$$H(p,q) = \mathcal{H}(p,q) + \mathcal{O}(\hbar; p, q) \qquad (4.7)$$

for many choices of the basic function $\eta(x)$. In such cases, the strictly classical action functional arises if the limit $\hbar \rightarrow 0$ is applied to (4.4) and (4.6).

Normally, showing that the quantum theory implies the classical theory is a somewhat convoluted exercise, seeking analog equations in both realms, such as commutators and Poisson brackets, and any convincing conclusion invariably requires macroscopic limits or $\hbar \rightarrow 0$. In obtaining the enhanced classical action functional directly from the quantum action functional and keeping $\hbar$ fixed, we have focused on a particular subset of variations, those known as canonical coherent states, which do not involve disturbing the system. But it should be appreciated that in fully varying the quantum action to derive the quantum equations of motion, we are implicitly varying as well, among many other states, the very subset that leads to the enhanced classical action functional. In other words, in the present derivation of the classical story we are simply restricting ourselves to a subset of the entire set of variations that describe the quantum story. *Consequently, we claim that "Classical Physics ⊂ Quantum Physics" in a very direct sense!*

The difference between canonical quantization and enhanced quantization may be briefly summarized: In enhanced quantization the proper classical theory is obtained from the quantum theory merely by restricting the variations to the coherent states, and there is no need whatsoever of initially starting with a classical theory, promoting some variables to operators, and confirming our quantum structure by regaining the classical theory by taking the limit $\hbar \to 0$, which is the rule in canonical quantization.

The introduction of classical canonical coordinate transformations—for example, $(p, q) \to (\tilde{p}, \tilde{q})$ such that $p\,dq = \tilde{p}\,d\tilde{q} + d\tilde{G}(\tilde{p}, \tilde{q})$, and for which we require that $\tilde{\eta}_{\tilde{p},\tilde{q}}(x) \equiv \eta_{p(\tilde{p},\tilde{q}),q(\tilde{p},\tilde{q})}(x) = \eta_{p,q}(x)$—leads to an enhanced classical action functional properly expressed in the new coordinates, given by

$$\int_0^T \int_{-\infty}^{\infty} \tilde{\eta}_{\tilde{p}(t),\tilde{q}(t)}(x)^* \left[i\hbar(\partial/\partial t) - \mathcal{H}(-i\hbar\partial/\partial x, x) \right] \tilde{\eta}_{\tilde{p}(t),\tilde{q}(t)}(x)\, dx\, dt$$

$$= \int_0^T \left[\tilde{p}(t)\,\dot{\tilde{q}}(t) + \dot{\tilde{G}}(\tilde{p}(t), \tilde{q}(t)) - \tilde{H}(\tilde{p}(t), \tilde{q}(t)) \right] dt\,, \tag{4.8}$$

but with *absolutely no change* of the quantum formalism—nor of the physics—whatsoever. Thus, in enhanced quantization, we have achieved full canonical-coordinate covariance of the enhanced classical action functional while at the same time maintaining the complete integrity of the underlying quantum theory. No corresponding relationship of complete and independent canonical coordinate transformation invariance exists within the canonical quantization formalism.

Stationary variation of the enhanced classical action functional leads to Hamilton's equations of motion based on the enhanced classical Hamiltonian $H(p, q)$. In this sense we have shown that a suitably restricted domain of the *quantum action functional*, consisting of a two-dimensional, continuously connected sheet of normalized functions, leads to an enhanced *canonical classical action functional*, with the benefit that $\hbar > 0$ still, and, in this way, we have achieved the goal of *embedding classical mechanics within quantum mechanics!* Moreover, the enhanced classical equations of motion may have $\mathcal{O}(\hbar; p, q)$ correction terms—the form of which would be dictated by (4.6)—that may be of interest in modifying the strictly classical solutions in interesting ways. An example of this behavior is offered below.

In a crude sense, and for a reasonable range of (p, q), elements of the set of states $\{\eta_{p,q}(x)\}$ act like the illumination from a flashlight used by a burglar in peering through the window of a deserted house on a pitch black night; indeed, the role of physically centering the function $\eta(x)$ is like ensuring the flashlight is aimed through the window and not at the brick wall. Changing the basic function $\eta(x)$ is like changing the orientation or the cone of illumination of the flashlight. Quantum mechanically, whatever the choice of $\eta(x)$, the set of functions $\{\eta_{p,q}(x)\}$, for all (p,q), span the space of all square-integrable functions, $L^2(\mathbb{R}) = \{\psi(x) : \int |\psi(x)|^2 \, dx < \infty\}$. In further analysis of the enhanced classical theory, however, some choices of $\eta(x)$ may be better than others.

A common choice for $\eta(x)$—and one that acts like a bright, narrow-beam flashlight—is given, with $\omega > 0$, by

$$\eta(x) = (\omega/\pi\hbar)^{1/4} e^{-\omega x^2/2\hbar} \tag{4.9}$$

for which $H(p,q)$ satisfies (4.7), i.e., $\mathcal{H}(p,q) = H(p,q)$ up to terms of order $\hbar$. This last property is *exactly* what is meant by having "Cartesian coordinates" [Dir58], although such coordinates can not originate from the classical phase space. However, Cartesian coordinates do have a natural origin from an enhanced quantization viewpoint. The set of allowed variational states $\{\eta_{p,q}(x)\}$—a set of canonical coherent states as noted earlier—forms a two-dimensional, continuously connected sheet of normalized functions within the set of normalized square-integrable functions, and a natural metric can be given for such functions. Since the overall phase of any wave function carries no physics, the (suitably scaled) square of the distance between two coherent-state rays is given by

$$d_R^2(p',q';p,q) \equiv (2\hbar) \min_\alpha \int_{-\infty}^{\infty} |\eta_{p',q'}(x) - e^{i\alpha} \eta_{p,q}(x)|^2 \, dx \,, \tag{4.10}$$

which for two infinitesimally close coherent-state rays becomes (this is also the Fubini-Study metric [Wiki-g])

$$d\sigma^2(p,q) \equiv (2\hbar)[\int |d\eta_{p,q}(x)|^2 \, dx - |\int \eta_{p,q}(x)^* \, d\eta_{p,q}(x) \, dx \,|^2] \,, \tag{4.11}$$

and for $\eta(x) = (\omega/\pi\hbar)^{1/4} \exp(-\omega x^2/2\hbar)$, (4.11) reduces to

$$d\sigma^2(p,q) = \omega^{-1} dp^2 + \omega \, dq^2 \,, \tag{4.12}$$

which ensures that p and q are indeed Cartesian coordinates. Although this metric originates in the quantum theory with the canonical coherent states, it may also be assigned to the classical phase space as well.

NOTE WELL: At this point WE HAVE RECREATED CONVENTIONAL CANONICAL QUANTIZATION in that we have identified canonical variables p and q that behave as Cartesian coordinates, and for which the quantum Hamiltonian is effectively

the same as the conventionally chosen one—particularly for clas-
sical Hamiltonians of the form $p^2/(2m) + V(q)$, and especially so if
ω in (4.9) is chosen very large. THUS, ENHANCED QUAN-
TIZATION CAN REPRODUCE CONVENTIONAL CANON-
ICAL QUANTIZATION—BUT IT HAS OTHER POSITIVE
FEATURES AS WELL.

Since enhanced quantization yields the same results as standard methods for many systems, there is no need to focus on those cases; instead, we
seek the new treasures that enhanced quantization has to offer.

4.2.2 *Enhanced affine quantization*

Classical canonical variables p and q fulfill the Poisson bracket $\{q, p\} = 1$,
which translates to the Heisenberg commutation rule $[x, -i\hbar(\partial/\partial x)] = i\hbar$;
these operators generate the two transformations that characterize the
canonical coherent states. Multiplying the Poisson bracket by q leads to
$\{q, pq\} = q$, which corresponds to $-(i\hbar/2)[x, x(\partial/\partial x) + (\partial/\partial x)x] = i\hbar x$ after both sides of the commutator are multiplied by x. This expression, *de-
rived from the Heisenberg commutation relation*, is known as an affine commutation relation between affine variables. While the operator $-i\hbar(\partial/\partial x)$
acts to generate *translations* of x, the operator $-(i\hbar/2)[x(\partial/\partial x) + (\partial/\partial x)x]$
acts to generate *dilations* of x. If one deals with a classical variable
$q > 0$ and its quantum analog $x > 0$ (both chosen dimensionless for
convenience), then the canonical coherent states are unsuitable and we
need a different set of coherent states. We choose a new basic func-
tion $\xi(x) \equiv M x^{\tilde{\beta}/\hbar - 1/2} e^{-\tilde{\beta} x/\hbar}$, $\tilde{\beta} > 0$ and $x > 0$, with M a nor-
malization factor, for which it also follows that $\int_0^\infty x|\xi(x)|^2\, dx = 1$ and
$\int_0^\infty \xi(x)^*[x(\partial/\partial x) + (\partial/\partial x)x]\xi(x)\, dx = 0$. We also introduce suitable
affine coherent states as $\xi_{p,q}(x) \equiv q^{-1/2} e^{ipx/\hbar} \xi(x/q)$, where $q > 0$ and
$-\infty < p < \infty$ [note: $\xi_{0,1}(x) \equiv \xi(x)$]. The physical meaning of the variables p and q follows from the relations

$$q = \int_0^\infty \xi_{p,q}(x)^* x \xi_{p,q}(x)\, dx \,, \tag{4.13}$$

$$pq = -\tfrac{1}{2} i\hbar \int_0^\infty \xi_{p,q}^*(x)[x(\partial/\partial x) + (\partial/\partial x)x)]\xi_{p,q}(x)\, dx \,. \tag{4.14}$$

The affine coherent states involve translation in Fourier space by p and
dilation—i.e., (de)magnification, partially realized by a magnifying glass—
in configuration space by q, and they describe a new, two-dimensional,
continuously connected sheet of normalized functions.

The quantum (Q) action functional on the half line $x > 0$ is given by

$$A'_Q = \int_0^T \int_0^\infty \phi(x,t)^* \left[i\hbar(\partial/\partial t) - \mathcal{H}(-i\hbar \partial/\partial x, x) \right] \phi(x,t) \, dx \, dt \, , \quad (4.15)$$

and a suitable stationary variation leads to Schrödinger's equation. Restricted (R) to the affine coherent states, we find that

$$
\begin{aligned}
A'_{Q(R)} &= \int_0^T \int_0^\infty \xi_{p(t),q(t)}(x)^* \left[i\hbar(\partial/\partial t) - \mathcal{H}(-i\hbar \partial/\partial x, x) \right] \xi_{p(t),q(t)}(x) \, dx \, dt \\
&= \int_0^T \left[-q(t)\dot{p}(t) - H(p(t), q(t)) \right] dt \, ,
\end{aligned}
\quad (4.16)
$$

where

$$
\begin{aligned}
&\int_0^\infty \xi_{p(t),q(t)}(x)^* \left[i\hbar(\partial/\partial t) \right] \xi_{p(t),q(t)}(x) \, dx \\
&= \int_0^\infty \xi(x)^* \{ -q(t)\dot{p}(t)\,x + \tfrac{1}{2}[q(t)^{-1}\dot{q}(t)]\,[x(\partial/\partial x) + (\partial/\partial x)x] \}\xi(x) \, dx \\
&= -q(t)\dot{p}(t)
\end{aligned}
\quad (4.17)
$$

and

$$
\begin{aligned}
H(p,q) &\equiv \int_0^\infty \xi_{p,q}(x)^* \mathcal{H}(-i\hbar\partial/\partial x, x) \xi_{p,q}(x) \, dx \\
&= \int_0^\infty \xi(x)^* \mathcal{H}(p - iq^{-1}\hbar\partial/\partial x, qx) \, \xi(x) \, dx \, .
\end{aligned}
\quad (4.18)
$$

Equations (4.16), (4.17), and (4.18) strongly suggest that they correspond to an enhanced *canonical classical action functional*. In other words, enhanced quantization has found a *different two-dimensional sheet of normalized functions* that nevertheless exhibits a *conventional canonical system for its enhanced classical behavior!* Invariance under canonical coordinate transformations follows along the same lines as before. For this system, "Cartesian coordinates" are *not* appropriate; instead, the geometry of the affine coherent-state rays leads to a (suitably scaled) Fubini-Study metric given by

$$
\begin{aligned}
d\sigma^2(p,q) &\equiv (2\hbar)[\int |d\xi_{p,q}(x)|^2 \, dx - |\int \xi_{p,q}(x)^* \, d\xi_{p,q}(x) \, dx|^2] \\
&= \tilde{\beta}^{-1} q^2 \, dp^2 + \tilde{\beta} q^{-2} \, dq^2 \, ,
\end{aligned}
\quad (4.19)
$$

which is a space of constant negative curvature: $-2/\tilde{\beta}$ (a Poincaré half plane [Wiki-h]), a space that is geodesically complete. As with the canonical case, this new geometry can be added to the classical phase space if so desired.

While we have focussed on the case where $x > 0$, it is also possible to consider $x < 0$, and even a *reducible representation* case where $|x| > 0$. In this latter situation, enhanced affine quantization can, effectively, replace enhanced canonical quantization. Examples involving reducible affine representations appear in Chaps. 8 and 9.

Example of enhanced affine quantization

Consider the classical action functional for a single degree of freedom given by [model #(1) in Chap. 1]

$$A_C = \int_0^T [-q\dot{p} - qp^2] \, dt , \qquad (4.20)$$

with the physical requirement that $q > 0$. The classical solutions for this example are given by

$$p(t) = p_0 (1 + p_0 t)^{-1} , \qquad q(t) = q_0 (1 + p_0 t)^2 , \qquad (4.21)$$

where (p_0, q_0) denote initial data at $t = 0$. Although $q(t)$ is never negative, *every solution* with positive energy, $E = q_0 p_0^2 > 0$, becomes singular since $q(-p_0^{-1}) = 0$.

We like to believe that quantization of singular classical systems may, sometimes, eliminate the singular behavior, and let us see if that can occur for the present system. Conventional canonical quantization is ambiguous up to terms of order $\hbar$, and that makes it difficult to decide on proper nonclassical corrections when $\hbar > 0$. In contrast, enhanced quantization always keeps $\hbar > 0$ and points to quite specific nonclassical corrections. Adopting enhanced affine quantization, the enhanced classical action functional becomes

$$A_{Q(R)} = \int_0^T \int_0^\infty \xi_{p(t),q(t)}(x)^* [i\hbar(\partial/\partial t) - (-i\hbar\partial/\partial x)x(-i\hbar\partial/\partial x)]$$
$$\times \xi_{p(t),q(t)}(x) \, dx \, dt$$
$$= \int_0^T [-q(t)\dot{p}(t) - q(t)p(t)^2 - Cq(t)^{-1}] \, dt , \qquad (4.22)$$

where $C \equiv \hbar^2 \int_0^\infty x|\xi'(x)|^2 \, dx > 0$. With our convention that q and x are dimensionless, it follows that the dimensions of C are those of $\hbar^2$. While the numerical value of C may depend on $\xi(x)$, the modification of the classical Hamiltonian has just one term proportional to $\hbar^2 q^{-1}$, which guarantees that the enhanced classical solutions do indeed eliminate the divergences encountered in the strictly classical solutions.

This model, which represents a toy model of gravity—where p stands for minus the Christoffel symbol and $q > 0$ stands for the metric with its signature requirements—is based on [K1A70], which also includes additional details. Recently, Fanuel and Zonetti [FaZ13] have employed affine quantization to study specific cosmological gravity models, and have concluded that conventional classical singularities are removed upon affine quantization. A version of their work is presented in the following chapter.

4.3   Enhanced Quantization: Take Two

What follows is similar in many ways to the previous section. What is different is an emphasis on an abstract version of quantum theory, which we will use almost exclusively from now on.

4.3.1   *Enhanced canonical quantization*

We use the language of one-dimensional classical phase-space mechanics with well-chosen canonical coordinates p and q, which, of course, fulfill the Poisson bracket $\{q, p\} = 1$. Likewise, we adopt the language of abstract quantum mechanics that includes basic Hermitian operators such as P and Q, for which $[Q, P] = i\hbar \mathbb{1}$, and Hilbert space vectors $|\psi\rangle$, their adjoints $\langle\psi|$, and their inner products such as $\langle\phi|\psi\rangle$, which is linear in the right-hand vector and anti-linear in the left-hand vector. The kinematical part of canonical quantization "promotes" $q \to Q$ and $p \to P$, for which $(aq + bp) \to (aQ + bP)$, etc., where a and b are fixed coefficients. This form of assignment extends to functions of the basic operators such as $H_c(p, q) \to \mathcal{H} = H_c(P, Q)$, at least to leading order in $\hbar$ in the right coordinates. If we interpret $H_c(p, q)$ as the classical Hamiltonian, it follows that the classical dynamical equations are given by $dq/dt = \partial H_c(p, q)/\partial p$ and $dp/dt = -\partial H_c(p, q)/\partial q$, with solutions determined by initial conditions (p_0, q_0) at $t = 0$. In turn, dynamics in the quantum theory may be expressed in the form $i\hbar \partial|\psi\rangle/\partial t = \mathcal{H}|\psi\rangle$, with a solution fixed by the initial condition $|\psi_0\rangle$ at $t = 0$. The two sets of dynamical equations of motion arise from separate classical (C) and quantum (Q) action functionals given, respectively, by

$$A_C = \int_0^T [p(t)\, \dot{q}(t) - H_c(p(t), q(t))]\, dt \, , \qquad (4.23)$$

$$A_Q = \int_0^T \langle\psi(t)|\, [i\hbar(\partial/\partial t) - \mathcal{H}(P, Q)]\, |\psi(t)\rangle\, dt \, , \qquad (4.24)$$

where we may assume that the vectors $|\psi(t)\rangle$ are normalized to unity. The formalisms of Heisenberg and Feynman are equivalent to the foregoing, and they also deal with distinct realms for classical and quantum mechanics. We now ask: Is there another way to quantize in which the classical and quantum realms coexist?

We initially chose basic quantum kinematical variables P and Q that were assumed to be *Hermitian* [e.g., $P^\dagger = P$ on the domain $\mathcal{D}(P) \subseteq \mathcal{D}(P^\dagger)$], but now we confine attention to those operators that are *self adjoint* [e.g., $P^\dagger = P$ on the domain $\mathcal{D}(P) = \mathcal{D}(P^\dagger)$]. This restriction is introduced because only self-adjoint operators can serve as generators of unitary one-parameter groups. Thus, we can introduce two, basic unitary groups defined by $U(q) \equiv \exp(-iqP/\hbar)$ and $V(p) \equiv \exp(ipQ/\hbar)$ that satisfy

$$U(q)V(p) = e^{-ipq/\hbar} V(p)U(q) , \qquad (4.25)$$

which guarantees [GoT03] a representation of P and Q, with a spectrum for each operator covering the whole real line, unitarily equivalent to a Schrödinger representation, provided the operators are irreducible. We choose a fiducial unit vector $|0\rangle$ that satisfies the relation $(Q + iP)|0\rangle = 0$, which implies that $\langle 0|P|0\rangle = \langle 0|Q|0\rangle = 0$, and use it to define a set of canonical coherent states

$$|p,q\rangle \equiv e^{-iqP/\hbar} e^{ipQ/\hbar} |0\rangle , \qquad (4.26)$$

for all $(p,q) \in \mathbb{R}^2$, which are jointly continuous in p and q; for convenience, we let P and Q (and thus p and q) have the same dimensions, namely, those of $\hbar^{1/2}$. It may be helpful to picture the coherent states as a continuous, two-dimensional sheet of unit vectors coursing through an infinite-dimensional space of unit Hilbert-space vectors.

With regard to Eq. (4.24), Schrödinger's equation results from stationary variation of A_Q over general histories of Hilbert space unit vectors $\{|\psi(t)\rangle\}_0^T$, modulo fixed end points. However, *macroscopic* observers studying a *microscopic* system cannot vary vector histories over such a wide range. Instead, they are confined to moving the system to a new position (q) or changing its velocity (p) [based on $\dot{q} = \partial H_c(p,q)/\partial p$]. Thus a macroscopic observer—or better a *classical observer*—is restricted to vary only the quantum states $|p(t), q(t)\rangle$, leading to a restricted (R) version of A_Q given by

$$A_{Q(R)} = \int_0^T \langle p(t), q(t)| [i\hbar(\partial/\partial t) - \mathcal{H}] |p(t), q(t)\rangle \, dt$$
$$= \int_0^T [p(t)\dot{q}(t) - H(p(t), q(t))] \, dt , \qquad (4.27)$$

in which $i\hbar\langle p,q| d|p,q\rangle = \langle 0|[(P + p\mathbb{1})\,dq - Q\,dp]|0\rangle = p\,dq$ and $H(p,q) \equiv \langle p,q|\mathcal{H}|p,q\rangle$. The restricted quantum action has the appearance of A_C, and, besides a new parameter $(\hbar)$, results in Hamiltonian classical mechanics apart from one issue. In standard classical mechanics, p and q represent *sharp* values, while in (4.27) the values of p and q are *mean* values since $p = \langle p,q|P|p,q\rangle$ and $q = \langle p,q|Q|p,q\rangle$, each variable having a standard deviation of $\sqrt{\hbar/2}$. However, we may question the assumption of sharp values for p and q since no one has ever measured their values to a precision, say, of one part in 10^{137}. Thus we conclude that the restricted quantum action functional IS the classical action functional, and, since $\hbar > 0$ still, we can claim that *the classical theory is a subset of the quantum theory*, meaning that *the classical and quantum theories coexist!* The equations of motion that follow from (4.27) will generally include the parameter $\hbar$. If we are interested in the idealized classical (c) Hamiltonian for which $\hbar = 0$, we may subsequently consider

$$H_c(p, q) \equiv \lim_{\hbar \to 0} H(p, q) = \lim_{\hbar \to 0} \langle p, q | \mathcal{H} | p, q \rangle , \tag{4.28}$$

but it is likely to be more interesting to retain $H(p, q)$ with $\hbar > 0$ and see what difference $\hbar$ makes, if any, to the usual classical story.

Changing coordinates by canonical coordinate transformations is sometimes problematic in traditional quantum theory. However, no such problems arise in the present formulation. Canonical coordinate transformations involve the change $(p, q) \to (\widetilde{p}, \widetilde{q})$ for which $\{\widetilde{q}, \widetilde{p}\} = 1$ as well as $p \, dq = \widetilde{p} \, d\widetilde{q} + d\widetilde{G}(\widetilde{p}, \widetilde{q})$, where $\widetilde{G}$ is called the generator of the canonical transformation. The coherent states $|p, q\rangle$ were chosen as a map from a particular point in phase space labeled by (p, q) to a unit vector in Hilbert space. We want the same result in the new coordinates, so we define a canonical coordinate transformation of the coherent states to be as a scalar, namely

$$|\widetilde{p}, \widetilde{q}\rangle \equiv |p(\widetilde{p}, \widetilde{q}), q(\widetilde{p}, \widetilde{q})\rangle = |p, q\rangle , \tag{4.29}$$

and thus the restricted quantum action functional becomes

$$\begin{aligned} A_{Q(R)} &= \int_0^T \langle \widetilde{p}(t), \widetilde{q}(t) | \left[i\hbar(\partial/\partial t) - \mathcal{H} \right] | \widetilde{p}(t), \widetilde{q}(t) \rangle \, dt \\ &= \int_0^T [\widetilde{p}(t)\dot{\widetilde{q}}(t) + \dot{\widetilde{G}}(\widetilde{p}(t), \widetilde{q}(t)) - \widetilde{H}(\widetilde{p}(t), \widetilde{q}(t))] \, dt , \end{aligned} \tag{4.30}$$

where $\widetilde{H}(\widetilde{p}, \widetilde{q}) \equiv H(p, q)$. Evidently this action functional leads to the proper form of Hamilton's equations after an arbitrary canonical transformation. Observe that *no* change of the quantum operators occurs from these canonical transformations.

The original coordinates we have chosen, i.e., (p, q), enter the unitary operators $U(q)$ and $V(p)$ in (4.26) as so-called canonical group coordinates [Coh61]. In these coordinates, and assuming that P and Q are irreducible, it follows that

$$\begin{aligned} H(p, q) &= \langle p, q | \mathcal{H}(P, Q) | p, q \rangle \\ &= \langle 0 | \mathcal{H}(P + p\mathbb{1}, Q + q\mathbb{1}) | 0 \rangle \\ &= \mathcal{H}(p, q) + \mathcal{O}(\hbar; p, q) . \end{aligned} \tag{4.31}$$

This expression confirms that in these coordinates $H_c(p,q) \to \mathcal{H} = H_c(P,Q)$ up to terms $\mathcal{O}(\hbar)$. As is well known [Dir58], in traditional quantum theory, the special property represented by (4.31) requires "Cartesian coordinates". Although that requirement is sometimes difficult to ascertain in canonical quantization, that is not the case in the proposed formalism. The (suitably scaled) Fubini-Study metric, which vanishes for vector variations that differ only in phase, is given for the canonical coherent states by

$$d\sigma^2(p,q) \equiv (2\hbar)\left[\|\,d|p,q\rangle\|^2 - |\langle p,q|\,d|p,q\rangle|^2\right]$$
$$= dp^2 + dq^2 \,, \tag{4.32}$$

which describes a flat, two-dimensional phase space in Cartesian coordinates. In different coordinates, e.g., $(\widetilde{p}, \widetilde{q})$, the metric still describes a flat space but it will generally no longer be expressed in Cartesian coordinates. Note carefully that this metric does not originate with the classical phase space but it describes the geometry of the sheet of coherent states in the Hilbert space of unit vectors. Of course, one can append such a metric to the classical phase space if one is so inclined.

Macroscopic effects of enhanced quantization

In enhanced quantization, the enhanced classical Hamiltonian $H(p,q)$, say for a classical Hamiltonian of the form $H_c(p,q) = p^2/2m + V(q)$, generally differs from the classical Hamiltonian such that $H(p,q) = H_c(p,q) + \mathcal{O}(\hbar; p,q)$. The additional $\mathcal{O}(\hbar; p,q)$ term may very well change the classical equations of motion and thus their solutions. Those changes depend on the chosen fiducial vector and the particular projection on Hilbert space vectors that choice entails. Can those terms introduce modifications that are apparent in the macroscopic world? In Chap. 6 we consider this question carefully and conclude that such modifications are confined to the quantum world and are negligible in the macroscopic world. For that reason we do not experience any 'classical uncertainty'—due to different choices of $|\eta\rangle$—in our everyday experience.

4.3.2 Enhanced affine quantization

We now discuss another variation on conventional quantization methods offered by the general principles of enhanced quantization. Starting with the usual classical theory, we note that multiplying the Poisson bracket

$1 = \{q, p\}$ by q leads to $q = q\{q, p\} = \{q, pq\}$, or $\{q, d\} = q$ where $d \equiv pq$. The two variables d and q form a Lie algebra and are worthy of consideration as a new pair of classical variables even though they are not canonical coordinates. It is also possible to restrict q to $q > 0$ or $q < 0$ consistent with d, a variable that acts to *dilate* q and not *translate* q as p does. Let us develop the quantum story featuring these variables.

Take the Heisenberg commutation relation $i\hbar\mathbb{1} = [Q, P]$ and multiply both sides by Q leads to $i\hbar Q = Q[Q, P] = [Q, QP] = [Q, \frac{1}{2}(PQ + QP)]$, or stated otherwise,

$$[Q, D] = i\hbar Q , \qquad\qquad D \equiv \tfrac{1}{2}(PQ + QP) . \qquad (4.33)$$

This is called the affine commutation relation and D and Q are called affine variables. Again, D acts to dilate Q while P acts to translate Q, just like the classical variables. Observe that the affine commutation relation has been *derived* from the canonical commutation relation, and thus it always exists when P and Q exist and fulfill the usual commutation relation. Even though the original P and Q are irreducible, the operators D and Q are reducible. Indeed, there are three inequivalent irreducible representations: one with $Q > 0$, one with $Q < 0$ and one with $Q = 0$, and all three involve representations that are self adjoint. The first two irreducible choices are the most interesting, and, for the present, we focus on the choice $Q > 0$. Conventional quantization techniques would suggest we start with the classical variables d and q and promote them to operators D and Q that fulfill the affine commutation relation. Although possible, that approach would seem to have little to do with canonical theories, either classical or quantum. However, if we adopt an enhanced quantization approach, that is no longer the case.

Clearly the dimensions of D are those of $\hbar$, and for convenience we treat Q as dimensionless (or replace Q by Q/q_0, with $q_0 > 0$, and use units where $q_0 = 1$). To build the affine coherent states based on the affine variables, we first must choose a fiducial vector. We choose the unit vector $|\tilde{\beta}\rangle$ as an extremal weight vector which is a solution of the equation

$$[(Q - 1) + (i/\tilde{\beta})D]|\tilde{\beta}\rangle = 0 , \qquad (4.34)$$

where $\tilde{\beta}$ has the units of $\hbar$, and it follows that $\langle\tilde{\beta}|Q|\tilde{\beta}\rangle = 1$ and $\langle\tilde{\beta}|D|\tilde{\beta}\rangle = 0$. The affine coherent states are now defined as

$$|p, q\rangle \equiv e^{ipQ/\hbar} e^{-i\ln(q)D/\hbar} |\tilde{\beta}\rangle , \qquad (4.35)$$

for $(p, q) \in \mathbb{R} \times \mathbb{R}^+$, i.e., $q > 0$. Despite a similar notation, all coherent states in this section are affine coherent states; indeed, the limited range of q is an implicit distinction between the affine and canonical coherent states.

It is noteworthy that the geometry of the affine coherent states is different from the geometry of the canonical coherent states. To see this we consider the rescaled Fubini-Study metric that holds for the affine coherent states, namely

$$d\sigma^2(p, q) \equiv (2\hbar)[\| d|p, q\rangle \|^2 - |\langle p, q| d|p, q\rangle|^2]$$
$$= \tilde{\beta}^{-1} q^2 dp^2 + \tilde{\beta} q^{-2} dq^2 . \qquad (4.36)$$

This metric describes a two-dimensional space of constant negative curvature: $-2/\tilde{\beta}$. Since $q > 0$, it follows that this (Poincaré) half plane is geodesically complete. Just as in the canonical case, however, this metric describes the geometry of the sheet of coherent states, and does not originate from the classical phase space.

We are now in position to examine the enhanced quantization of such a system. Although the basic affine variables are D and Q, it is sometimes convenient to use P as well, even though it cannot be made self adjoint. Thus, the Hamiltonian operators $\mathcal{H}$ for such variables may be taken as $\mathcal{H}'(D, Q)$ or as $\mathcal{H}(P, Q) \equiv \mathcal{H}'(D, Q)$. We start with the quantum action functional for affine variables given by

$$A'_Q = \int_0^T \langle \psi(t)| [i\hbar(\partial/\partial t) - \mathcal{H}] |\psi(t)\rangle \, dt , \qquad (4.37)$$

which leads to the same formal Schrödinger's equation under general stationary variations. But what classical system does this correspond to? As part of enhanced quantization, the answer arises when we restrict the domain of the quantum action functional to the affine coherent states,

$$A'_{Q(R)} = \int_0^T \langle p(t), q(t)| [i\hbar(\partial/\partial t) - \mathcal{H}] |p(t), q(t)\rangle \, dt$$
$$= \int_0^T [-q(t)\dot{p}(t) - H(p(t), q(t))] \, dt , \qquad (4.38)$$

where $i\hbar \langle p, q| d|p, q\rangle = \langle \tilde{\beta}| (-qQ \, dp + D \, d\ln(q))|\tilde{\beta}\rangle = -q \, dp$ and $H(p, q) \equiv \langle p, q|\mathcal{H}|p, q\rangle$. In short, *the classical limit of enhanced affine quantization is a canonical theory!* Thus, if we deal with a classical system for which the physics requires that $q > 0$, then enhanced quantization, which requires self-adjoint generators, cannot proceed with the canonical variables P and Q (because P is not self adjoint), but it must use the self-adjoint affine variables D and Q. To emphasize the canonical nature of the enhanced classical theory, we note that we can again introduce canonical coordinate transformations such that $|\tilde{p}, \tilde{q}\rangle \equiv |p(\tilde{p}, \tilde{q}), q(\tilde{p}, \tilde{q})\rangle = |p, q\rangle$, leading, as before, to an expression for $A'_{Q(R)}$ involving the new coordinates $(\tilde{p}, \tilde{q})$, but without any change of the quantum story whatsoever.

To complete the story we need to give further details regarding the dynamics. In particular, we note that

$$H(p,q) = \langle \tilde{\beta} | \mathcal{H}'(D + pqQ, qQ) | \tilde{\beta} \rangle = \langle \tilde{\beta} | \mathcal{H}(P/q + p\mathbb{1}, qQ) | \tilde{\beta} \rangle . \quad (4.39)$$

In order to proceed further we let $|x\rangle$, $x > 0$, where $Q|x\rangle = x|x\rangle$, be a basis in which $Q \to x$ and $D \to -i\hbar(x\partial/\partial x + \frac{1}{2})$. With M a normalization factor, it follows that

$$\langle x | \tilde{\beta} \rangle = M \, x^{\tilde{\beta}/\hbar - 1/2} \, e^{-(\tilde{\beta}/\hbar) x} , \quad (4.40)$$

for which $\langle p,q | Q^n | p,q \rangle = \langle \tilde{\beta} | (qQ)^n | \tilde{\beta} \rangle = q^n (1 + \mathcal{O}(\hbar))$ and $\langle p,q | P^2 | p,q \rangle = \langle \tilde{\beta} | (P/q + p\mathbb{1})^2 | \tilde{\beta} \rangle = p^2 + C_2/q^2$, where $C_2 \equiv \langle \tilde{\beta} | P^2 | \tilde{\beta} \rangle \propto \hbar^2$ and $\langle \tilde{\beta} | P | \tilde{\beta} \rangle = 0$.

Example of enhanced affine quantization

As our example let us consider the classical, one-dimensional 'Hydrogen atom' [model #(2) in Chap. 1] with $H_c(p,q) = p^2/(2m) - e^2/q$, $q > 0$, which actually leads to singularities since, generally, $q(t) \to 0$ in a finite time. On the basis of an affine quantization, however, the enhanced classical Hamiltonian $H(p,q) = p^2/(2m) - C_1/q + C_2/(2mq^2)$, where $C_1 = e^2 + \mathcal{O}(\hbar)$, but, more importantly, $C_2/(2m) \simeq (\hbar^2/me^2) C_1$, and the coefficient here is recognized as the Bohr radius! Thus, for this model, all conventional classical singularities are removed by quantum corrections to the classical theory using $H(p,q)$. However, if we first let $\hbar \to 0$, leading to $H_c(p,q)$, the singularities will return. In addition, one may use a *reducible representation* of the affine group in which $|Q| > 0$ (and thus $|q| > 0$) to describe a similar problem in which the classical potential becomes $-e^2/|q|$.

It is noteworthy that simple models of cosmic singularities of the gravitational field, which exhibit classical singularities, have been shown to have classical bounces instead, based on enhanced affine quantization [FaZ13]; these models are summarized in the following chapter.

4.4   Enhanced Spin Quantization

For completeness we ask whether there is still another pair of quantum operators that can also serve as partners for an enhanced quantization of classical phase space. Indeed, there is another family, and we briefly outline its properties. Consider a set of three spin operators which satisfy $[S_1, S_2] = i\hbar S_3$, plus cyclic permutations. We choose an irreducible representation where $\Sigma_{j=1}^3 S_j^2 = \hbar^2 s(s+1) \mathbb{1}_s$, the spin $s \in \{1/2, 1, 3/2, 2, \ldots\}$, and the representation space is a $(2s+1)$-dimensional Hilbert space. Normalized eigenvectors and their eigenvalues for S_3 are given by $S_3 |s, m\rangle = m\hbar |s, m\rangle$ for $-s \le m \le s$. We choose $|s, s\rangle$ as our fiducial vector, which is also an extremal weight vector since $(S_1 + i S_2)|s, s\rangle = 0$. As spin coherent states we choose [KlS85]

$$|\theta, \phi\rangle \equiv e^{-i\phi S_3/\hbar} e^{-i\theta S_2/\hbar} |s, s\rangle , \qquad (4.41)$$

for $0 \le \theta \le \pi$ and $-\pi < \phi \le \pi$; they may also be described by

$$|p, q\rangle \equiv e^{-i(q/(s\hbar)^{1/2}) S_3/\hbar} e^{-i\cos^{-1}(p/(s\hbar)^{1/2}) S_2/\hbar} |s, s\rangle , \quad (4.42)$$

where $-(s\hbar)^{1/2} \le p \le (s\hbar)^{1/2}$ and $-\pi(s\hbar)^{1/2} < q \le \pi(s\hbar)^{1/2}$. The quantum action functional is

$$A_Q'' = \int_0^T \langle \psi(t)| [i\hbar(\partial/\partial t) - \mathcal{H}] |\psi(t)\rangle \, dt , \qquad (4.43)$$

and with $|\psi(t)\rangle \to |\theta(t), \phi(t)\rangle$, it becomes

$$\begin{aligned} A_{Q(R)}'' &= \int_0^T [s\hbar \cos(\theta)\, \dot{\phi}(t) - H(\theta(t), \phi(t))] \, dt \\ &= \int_0^T [p(t)\, \dot{q}(t) - H(p(t), q(t))] \, dt . \end{aligned} \qquad (4.44)$$

Canonical transformations follow the same pattern as before. The rescaled Fubini-Study metric on the spin coherent-state sheet in the $(2s+1)$-dimensional Hilbert space is given by

$$\begin{aligned} d\sigma^2 &\equiv (2\hbar) [\| d|\theta, \phi\rangle \|^2 - |\langle \theta, \phi| d|\theta, \phi\rangle|^2] \\ &= (s\hbar) [d\theta^2 + \sin(\theta)^2 \, d\phi^2] \\ &= [1 - p^2/(s\hbar)]^{-1} dp^2 + [1 - p^2/(s\hbar)] dq^2 , \end{aligned} \qquad (4.45)$$

corresponding to the two-dimensional surface of a three-dimensional sphere of radius $(s\hbar)^{1/2}$ and area of $4\pi(s\hbar)$. If we let $\hbar \to 0$, the whole story in this section vanishes; but in the real world, the only realm that really counts, $\hbar > 0$!

4.5 Special Situations

4.5.1 *Enhanced canonical quantization on a circle*

Quantization of a system with a periodic position variable q, such that, symbolically, physics$(q + 1) =$ physics(q), is a special case. First observe, that q (or Q) may be regarded as dimensionless and thus p (or P) has the dimensions of $\hbar$. As shown in Chap. 3, the self-adjoint momentum operator P has many, inequivalent incarnations: P_α, $0 \le \alpha < 2\pi$, with a spectrum $\mathsf{spec}(P_\alpha) = (\alpha + 2\pi n)\hbar$, $n \in \mathbb{Z} \equiv \{0, \pm 1, \pm 2, \dots\}$. The enhanced quantization of such a system depends on α, but the differences for different value of α all disappear in the limit that $\hbar \to 0$. In what follows we are guided by [BeK13].

The coherent states for this problem are chosen as

$$|p,q\rangle = e^{-iqP_\alpha/\hbar} e^{ip(Q - 1/2)/\hbar} |\eta\rangle \,, \qquad (4.46)$$

where $Q|x\rangle = x|x\rangle$ and $|\eta\rangle$ is chosen—with M_r fixed by normalization and $c \equiv 1/3$ (although any c where $0 < c < 1/2$ would do just as well)—such that $\langle x|\eta\rangle = M_r e^{i\alpha x}[1 - c - c\cos(2\pi x)]^{r/\hbar}$ consistent with the requirement that $\psi(x + 1) = e^{i\alpha}\psi(x)$ for all elements of the Hilbert space. In view of the spectrum of P_α, it follows that $|p, q + 1\rangle = e^{-i\alpha}|p, q\rangle$ leading to $\langle p, q + 1|\psi\rangle = e^{i\alpha}\langle p, q|\psi\rangle$, which reflects the proper periodicity.

Periodic potentials are represented by

$$V(Q - 1/2) \equiv W(\cos(2\pi Q), \sin(2\pi Q)) \,, \qquad (4.47)$$

and thus a common Hamiltonian operator (with $m = 1$) would be $\mathcal{H}_\alpha = \frac{1}{2}P_\alpha^2 + V(Q - 1/2)$. The quantum action functional is given by

$$A_Q = \int_0^T \langle \psi(t)|\,[i\hbar(\partial/\partial t) - \mathcal{H}_\alpha]\,|\psi(t)\rangle\, dt \,, \qquad (4.48)$$

and the restricted quantum action functional becomes

$$\begin{aligned} A_{Q(R)} &= \int_0^T \langle p(t), q(t)|\,[i\hbar(\partial/\partial t) - \mathcal{H}_\alpha]\,|p(t), q(t)\rangle\, dt \\ &= \int_0^T [p(t)\dot{q}(t) - H_\alpha(p(t), q(t))]\, dt \,. \end{aligned} \qquad (4.49)$$

Here, we have used

$$i\hbar \langle p, q|\,(\partial/\partial t)|p, q\rangle = \langle \eta|[(P_\alpha + p)\dot{q} - \dot{p}(Q - 1/2)]\eta\rangle = p\dot{q} + \hbar\alpha\dot{q} \,, \quad (4.50)$$

the last term of which can be dropped, as well as

$$\begin{aligned} H_\alpha(p, q) &= \langle \eta|\,[\tfrac{1}{2}(P_\alpha + p)^2 + V(Q + q - 1/2)]\,|\eta\rangle \\ &= \tfrac{1}{2}p^2 + V(q) + \mathcal{O}(\hbar; p, q) \,, \end{aligned} \qquad (4.51)$$

where, using $\langle\!\langle (\cdot) \rangle\!\rangle \equiv \langle \eta|(\cdot)|\eta\rangle$,

$$\mathcal{O}(\hbar; p, q) = \hbar\alpha p + \tfrac{1}{2}\langle\!\langle P_\alpha^2 \rangle\!\rangle + \langle\!\langle [V(Q + q - 1/2) - V(q)] \rangle\!\rangle \,. \qquad (4.52)$$

Our choice of $|\eta\rangle$ is such that as the ratio $r/\hbar$ becomes large, the function $\eta(x) = M_r e^{i\alpha x}[1 - c - c\cos(2\pi x)]^{r/\hbar}$ is highly peaked about $x = 1/2$ with deviations from $1/2$ involving $\hbar$. This remark holds because, for large $r/\hbar$ and using $c = 1/3$,

$$M_r[1 - c - c\cos(2\pi x)]^{r/\hbar} \simeq M_r e^{-4r\pi^2(x - 1/2)^2/6\hbar} . \qquad (4.53)$$

In short, for large values of $r/\hbar$, this fiducial vector acts rather like a Gaussian fiducial vector with a large variance parameter, leading to $\langle (Q - 1/2)^2 \rangle = \mathcal{O}(\hbar)$, etc.

The enhanced classical solutions are largely unchanged by α which only enters into the first term in $\mathcal{O}(\hbar; p, q)$ via the term $\hbar\alpha p$. The second term is a constant and does not contribute to the classical solutions, and the third term is independent of α, but it does modify the potential. However, in the true classical limit in which $\hbar \to 0$, the classical Hamiltonian $H_c(p, q) = \lim_{\hbar \to 0} H_\alpha(p, q) = \frac{1}{2}p^2 + V(q)$ and no trace of α remains.

4.5.2 *Infinite square-well potentials*

One of the favorite examples in elementary quantum mechanics is a one-dimensional system with a potential $V(x) = 0$ for, say, $0 < x < 1$, while $V(x) = \infty$ when $x \leq 0$ and $x \geq 1$. Such an example is said to have an infinite square-well potential. For the Hamiltonian operator $\mathcal{H} = -\frac{1}{2}\hbar^2 \partial^2/\partial x^2 + V(x)$ (with $m = 1$), this potential forces all eigenfunctions to vanish at $x = 0$ and $x = 1$ since they must be continuous and they must be identically zero outside the interval $(0, 1)$. The eigenfunctions are given by $\psi_n(x) = \sqrt{2}\sin(n\pi x)$, with the energy eigenvalues $E_n = \frac{1}{2}\hbar^2 n^2 \pi^2$, in each case with $n \in \mathbb{N}^+ = \{1, 2, 3, ...\}$. How should we address this problem by means of enhanced quantization?

In the present case, the operator P has the domain $\mathcal{D}(P) = \{\psi(x) : \int_0^1 [|\psi'(x)|^2 + |\psi(x)|^2] \, dx < \infty, \psi(0) = \psi(1) = 0\}$, and the domain of $P^\dagger$ is $\mathcal{D}(P^\dagger) = \{\phi(x) : \int_0^1 [|\phi'(x)|^2 + |\phi(x)|^2] \, dx < \infty\}$. There is no extension to make P into a self-adjoint operator. In our Hamiltonian, the term "P^2" should be interpreted as $P^\dagger P$, which is a proper self-adjoint operator with a real spectrum. Without a self-adjoint P we cannot define coherent states. In order to overcome this problem, we first need to regularize our system.

We modify the potential $V(x)$ and replace it with $V_{(N)}(x) = 0$ for $0 < x < 1$, as before, but now $V_{(N)}(x) = N, 0 < N < \infty$, for $x \leq 0$ and $x \geq 1$. In this case the problem now lives on the Hilbert space $L^2(\mathbb{R})$, and therefore we can introduce the usual self-adjoint canonical operators P and Q both of which have spectrum on the whole real line. In this case our regularized Hamiltonian operator becomes $\mathcal{H}_{(N)} = \frac{1}{2}P^2 + V_{(N)}(Q)$, and we can analyze this problem as an enhanced canonical system using traditional coherent states—except for one aspect. Since we ultimately wish to remove the regularization when $N \to \infty$, it would be helpful if the fiducial vector is already supported on the interval $(0,1)$. Thus, in this case we choose $\eta(x) = \sqrt{2} \sin(\pi x)$, for $0 < x < 1$, and $\eta(x) = 0$ outside the interval $(0,1)$. Armed with these canonical coherent states, we construct the restricted quantum action functional given by

$$A_{Q(R)} = \int_0^T \langle p(t), q(t)| \, [i\hbar(\partial/\partial t) - \mathcal{H}_{(N)}] \, |p(t), q(t)\rangle \, dt$$
$$= \int_0^T [p(t)\dot{q}(t) - H_{(N)}(p(t), q(t))] \, dt \, , \tag{4.54}$$

where the enhanced classical Hamiltonian is given by

$$H_{(N)}(p, q) = \tfrac{1}{2}p^2 + \tfrac{1}{2}\| P|\eta\rangle \|^2 + V_{(N)}(q) = \tfrac{1}{2}p^2 + \tfrac{1}{2}\hbar^2\pi^2 + V_{(N)}(q) \, . \tag{4.55}$$

The classical solution, $q(t) = q_0 + p_0 t$, is limited to the region $(0,1)$ unless the energy is sufficiently large, namely, $\tfrac{1}{2}p_0^2 = \tfrac{1}{2}p_{0\,(N)}^2 + N$, with $p_{0\,(N)} \neq 0$, in which case it moves outside the interval $(0,1)$ with momentum $p_{0\,(N)}$. To achieve our original problem, we now take the limit $N \to \infty$. In this case the classical solution is always restricted to the interval $(0,1)$ bouncing back and forth between the hard walls with a motion corresponding either to $q(t) = q_0 + p_0 t$ or $q(t) = q_0 - p_0 t$ for alternating intervals of time.

In summary, this example is obtained as the limit of a series of conventional enhanced quantization systems as the height of the walls tends toward infinity.

4.6   Reducible Representations

So far we have implicitly assumed the basic kinematical operator representations are irreducible because that is what conventional canonical quantization requires. However, enhanced quantization does *not* require that assumption. We start our analysis by allowing for reducible representations of the canonical operators P and Q. Since it is known that all irreducible representations of these operators (in the Weyl, i.e., exponential, form) are equivalent up to unitary transformations, we can accommodate all possible reducible representations simply by repeating the standard representation. Thus we choose a direct sum of Hilbert spaces of the form $\widetilde{\mathfrak{H}} = \oplus_j \mathfrak{H}_j$, where each $\mathfrak{H}_j$ is a copy of the prototypical Hilbert space $\mathfrak{H}$, and the sum may be finite or infinite (with care taken for absolute convergence of certain sums when necessary). In a related language, we may describe the same situation as a direct product of Hilbert spaces, namely, $\widetilde{\mathfrak{H}} = \mathfrak{H} \otimes \mathfrak{H}_N$, where N denotes how many copies we have. Elements of either form may be given by $|\widetilde{\psi}\rangle = \oplus_j |\psi_j\rangle$. In this Hilbert space, the basic kinematical operators become $\widetilde{Q} = \oplus_j Q_j$ and $\widetilde{P} = \oplus_j P_j$—namely, they are block diagonal and each P_j and Q_j is a copy of the prototypical operators P and Q. They may also be defined as $\widetilde{Q} = Q \otimes \mathbb{1}_N$ and $\widetilde{P} = P \otimes \mathbb{1}_N$ In this case, the Heisenberg commutation relation becomes

$$[\widetilde{Q}, \widetilde{P}] = \oplus_j [Q_j, P_j] = i\hbar \oplus_j \mathbb{1}_j = i\hbar \widetilde{\mathbb{1}} , \qquad (4.56)$$

or

$$[\widetilde{Q}, \widetilde{P}] = [Q \otimes \mathbb{1}_N, P \otimes \mathbb{1}_N] = i\hbar \mathbb{1} \otimes \mathbb{1}_N = i\hbar \widetilde{\mathbb{1}} , \qquad (4.57)$$

as required. Hilbert space vectors in $\widetilde{\mathfrak{H}}$ are given by $|\widetilde{\psi}\rangle = \oplus |\psi_j\rangle$ with an inner product given by $\langle \widetilde{\phi} | \widetilde{\psi} \rangle \equiv \Sigma_j \langle \phi_j | \psi_j \rangle$, at which point absolute convergence of the sum becomes an issue when there are an infinite number of elements.

The Hamiltonian operator is defined as $\widetilde{\mathcal{H}} = \widetilde{\mathcal{H}}(\widetilde{P}, \widetilde{Q}) = \oplus_j \mathcal{H}(P_j, Q_j) = \mathcal{H}(P, Q) \otimes \mathbb{1}_N$, and like the basic operators $\widetilde{P}$ and $\widetilde{Q}$, this Hamiltonian is also block diagonal in form. We are now in a position to introduce the quantum action functional

$$A_Q = \int_0^T \langle \widetilde{\psi}(t) | [i\hbar(\partial/\partial t) - \widetilde{\mathcal{H}}(\widetilde{P}, \widetilde{Q})] | \widetilde{\psi}(t) \rangle \, dt , \qquad (4.58)$$

which, under general stationary variation of each element $|\psi_j(t)\rangle$ of $|\widetilde{\psi}(t)\rangle$, subject to the overall requirement that $\langle \widetilde{\psi}(t) | \widetilde{\psi}(t) \rangle = 1$, along with $\delta|\psi_j(0)\rangle = \delta|\psi_j(T)\rangle = 0$ for all j, leads to the abstract Schrödinger equation given by

$$i\hbar \, \partial |\widetilde{\psi}(t)\rangle / \partial t = \widetilde{\mathcal{H}}(\widetilde{P}, \widetilde{Q}) |\widetilde{\psi}(t)\rangle , \qquad (4.59)$$

or, equivalently,

$$i\hbar \, \partial |\psi_j(t)\rangle / \partial t = \mathcal{H}(P, Q) |\psi_j(t)\rangle \qquad (4.60)$$

for all j. What is gained by this analysis is the variety that comes from the initial state given by $|\widetilde{\psi}(0)\rangle = \oplus_j |\psi_j(0)\rangle$ wherein, apart from an overall normalization condition, the separate elements may be specified independently. To eliminate any unnecessary redundancy in the set of vectors $\{|\psi_j(0)\rangle\}$, it is useful to choose this subset of nonzero vectors to be mutually orthogonal $\langle \psi_j(0) | \psi_k(0) \rangle = 0$, when $j \neq k$; of course, this set of vectors cannot be orthonormal since we require that $\Sigma_j \langle \psi_j(0) | \psi_j(0) \rangle = \langle \widetilde{\psi}(0) | \widetilde{\psi}(0) \rangle \equiv 1$.

A second equation follows from the stationary variation of the action functional that amounts to the conjugate Schrödinger equation, namely

$$-i\hbar\,\partial\langle\psi_k(t)|/\partial t = \langle\psi_k(t)|\mathcal{H}(P,Q) \tag{4.61}$$

since $\mathcal{H}$ is a Hermitian operator. By adding a suitable factor to each equation, we can combine Eqs. (4.60) and (4.61) to yield

$$i\hbar\,\partial\{|\psi_j(t)\rangle\,\langle\psi_k(t)|\}/\partial t = \mathcal{H}(P,Q)\{|\psi_j(t)\rangle\,\langle\psi_k(t)|\}$$
$$-\{|\psi_j(t)\rangle\,\langle\psi_k(t)|\}\,\mathcal{H}(P,Q)\,. \tag{4.62}$$

Each such equation lies in the j-k block in $\mathcal{H}_N$. As the next step, we take the partial trace over $\mathcal{H}_N$, leading to a sum over the diagonal blocks, given by

$$i\hbar\,\partial\{\Sigma_j\,|\psi_j(t)\rangle\,\langle\psi_j(t)|\}/\partial t = \mathcal{H}(P,Q)\,\{\Sigma_j|\psi_j(t)\rangle\,\langle\psi_j(t)|\}$$
$$-\{\Sigma_j\,|\psi_j(t)\rangle\,\langle\psi_j(t)|\}\,\mathcal{H}(P,Q)\,. \tag{4.63}$$

Lastly, we introduce

$$\rho(t) \equiv \Sigma_j\,|\psi_j(t)\rangle\,\langle\psi_j(t)| \equiv \Sigma_j p_j|\phi_j(t)\rangle\,\langle\phi_j(t)|\,, \tag{4.64}$$

where each $|\phi_j(t)\rangle$ denotes an element of a complete orthonormal set of vectors, each vector proportional to $|\psi_j(t)\rangle$, and $\langle\psi_j(t)|\psi_j(t)\rangle \equiv p_j \geq 0$ for all j as well as $\Sigma_j p_j = 1$. In terms of $\rho(t)$, (4.63) becomes

$$i\hbar\,\partial\rho(t)/\partial t = [\mathcal{H}(P,Q),\rho(t)]\,, \tag{4.65}$$

an operator equation on the Hilbert space $\mathcal{H}$, and which is the usual equation of motion for the density matrix (or density operator) $\rho(t)$, with the properties that $\rho(t) \geq 0$ and on which we impose the normalization that $\mathrm{Tr}(\rho(t)) \equiv 1$, a restriction that is consistent with (4.64). In this fashion, the quantum density operator is obtained.

4.6.1 *Restricted quantum action functional*

To analyze the enhanced classical theory that emerges from the present story, we first introduce suitable, normalized, canonical coherent states

$$|p,q\rangle \equiv e^{-iq\widetilde{P}/\hbar}e^{ip\widetilde{Q}/\hbar}\,|\widetilde{\eta}\rangle\,, \tag{4.66}$$

which are vectors in the Hilbert space $\widetilde{\mathfrak{H}}$. Observe that only *one* p and *one* q is involved, each of which multiplies all components of the reducible operators $\widetilde{P}$ and $\widetilde{Q}$. Here, the fiducial vector $|\widetilde{\eta}\rangle = \oplus_j |\eta_j\rangle$ involves generally different vectors in the various irreducible components. The restricted quantum action functional then becomes

$$A_{Q(R)} = \int_0^T \langle p(t), q(t)| \left[i\hbar(\partial/\partial t) - \widetilde{\mathcal{H}} \right] |p(t), q(t)\rangle \, dt$$
$$= \int_0^T \left[p(t)\dot{q}(t) - H(p(t), q(t)) \right] dt \,, \tag{4.67}$$

where

$$\begin{aligned} H(p, q) &\equiv \langle p, q| \widetilde{\mathcal{H}}(\widetilde{P}, \widetilde{Q})|p, q\rangle \\ &= \langle \widetilde{\eta}|| \widetilde{\mathcal{H}}(p + \widetilde{P}, q + \widetilde{Q})|\widetilde{\eta}\rangle \\ &= \Sigma_j \langle \eta_j| \mathcal{H}(p + P, q + Q)|\eta_j\rangle \\ &= \mathcal{H}(p, q) + \Sigma_j O_j(\hbar; p, q) \,. \end{aligned} \tag{4.68}$$

The most important feature of this exercise is that the restricted quantum action functional for this reducible representation of the canonical operators essentially leads to a *conventional enhanced classical canonical theory*. Note that the appearance of a quantum density operator does not imply the existence of a classical density function, contrary to what might have been expected; this is a consequence of the different classical/quantum connection of enhanced quantization. In any case, this example establishes the general principle that reducible as well as irreducible representations of the canonical operators both lead to acceptable enhanced classical canonical systems!

4.7 Comments

The foregoing discussion of a few simple mechanical examples shows how enhanced quantization techniques offer a new outlook on the very process of quantization itself. The principal motivating force behind enhanced quantization is to develop a formalism that lets both the classical and quantum theories coexist. This requires rejecting the usual assignment of classical and quantum variables by the promotion of one into the other. Instead, using enhanced quantization, in which the classical and quantum systems always coexist, we have seen that enhanced canonical quantization completely recovers the good conventional canonical quantization results and does so by explaining the origin of the favored Cartesian coordinates with the added benefit that the usual difficulties surrounding invariance under arbitrary canonical coordinate transformations disappear. Although we have principally focused on an enhanced affine quantization for which $q > 0$, and thus $Q > 0$—and even that domain can be readily changed to $q > -\gamma$ and $Q > -\gamma\mathbb{1}$—it is also possible to study affine models for which both q and the spectrum of Q run over the entire real line almost everywhere, a result that lets enhanced affine quantization compete with canonical quantization itself, as we will see in later chapters. The extension of enhanced quantization to multiple degrees of freedom is straightforward, and several interesting examples appear in later chapters. In most cases, the features needed to attack such problems are natural generalizations of the various topics we have considered in the present chapter.

Chapter 5

Enhanced Affine Quantization and the Initial Cosmological Singularity

WHAT TO LOOK FOR

The affine coherent-state quantization procedure of enhanced quantization is applied to the case of an FLRW universe in the presence of a cosmological constant. The quantum corrections alter the dynamics of the system in the semiclassical regime, providing a potential barrier term which avoids all classical singularities, as already suggested for other models. Furthermore—and surprisingly so—the quantum corrections may also be responsible for enhancing an accelerated cosmic expansion.

5.1 Introduction

This chapter draws heavily on an article with a similar title by M. Fanuel and S. Zonetti [FaZ13]. Although the text and format have sometimes been changed to fit the overall style of this monograph, and some aspects of their work have been omitted, this version still attempts to retain the spirit and emphasis of the original paper. The authors are thanked for permission to use their research in this manner. Please consult the original article for their own prose as well as additional insight into the topics discussed below.

****

The initial singularity problem is a long standing one in modern cosmology. It is often believed that the effects of quantum gravity should provide an answer to this question. Popular theories for quantum gravity are loop quantum gravity and superstring theories, but alternative approaches deserve to be considered as well. Among them, affine quantization has recently been put forward in order to quantize gravity [Kl12a; Kl12b; Kl12c], but it has also been studied previously in [IK84a; IK84b]. This approach was also used to study a strong coupling limit of gravity in [Pil82; Pil83; FrP85]. In this chapter, the affine coherent-state quantization program is used to study the dynamics of the scale factor in the FLRW (Friedman-Lemaître-Robertson-Walker) model for a cosmology including a cosmological constant. We analyze the enhanced classical behavior with help from the weak correspondence principle [Kla67], and the quantum corrections provide a natural potential barrier term.

5.2   The Classical Model

We choose the classical action functional given by

$$A_C = \alpha \int \{-\tfrac{1}{2} N(t)^{-2} a \dot{a}^2 - \tfrac{1}{6} \Lambda a^3 + \tfrac{1}{2} k a\} N(t) \, dt \, , \qquad (5.1)$$

where $a = a(t)$ is the scale factor, Λ is the cosmological constant, k is a geometric factor, and the overall factor α ensures that the action has the right dimensions. In the following we will set $\alpha = 1$ for simplicity. The arbitrariness in choosing a time coordinate t is emphasized by the presence of the lapse function $N(t)$.

Because of the constraint that the Hamiltonian vanishes, it is clear that the classical solutions to this model depend on the value of the factor k ($k = 0, \pm 1$) and the sign of the cosmological constant. In particular, de Sitter-like solutions ($\Lambda > 0$) are available for all values of k, while anti-de Sitter-like solutions ($\Lambda < 0$) are only possible with $k = -1$. A vanishing cosmological constant, on the other hand, does not allow a solution for $k = 1$.

5.2.1   *Hamiltonian formulation*

In what follows we will relabel $a(t)$ by $q(t)$ to make contact with the usual phase-space notation. Given the Lagrangian density in (5.1), the corresponding classical Hamiltonian, in the gauge $N(t) = 1$ (used hereafter), reads [model #(3) in Chap. 1]

$$H_c(p, q) = -\tfrac{1}{2} p^2 q^{-1} - \tfrac{1}{2} k q + \tfrac{1}{6} \Lambda q^3 \, , \qquad (5.2)$$

where p is the conjugate momentum for q. The Hamiltonian is constrained to vanish as an effect of coordinate invariance. The symplectic structure is given by the Poisson bracket $\{q, p\} = 1$. The configuration space variable

q is constrained to stay strictly positive: $q > 0$. Insisting that the quantum operator $Q > 0$ leads to difficulties with the conjugate operator P, which can not be made self adjoint. Thus the use of canonical operators is not suitable, and the use of affine operators is a natural choice.

5.3   Affine Coherent-State Quantization

5.3.1   *Construction of affine coherent states*

Some time ago, affine kinematical variables have been suggested for quantum gravity in the ADM (Arnowitt-Deser-Misner) formulation [Kla06; IK84a; IK84b]. These variables arise already in the group of transformations "$x \to x' = ax + b$", $a > 0$. For gravity, the major advantage they have is that they automatically implement the condition of positive definiteness of the spatial metric (see Chap. 10). In the model at hand, we have a similar constraint on the scale factor: $q > 0$. In order to define affine coherent states, we initially recall that the classical affine variables (q, d), $d \equiv pq$, which have the Poisson bracket $\{q, d\} = q$, have the feature that d *dilates* q instead of *translating* it like p does. By analogy, we introduce operators Q and D that satisfy the *affine commutation relation*, $[Q, D] = i\hbar Q$, and D has the virtue of dilating Q rather than translating it as P does. The *affine coherent states*, defined as follows

$$|p, q\rangle = e^{ipQ/\hbar} e^{-i\ln(q/\mu)D/\hbar} |\eta\rangle , \qquad (5.3)$$

for $(p, q) \in \mathbb{R} \times \mathbb{R}^+$, form an overcomplete basis of the Hilbert space, where μ is a scale factor with the dimensions of length; hereafter we choose units so that $\mu = 1$. We choose the fiducial vector $|\eta\rangle$ to fulfill the relation

$$[(Q - 1) + iD/(\beta\hbar)]|\eta\rangle = 0 , \qquad (5.4)$$

with β as a free dimensionless parameter; we also consider $\widetilde{\beta} (\equiv \beta\hbar)$ and $\hbar$ as an alternative pair of separate variables. It follows that

$$\langle\eta|Q|\eta\rangle = 1 , \qquad \langle\eta|D|\eta\rangle = 0 . \qquad (5.5)$$

It is noteworthy that Eq. (5.4) is chosen by analogy with the canonical coherent-state construction and provides a differential equation for the wave function of the fiducial state. The state $|\eta\rangle$ satisfies $0 < \langle\eta|Q^{-1}|\eta\rangle < \infty$ for $\beta > 3/2$, a value that will be needed to include the affine coherent states in the domain of the Hamiltonian operator. The associated coherent states (5.3) therefore admit a resolution of unity as given by

$$\int |p, q\rangle\langle p, q| \, dp \, dq/(2\pi\hbar C) = \mathbb{1} , \qquad (5.6)$$

where $C \equiv \langle \eta | Q^{-1} | \eta \rangle$.

We may now proceed to an affine quantization of the Hamiltonian formulation. In terms of the affine variables q and $d = pq$ the classical Hamiltonian takes the form

$$H_c(p, q) = -\tfrac{1}{2} d^2 q^{-3} - \tfrac{1}{2} kq + \tfrac{1}{6} \Lambda q^3 \,. \tag{5.7}$$

Next, we may apply the affine quantization principle: $q \to Q$ and $d \to D$, where $[Q, D] = i\hbar Q$. The operators D and Q are conveniently represented in x-space by

$$Df(x) = -i\hbar(x\partial_x + 1/2)f(x) = -i\hbar \, x^{1/2}\partial_x(x^{1/2}f(x)) \,, \tag{5.8}$$

$$Qf(x) = xf(x) \,, \tag{5.9}$$

and the interpretation of the algebra in terms of dilations is now completely intuitive. For the sake of consistency with the coherent-state definition (5.3), the operator D represented above must be self-adjoint, while there is no difficulty to define the operator Q. The self-adjoint operators Q and D appropriately realize the affine commutation relation, and the representation theory of such operators guarantees the existence of a unitary irreducible representation in which $Q > 0$ [AsK68]. The fiducial vector is then described by the wave function

$$\langle x | \eta \rangle = Nx^{\beta - 1/2} \exp(-\beta x) \,, \tag{5.10}$$

with $N = [2^{-2\beta}\beta^{-2\beta}\Gamma(2\beta)]^{-1/2}$.

The existence of β can be understood as an artifact of the representation. Different values of β lead to different representations of the same physical states. However, it is possible to see that there is a lower bound on the value of β: if we require the matrix element $\langle \eta | Q^{-1}DQ^{-1}DQ^{-1} \cdots | \eta \rangle$ [containing a number n (resp. $n-1$) of Q^{-1} (resp. D) operators] and $\langle \eta | Q^{-n} | \eta \rangle$ to be finite, we are forced to have $\beta > n/2$. We emphasize that this lower bound on the value of β is dictated by mathematical consistency and not by physical arguments. Besides this constraint, there are no other requirements on β at this stage; hence β will be considered as a free parameter. We will find that the specific value of β ($> 3/2$) is irrelevant in determining the qualitative cosmological behavior in the semiclassical regime.

5.3.2   *Quantization and the semiclassical regime*

We proceed to quantize the classical Hamiltonian (5.7) by choosing the quantum Hamiltonian as

$$\mathcal{H}'(D, Q) = -\tfrac{1}{2} Q^{-1}DQ^{-1}DQ^{-1} - \tfrac{1}{2} kQ + \tfrac{1}{6} \Lambda Q^3 \,. \tag{5.11}$$

In order to have a self-adjoint operator, the conditions on the domain of $K = Q^{-1}DQ^{-1}DQ^{-1}$ require that the boundary term in the following equation

$$\langle\phi|K\psi\rangle - \langle K^\dagger\phi|\psi\rangle = \hbar^2 \int_0^\infty dx\, \partial_x[\psi(x)\frac{1}{x}\partial_x\phi^*(x) - \phi^*(x)\frac{1}{x}\partial_x\psi(x)]\,,$$

(5.12)

vanishes. To ensure that $\psi(x) \in \mathcal{D}(K)$, the domain of K, $\psi(x)$ must satisfy

$$\lim_{x\to 0} x^{-1}\psi(x) = \lim_{x\to+\infty} x^{-1}\psi(x) = 0\,.$$

(5.13)

The functions $\phi(x)$ in $\mathcal{D}(K^\dagger)$, as may be seen from (5.12), must satisfy the same conditions (5.13). As a result, the domains of K and $K^\dagger$ coincide. Let us point out that the affine coherent states also belong to the domain of K whenever $\beta > 3/2$, and the wave function of an affine coherent state is given by

$$\langle x|p,q\rangle = N x^{-1/2}(x/q)^\beta\, e^{-\beta x/q}\, e^{ipx/\hbar}\,,$$

(5.14)

which satisfies (5.13) when $\beta > 3/2$. The Hilbert space of states and the domain of the relevant operators being properly identified, we may now use the coherent states to understand the dynamics.

Since the notion of geometry is difficult to interpret in a purely quantum theory of gravity, we consider a semiclassical quantity that could emerge from the quantum theory and be interpreted in a geometrical context. The enhanced classical Hamiltonian provides such a description by the Weak Correspondence Principle, which takes the form

$$H(p,q) = \langle p,q|\mathcal{H}'(D,Q)|p,q\rangle\,.$$

(5.15)

Intuitively, we appreciate that classical and quantum mechanics must co-exist as they do in the physical world. Thus, the weak correspondence principle allows us to consider quantum effects in a classical description of the world where we know that $\hbar$ takes a non-vanishing finite value. The fundamental reason why (5.15) is believed to incorporate proper quantum corrections is that it originates (as was discussed in the previous chapter) from the variational principle implementing the Schrödinger equation

$$A_Q = \int_0^T \langle\psi(t)|[i\hbar(\partial/\partial t) - \mathcal{H}'(D,Q)]|\psi(t)\rangle\, dt\,,$$

(5.16)

but in which the "restricted" quantum action functional can vary only over the set of (affine) coherent states $|p(t), q(t)\rangle$ rather than a general set of quantum states. Because of their semiclassical features, we believe that the coherent states may be the only states accessible to a classical observer. Consequently, this restricted quantum action principle

$$A_{Q(R)} = \int \langle p(t), q(t)|\, [i\hbar(\partial/\partial t) - \mathcal{H}'(D,Q)]\, |p(t), q(t)\rangle\, dt$$
$$= \int \left[-q(t)\dot{p}(t) - H(p,q) \right] dt \,, \tag{5.17}$$

gives a motivation for considering the equations of motion of $H(p,q)$ as a meaningful semiclassical approximation of the dynamics of the quantum system [Kl12c]. Finally, we emphasize here the noteworthy result that, starting from an *affine* quantized theory, the restricted action leads to an enhanced canonical theory (5.17). Making use of

$$\langle p, q|\mathcal{H}'(D,Q)|p,q\rangle = \langle \eta|\mathcal{H}'(D + pqQ, qQ)|\eta\rangle \,, \tag{5.18}$$

based on our choice that $\mu = 1$, we obtain

$$H(p,q) = -\tfrac{1}{2}q^{-3}\left\langle Q^{-1}DQ^{-1}DQ^{-1}\right\rangle - \tfrac{1}{2}p^2 q^{-1}\left\langle Q^{-1}\right\rangle$$
$$-\tfrac{1}{2}kq\langle Q\rangle + \tfrac{1}{6}\Lambda q^3\left\langle Q^3\right\rangle \,. \tag{5.19}$$

The required matrix elements can easily be calculated using (5.10) and (5.8), and (with $\mu = 1$ still) are given by

$$\left\langle Q^{-1}DQ^{-1}DQ^{-1}\right\rangle = \gamma \qquad \text{with } \beta > 3/2 \,, \tag{5.20}$$

$$\left\langle Q^{-1}\right\rangle = Z \qquad \text{with } \beta > 3/2, \tag{5.21}$$

$$\langle Q\rangle = 1 \qquad \text{with } \beta > 3/2, \tag{5.22}$$

$$\left\langle Q^3\right\rangle = \delta \qquad \text{with } \beta > 3/2 \,, \tag{5.23}$$

where

$$\gamma \equiv \frac{2\beta^3 \hbar^2}{(3 + 4(-2 + \beta)\beta)} > 0 \,, \tag{5.24}$$

$$Z \equiv \frac{2\beta}{(2\beta - 1)} > 1 \,, \tag{5.25}$$

$$\delta \equiv \frac{(1 + \beta)(1 + 2\beta)}{2\beta^2} > 1 \,. \tag{5.26}$$

Observe that these quantities are independent of the scale μ (if we had included it), and finiteness of these matrix elements requires a lower bound on the value of β (and $\beta > 3/2$ is sufficient). We emphasize that once β is chosen so that all matrix elements are finite, the value of $\gamma = ||Q^{-1/2}DQ^{-1}|\eta\rangle||^2$ *is never negative*. We stress that the value of the matrix elements given above should only be evaluated for $\beta > 3/2$ since other values of β can lead to inconsistent results. To summarize, we find that the enhanced Hamiltonian takes the form

$$H(p,q) = -\tfrac{1}{2}Z p(t)^2 q(t)^{-1} - \tfrac{1}{2}\gamma q(t)^{-3} + \tfrac{1}{6}\delta\Lambda q(t)^3 - \tfrac{1}{2}kq(t) \,. \tag{5.27}$$

Once again coordinate invariance requires that $H(p,q) = 0$ be enforced by the dynamics. The classical limit (5.7) of the enhanced Hamiltonian is readily reproduced by taking simultaneously $\hbar \to 0$ and $\beta \to \infty$, while their product $\hbar\beta \equiv \tilde{\beta}$ is fixed and finite. In this way, it follows that $Z \to 1$ and $\delta \to 1$, while $\gamma \to 0$.

5.3.3 *Qualitative analysis of the dynamics*

Interestingly, the quantum corrections generate one unique new dynamical term in the Hamiltonian, proportional to $\hbar^2 q^{-3}$. This contribution will naturally affect the dynamics for small values of the scale factor q, and indeed it will ensure that $q(t) > 0$ so no singularity at the origin will ever arise. As is known in General Relativity, the large scale gravitational dynamics, i.e., $q \gg 1$, is dominated by the cosmological constant term, $\Lambda > 0$, which generates a repulsive force and determines an accelerated expansion, while $\Lambda < 0$ is responsible for an attractive force that, for example, can slow down cosmic expansion. In the same way, the small scale dynamics, i.e., $q \ll 1$, will be dominated by the second term proportional to γ. This quantity is always positive for $\beta > 3/2$, and would be positive for any acceptable fiducial vector as well. Hence, it behaves as a small-scale equivalent of a positive cosmological constant, generating a repulsive force when the universe contracts. Furthermore the large-scale behavior is also modified: the constant δ, defined in (5.26), multiplies Λ and it is strictly greater than 1 for finite β, so that the effects of the cosmological constant are *amplified* for finite β since the effective cosmological constant is $\delta\Lambda$. Finally, we can see also that $Z > \delta > 1$ for all β. Therefore, even if different β's label distinct but similar enhanced classical theories, the qualitative effects, such as the avoidance of the classical singularity and the increased expansion rate, are universal.

****

At this point the original article develops numerical solutions for the equations of motion and presents figures to illustrate certain aspects of the solutions. For these features, please consult the original article [FaZ13].

5.4 Comments

In this chapter, the possibility of applying enhanced affine quantization procedures to a model of FLRW cosmology in the presence of a cosmological constant has been studied. Focus was on a semiclassical regime determined by the Weak Correspondence Principle [Kla67], which is consistent with the principles of enhanced quantization.

It was found that the additional terms and multiplicative constants arising from the quantization of the dilation algebra profoundly change the dynamics, and do so largely independent of the specific value of the parameter $\beta > 3/2$, which labels different enhanced classical theories. A change of β, or indeed a general change of the fiducial vector that still satisfies the required domain conditions, yields different two-dimensional projections of the infinite-dimensional Hilbert space, and it would be a valuable exercise to see if the general qualitative modifications introduced by the quantum terms continue to support the qualitative features emphasized here. Specifically, we have found that the large-scale dynamics is modified by an increased absolute value of the cosmological constant and the small-scale dynamics is affected by a potential barrier generated by quantum corrections. In the case of an open de Sitter universe, which already at the classical level exhibits no singularity and expands eternally, expansion is accelerated by a combination of small- and large-scale effects; the possibility of a connection with Dark Energy is worth investigating. More interestingly, all cases that possess a classical singularity when $q \to 0$, exhibit a nonsingular behavior in the semiclassical regime and enter an expansion phase after reaching a minimal length at which point the quantum dynamics is dominant. In the case of a closed anti-de Sitter universe, the scale factor enters an infinite cycle of expansions and contractions. These results, although limited to semiclassical considerations, provide additional support to the proposal of applying the affine quantization procedure in the approach of quantum gravity and quantum cosmology. [Remark: Preliminary work in this direction is the topic of Chap. 10 in this monograph.] Further investigations should be put forward to fully understand the role of affine coherent states and the potential of this approach: for instance, as already noted, it would be interesting to see whether alternative choices for the fiducial vector provide a similar behavior, and alternative ordering prescriptions can also be employed and their consistency should be analyzed.

PART 3

Enhanced Quantization of Some Complex Systems

Chapter 6

Examples of Enhanced Quantization: Bosons, Fermions, and Anyons

WHAT TO LOOK FOR

Fiducial vectors are generally chosen as Gaussian functions, partly because of their analytical simplicity as well as the fact that they are the only vectors that saturate the minimum uncertainty product $\triangle P \cdot \triangle Q = \frac{1}{2}\hbar$, and thus they are regarded as 'closest to the classical realm'. However, there are systems for which the fiducial vector can not be Gaussian, but must fulfill other criteria. Focussing on various specific systems, this chapter is devoted to examining the role of proper fiducial vectors and how they enter into enhanced quantization.

6.1 Many Degrees of Freedom

The passage from one, or a few, degrees of freedom to many degrees of freedom is basically quite easy and yet there are often special issues to be dealt with. In this short, introductory section we describe the 'easy' part saving the 'special issues' and further details to the following sections. Here, we focus entirely on the enhanced canonical and enhanced affine stories. In large measure, this chapter follows the article [AdK14].

In brief, for N degrees of freedom, where $N < \infty$ or $N = \infty$, there is a Hamiltonian operator $\mathcal{H}$ that is assumed to be self adjoint, as well as the existence of a set of N kinematic, self-adjoint *canonical operators*, namely, Q_n and P_n, $1 \leq n \leq N$, which satisfy the only nonvanishing commutators $[Q_m, P_n] = i\hbar\delta_{m,n}\mathbb{1}$ for all m and n—or—instead, there is a set of N kinematic, self-adjoint *affine operators*, namely, Q_n and D_n, $1 \leq n \leq N$, which satisfy the only nonvanishing commutators $[Q_m, D_n] = i\hbar\delta_{m,n} Q_m$

for all m and n. These various operators act on vectors $|\psi\rangle \in \mathfrak{H}$, where $\mathfrak{H}$ is a separable Hilbert space for all $N \leq \infty$, i.e., there exist complete, discrete, orthonormal bases. The quantum action functional is given, again, by

$$A_Q = \int_0^T \langle\psi(t)|\,[i\hbar\,(\partial/\partial t) - \mathcal{H}]\,|\psi(t)\rangle\,dt\,, \tag{6.1}$$

and, as usual, stationary variation of A_Q leads to Schrödinger's equation

$$i\hbar\,\partial|\psi(t)\rangle/\partial t = \mathcal{H}\,|\psi(t)\rangle \tag{6.2}$$

and its adjoint.

To obtain the enhanced classical action functional, we first appeal to canonical coherent states, which are constructed in analogous fashion. To simplify notation here, we let $p = \{p_1, p_2, \ldots\}$ and $q = \{q_1, q_2, \ldots\}$, which leads us to

$$|p, q\rangle \equiv \{\Pi_{n=1}^N e^{-iq_n P_n/\hbar}\, e^{ip_n Q_n/\hbar}\}\, |\eta\rangle\,, \tag{6.3}$$

where $|p, q\rangle \in \mathfrak{H}$, and, in particular, $|\eta\rangle \in \mathfrak{H}$. As we shall see, special attention must be paid to the choice of the fiducial vector $|\eta\rangle$ in some of the cases that we study. Armed with these coherent states, the reduced quantum action becomes

$$A_{Q(R)} = \int_0^T \langle p(t), q(t)|\,[i\hbar\,(\partial/\partial t) - \mathcal{H}]\,|p(t), q(t)\rangle\,dt$$
$$= \int_0^T [p(t) \cdot \dot{q}(t) - H(p(t), q(t))]\,dt\,, \tag{6.4}$$

where, as usual, $p \cdot \dot{q} \equiv \Sigma_{n=1}^N p_n \dot{q}_n$, with absolute convergence understood whenever $N = \infty$. By definition, the enhanced classical Hamiltonian is given by the weak correspondence principle,

$$H(p, q) \equiv \langle p, q|\mathcal{H}|p, q\rangle\,. \tag{6.5}$$

Instead, if we choose affine kinematical operators, we must first introduce affine coherent states given by

$$|p, q\rangle \equiv \{\Pi_{n=1}^N e^{ip_n Q_n/\hbar}\, e^{-i\ln(q_n)\,D_n/\hbar}\}\, |\xi\rangle\,. \tag{6.6}$$

Here we have chosen to illustrate a case for which $q_n > 0$ and the spectrum of $Q_n > 0$ as well. The choice of the fiducial vector $|\xi\rangle$ is again an important part of the discussion. There are many variations possible in the choice made for affine coherent states. For example, some or all of the coordinates may instead satisfy $q_n < 0$, or even $|q_n| > 0$, with a corresponding change of the spectrum of Q_n in each such case. Sticking with the chosen example, the reduced quantum action functional becomes

$$A_{Q(R)} = \int_0^T \langle p(t), q(t)|\,[i\hbar\,(\partial/\partial t) - \mathcal{H}]\,|p(t), q(t)\rangle\,dt$$
$$= \int_0^T [-q(t) \cdot \dot{p}(t) - H(p(t), q(t))]\,dt\,, \tag{6.7}$$

with $q \cdot \dot{p} \equiv \Sigma_{n=1}^{N} q_n \dot{p}_n$, again with absolute convergence understood whenever $N = \infty$. By definition, the enhanced classical Hamiltonian is again given by

$$H(p, q) \equiv \langle p, q | \mathcal{H} | p, q \rangle \,, \tag{6.8}$$

this time with the affine coherent states understood in place of the canonical coherent states.

Of course, there is always the possibility that some of the degrees of freedom are canonical while others are affine in nature; however, we choose not to consider such hybrid examples. If that situation arises, it would seem to be preferable to admit two distinct arrays, one which is strictly canonical, the other strictly affine.

6.1.1 *Two degrees of freedom*

For two degrees of freedom, we deal with classical variables p_1, p_2 and q_1, q_2, canonical operators P_1, P_2 and Q_1, Q_2, and conventional Hamiltonians built from these irreducible operators, $\mathcal{H} = \mathcal{H}(P_1, P_2; Q_1, Q_2)$. This formalism may be interpreted in two distinct ways. On the one hand, it is possible to imagine that this system takes place on two separate real lines—diagonalization of both Q_1 and Q_2—as would be the case for two separate, one-dimensional particles interacting with each other. On the other hand, we may interpret this system as one particle moving in a two-dimensional configuration space. However, there is a special situation when the Hamiltonian is symmetric by which we mean that

$$\mathcal{H}(P_2, P_1; Q_2, Q_1) \equiv \mathcal{H}(P_1, P_2; Q_1, Q_2) \,, \tag{6.9}$$

equivalent to an interchange between the two separate particles envisioned above. The Hilbert space of square-integrable functions may be defined by $\psi(x_1, x_2) \equiv \langle x_1, x_2 | \psi \rangle$, where $(aQ_1 + bQ_2) | x_1, x_2 \rangle = (ax_1 + bx_2) | x_1, x_2 \rangle$, etc. It is useful to divide the functions $\psi(x_1, x_2)$ into even and odd components so that

$$\psi(x_1, x_2) = \tfrac{1}{2}[\psi(x_1, x_2) + \psi(x_2, x_1)] + \tfrac{1}{2}[\psi(x_1, x_2) - \psi(x_2, x_1)]$$
$$\equiv \psi_+(x_1, x_2) + \psi_-(x_1, x_2) \,, \tag{6.10}$$

where, clearly, $\psi_+(x_2, x_1) \equiv \psi_+(x_1, x_2)$ and $\psi_-(x_2, x_1) \equiv -\psi_-(x_1, x_2)$. It is also useful to introduce abstract projection operators $\mathbb{E}_+$ and $\mathbb{E}_-$, which satisfy $\mathbb{E}_+ \mathbb{E}_- = 0$ and $\mathbb{E}_+ + \mathbb{E}_- = \mathbb{1}$, and use them to divide the set of all Hilbert space vectors into even and odd sectors as in

$$|\psi\rangle \equiv \mathbb{E}_+ |\psi\rangle + \mathbb{E}_- |\psi\rangle \equiv |\psi_+\rangle + |\psi_-\rangle \,. \tag{6.11}$$

The symmetry assumed regarding the Hamiltonian operator means that the temporal evolution of a symmetric wave function remains symmetric, and likewise, the temporal evolution of an antisymmetric wave function remains antisymmetric. In this sense, the Hilbert space of square integrable functions is divided for all time into either the set of symmetric functions $\psi_+(x_1, x_2, t)$, and/or the set of antisymmetric wave functions $\psi_-(x_1, x_2, t)$. Nature makes use of this separation, and the even sector (symmetric functions) refers to *bosons* while the odd sector (antisymmetric functions) refers to *fermions*. Thus, the same two-particle Hamiltonian can be quantized either as two bosons or two fermions.

Let us see what enhanced quantization has to say about this situation. The quantum action principle may be written in the form

$$A_Q = \int_0^T \langle \psi(t) | [i\hbar(\partial/\partial t) - \mathcal{H}] | \psi(t) \rangle \, dt$$
$$\equiv \int_0^T \langle \psi_+(t) | [i\hbar(\partial/\partial t) - \mathcal{H}] | \psi_+(t) \rangle \, dt \qquad (6.12)$$
$$+ \int_0^T \langle \psi_-(t) | [i\hbar(\partial/\partial t) - \mathcal{H}] | \psi_-(t) \rangle \, dt \, ,$$

thanks to the symmetry of the Hamiltonian.

The next step is to introduce the coherent states and examine the restricted quantum action functional. We introduce two fiducial vectors, $|\eta_+\rangle$ and $|\eta_-\rangle$, which live in the even and odd subspaces, respectively. We also introduce $|\eta\rangle = |\eta_+\rangle \oplus |\eta_-\rangle$ normalized such that $\langle \eta | \eta \rangle = \langle \eta_+ | \eta_+ \rangle + \langle \eta_- | \eta_- \rangle \equiv c_+ + c_- = 1$. The canonical coherent states themselves are given by

$$|p_1, q_1; p_2, q_2\rangle \equiv e^{-i(q_1 P_1 + q_2 P_2)/\hbar} \, e^{i(p_1 Q_1 + p_2 Q_2)/\hbar} \, |\eta\rangle$$
$$\equiv |p_1, q_1; p_2, q_{2+}\rangle \oplus |p_1, q_1; p_2, q_{2-}\rangle \, , \qquad (6.13)$$

and it follows that

$$|p_2, q_2; p_1, q_{1+}\rangle \equiv |p_1, q_1; p_2, q_{2+}\rangle$$
$$|p_2, q_2; p_1, q_{1-}\rangle \equiv -|p_1, q_1; p_2, q_{2-}\rangle \, . \qquad (6.14)$$

Introducing the canonical coherent states as vectors in the restricted quantum action functional leads to

$$A_{Q(R)} \qquad (6.15)$$
$$= \int_0^T \langle p_1(t), q_1(t); p_2(t), q_2(t) | [i\hbar(\partial/\partial t) - \mathcal{H}] | p_1(t), q_1(t); p_2(t), q_2(t) \rangle \, dt$$
$$= \int_0^T \langle p_1(t), q_1(t); p_2(t), q_2(t)_+ | [i\hbar(\partial/\partial t) - \mathcal{H}] | p_1(t), q_1(t); p_2(t), q_2(t)_+ \rangle \, dt$$
$$+ \int_0^T \langle p_1(t), q_1(t); p_2(t), q_2(t)_- | [i\hbar(\partial/\partial t) - \mathcal{H}] | p_1(t), q_1(t); p_2(t), q_2(t)_- \rangle \, dt \, ,$$

which then leads to the expression

$$A_{Q(R)} = \int_0^T [p_1 \dot{q}_1 + p_2 \dot{q}_2 - H(p_1, q_1; p_2, q_2)] \, dt \, , \qquad (6.16)$$

where

$$H(p_1, q_1; p_2, q_2) \equiv c_+ H_+(p_1, q_1; p_2, q_2) + c_- H_-(p_1, q_1; p_2, q_2) \qquad (6.17)$$
$$= \mathcal{H}(p_1, q_1; p_2, q_2) + c_+ \mathcal{O}_+(\hbar; p_1, q_1; p_2, q_2) + c_- \mathcal{O}_-(\hbar; p_1, q_1; p_2, q_2) ,$$

meaning that, apart from normalization, the distinctions between H_+ and H_- arise only in the $\mathcal{O}(\hbar)$ corrections due to the difference between $|\eta_+\rangle$ and $|\eta_-\rangle$.

To study this situation further it is useful to consider the two cases separately where only even or odd wave functions are allowed. Thus we consider $|\eta_\pm\rangle$—and therefore $|\eta_\mp\rangle = 0$—given by

$$H_\pm(p_1, q_1; p_2, q_2) = \langle \eta_\pm | \mathcal{H}(P_1 + p_1, Q_1 + q_1; P_2 + p_2, Q_2 + q_2) | \eta_\pm \rangle$$
$$= \mathcal{H}(p_1, q_1; p_2, q_2) + \mathcal{O}_\pm(\hbar; p_1, q_1; p_2, q_2) . \qquad (6.18)$$

To illustrate the quantum correction we suppose that

$$\mathcal{H}(P_1, Q_1; P_2, Q_2) = \tfrac{1}{2}[P_1^2 + P_2^2] + Q_1^4 + Q_2^4 + Q_1^2 Q_2^2 , \qquad (6.19)$$

in which case, and assuming odd operator powers do not contribute, it follows that

$$\mathcal{O}_\pm(\hbar; p_1, q_1; p_2, q_2) = 6\,[q_1^2 \langle Q_1^2 \rangle + q_2^2 \langle Q_2^2 \rangle] + q_1^2 \langle Q_2^2 \rangle + q_2^2 \langle Q_1^2 \rangle$$
$$+ \langle \{ \tfrac{1}{2}[P_1^2 + P_2^2] + Q_1^4 + Q_2^4 + Q_1^2 Q_2^2 \} \rangle , \quad (6.20)$$

where $\langle (\cdot) \rangle \equiv \langle \eta_\pm | (\cdot) | \eta_\pm \rangle$, as the case may be.

6.2   Choosing the Fiducial Vector

6.2.1   *General properties*

Coherent states are often defined with the aid of a group, and in so doing the necessary set of group representatives acting on a fixed, fiducial vector defines the coherent states. As an example, consider the set of canonical coherent states given by

$$|p, q; \eta\rangle \equiv e^{-iqP/\hbar} e^{ipQ/\hbar} |\eta\rangle . \qquad (6.21)$$

Normally, the choice of the normalized fiducial vector $|\eta\rangle$ is left implicit, but on this occasion, since we are deciding on how to choose a "good"—or the "best"—fiducial vector, we include it explicitly in the previous equation. The transformation of the fiducial vector to make coherent states involves unitary transformations by the given expressions, which means that both P and Q must be self-adjoint operators so they may generate unitary operators. The real variables p and q each generate one-parameter groups expressed in so-called canonical group coordinates of the second kind [Coh61]. Since such unitary operators are strongly continuous in their parameters, e.g., $\| [e^{ipQ/\hbar} - 1] |\psi\rangle \| \to 0$ as $p \to 0$ for all $|\psi\rangle \in \mathfrak{H}$, it follows that coherent states are strongly continuous in their parameters for any $|\eta\rangle$. This property ensures the continuity of the coherent-state representation for any abstract vector $|\psi\rangle$ given by $\psi(p, q; \eta) \equiv \langle p, q; \eta | \psi \rangle$ for any $|\eta\rangle$.

Besides continuity, the other basic feature of coherent states is a resolution of unity by an integral over the entire phase space of coherent-state projection operators that involves an absolutely continuous measure with a suitable positive weighting, which, for the example under consideration, is given by

$$\mathbb{1} = \int |p, q; \eta\rangle \langle p, q; \eta| \, d\mu(p, q) \,, \qquad d\mu(p, q) \equiv dp \, dq/(2\pi\hbar) \,, \quad (6.22)$$

a relation that holds weakly (as well as strongly) for any choice of the normalized fiducial vector $|\eta\rangle$. [Remark: For other sets of putative coherent states, it sometimes happens that the resolution of unity fails; in this case we deal with so-called "weak coherent states" [KlS85]. When that occurs, it is useful to let the inner product of weak coherent states serve as a reproducing kernel and use it to generate a reproducing kernel Hilbert space [Aro43]; see also Sec. 10.10 later in this mongraph.]

6.2.2   *Choosing the fiducial vector: Take one*

In enhanced quantization, such as discussed in Chap. 4 of this monograph, the restricted quantum action functional is given by

$$A_{Q(R)} = \int_0^T \langle p, q; \eta | [i\hbar(\partial/\partial t) - \mathcal{H}(P, Q)] | p, q; \eta \rangle \, dt$$
$$= \int_0^T [p(t)\dot{q}(t) - H(p(t), q(t))] \, dt \,, \qquad (6.23)$$

and thus it is necessary that the the coherent states are in the domain of the Hamiltonian operator $\mathcal{H}$, which, for an unbounded Hamiltonian operator, will already induce a certain restriction on $|\eta\rangle$. Additionally, in giving physical meaning to the variables p and q, it is useful to impose 'physical centering', i.e., $\langle P \rangle = \langle Q \rangle = 0$, wherein we have again used the shorthand that $\langle (\cdot) \rangle \equiv \langle \eta | (\cdot) | \eta \rangle$. The virtue of physical centering becomes clear when we note that it leads to

$$\langle p, q; \eta | P | p, q; \eta \rangle = p \,, \qquad \langle p, q; \eta | Q | p, q; \eta \rangle = q \,, \qquad (6.24)$$

yielding a natural physical interpretation of the parameters p and q.

The enhanced classical Hamiltonian $H(p,q)$ (with $\hbar > 0$) differs from the classical Hamiltonian $H_c(p,q) = \lim_{\hbar \to 0} H(p,q)$ by terms of order $\hbar$. As an example, let $\mathcal{H}(P,Q) = P^2 + Q^2 + Q^4$, in which case

$$H(p,q) = p^2 + q^2 + q^4 + 6q^2\langle Q^2 \rangle + \langle [P^2 + Q^2 + Q^4] \rangle , \qquad (6.25)$$

assuming for simplicity that odd expectations vanish. Apart from a constant, there is the term $6q^2\langle Q^2 \rangle$ which will modify the usual equations of motion and their solution. If we choose $\langle x|\eta \rangle \propto \exp(-\omega x^2/2\hbar)$, where ω is a macroscopic parameter, it follows that $\langle Q^2 \rangle \propto \hbar$, meaning that the correction is important only when p and q are "quantum-sized" themselves. That limitation makes good sense since then the enhanced classical description is effectively unchanged for macroscopic motion, only showing quantum "uncertainty", i.e., dependence on ω, for quantum-sized motion. Alternatively, we could in principle choose $\langle x|\eta \rangle \propto \exp(-ax^2/2)$, where, say, $a = 10^{-137} cm^{-2}$ and independent of $\hbar$, which means that the additional term would modify the quadratic term by a potentially huge amount even for macroscopic motion. Such a choice is mathematically possible, just as studying a simple harmonic oscillator with displacements on a planetary scale or energies equivalent to the mass energy of the Earth are mathematically possible. However, these are unphysical applications of the mathematical description of a simple harmonic oscillator. In a similar story, although it is mathematically possible to choose fiducial vectors so that $\langle Q^2 \rangle$ leads to macroscopic modifications, such a fiducial vector could never be physically realized. Thus we conclude that it is logical to choose the fiducial vector supported largely on a "quantum-sized" region. **This fact means that the quantum corrections term $\mathcal{O}(\hbar; p, q)$ is negligible for systems involving macroscopic motion, and as such, the solutions to the related equations of motion have all the features of deterministic variables as commonly assumed about classical dynamics.**

However, that still leaves open many possibilities for the fiducial vector. Indeed, Truong [Tru75] has showed that choosing the ground state of a quartic Hamiltonian as the fiducial vector can recast Schrödinger's equation into a new form that offers novel options for analysis.

6.2.3   *Choosing the fiducial vector: Take two*

It is popular to choose the fiducial vector to be a "quantum-sized" Gaussian; indeed, we have done so earlier in this chapter. This choice often leads to fairly simple analytic expressions, and sometimes plausible arguments can be advanced that even help choose the variance parameter for such a vector [BGY13]. However, it is important to understand that such a choice is *not* suitable in all cases. Let us reexamine the old discussion about *"The rest of the universe"* [Fey72]. A single system seldom exists in isolation; instead, it is surrounded by other systems. If we can imagine one specific system, then it is possible to imagine N independent, identical systems, and even infinitely many such systems, i.e., $N = \infty$. A system may involve several degrees of freedom; however, for clarity our basic system is chosen to have a single degree of freedom.

For example, consider the operator $\mathcal{H}_{(N)} \equiv \Sigma_{n=1}^{N} \mathcal{H}(P_n, Q_n)$, which represents N independent, identical versions of the "original" Hamiltonian $\mathcal{H}(P_1, Q_1)$ involving independent operator pairs (P_n, Q_n). The first system (for P_1 and Q_1) is chosen as the "physical" one, while the other sub-systems are "spectator" systems. The coherent states we choose are only for the physical system, i.e., only for the *first* operator pair, specifically

$$|p, q; \eta\rangle = e^{-iqP_1/\hbar} e^{ipQ_1/\hbar} |\eta\rangle , \qquad |\eta\rangle \equiv \otimes_{n=1}^{N} |\eta_n\rangle ; \qquad (6.26)$$

note that these 'coherent states' do not span the entire Hilbert space. To preserve the equivalence of all subsystems we choose the $|\eta_n\rangle$ to be identical to one another, i.e., the same single-system fiducial vector. In this case it follows that

$$A_{Q(R)} = \int_0^T [p(t)\dot{q}(t) - H_{(N)}(p(t), q(t))] \, dt , \qquad (6.27)$$

where, in the present case,

$$\begin{aligned} H_{(N)}(p, q) &= H(p, q) + \Sigma_{n=2}^{N} \langle \mathcal{H}(P_n, Q_n)\rangle \\ &= H(p, q) + (N - 1) \langle \mathcal{H}(P_2, Q_2)\rangle , \end{aligned} \qquad (6.28)$$

which differs by a constant from the single system story. So long as $N < \infty$, that constant is finite and not important. But, as $N \to \infty$, and we eventually deal with an infinite number of spectator systems, it becomes important that this constant must be zero. At this point in the discussion we restrict attention to quantum systems that have a non-negative spectrum and a unique ground state, which we choose as $|\eta\rangle$, with an energy eigenvalue adjusted to be zero. If that is the case, then the added constant in (6.28) vanishes for all N including $N = \infty$. It may be argued that we could choose $|\eta\rangle$, say, as the first excited state for each subsystem and subtract that energy to obtain a zero. However, that would imply, for $N = \infty$, that there were infinitely many energy levels with $-\infty$ for their energy value, which is clearly an unphysical situation. The remedy for that situation is to insist that the fiducial vector be chosen as the unique ground state of each system with an energy value adjusted to vanish. This choice applies to the physical system, and to all of the spectator systems as well. Thus we can imagine any one of the N identical systems being the physical one and the remaining $N - 1$ systems as spectators.

Moreover, we can imagine there are many other spectator systems different from the physical one we have chosen. For example, suppose there is another multiple-system type, with a Hamiltonian $\widetilde{\mathcal{H}}_{(\widetilde{N})} = \Sigma_{\widetilde{n}=1}^{\widetilde{N}} \widetilde{\mathcal{H}}(P_{\widetilde{n}}, Q_{\widetilde{n}})$, that is also present. This operator, too, is assumed to have a unique ground state with a zero-energy eigenvalue. Thus, this new Hamiltonian could be present in the overall Hamiltonian but it would contribute nothing to the restricted quantum action functional because coherent states for it have not been "turned on". Indeed there could be many such systems, even infinitely many such new (sub)systems. This argument can be carried to yet new families of spectator systems all of which are there, just "resting", or "hibernating", in their own ground state, and contributing nothing to the restricted quantum action functional. In this fashion we have found how to include "the rest of the universe" in such a way that it makes no contribution whatsoever, just as if we had ignored it altogether at the beginning of the story.

This desirable property requires that we choose the fiducial vector as the *unique ground state of the system under consideration adjusted to have zero energy*, which for this esoteric exercise of dealing with the surroundings proved extremely convenient if not absolutely necessary. This choice of fiducial vector also eliminates any nonsense regarding intrusion of the micro world into the macro world as we discussed above. Of course, concerns about the intrusion of the surroundings are not always necessary, and thus it is acceptable to consider other fiducial vectors that are "close" to the ground state in some unspecified way, if one so desires.

Finally, we need to comment on other model systems that have Hamiltonians, which (i), near a lower bound, have a continuous spectrum, or (ii) instead have a spectrum that is unbounded below. These are interesting mathematical models, but it it is difficult to find any real physical systems that have such features.

Having shown how we can, if necessary, deal with the rest of the universe, we revert to simple systems without concerning ourselves with such big issues. Thus, in what follows, we allow ourselves to consider a variety of useful fiducial vectors, particularly those where a Gaussian form is not appropriate to describe certain physical systems, as in the case of fermions and anyons.

6.3   Bosons, Fermions, and Anyons

In this section we confine ourselves to $2 + 1$ spacetime dimensions so that anyons can be treated as well as bosons and fermions. Although the results and calculations for bosons and fermions are limited to $2 + 1$ dimensions, generalizations to arbitrary spacetime dimensions for them are straightforward.

Charged particles orbiting around a magnetic flux tube and interacting with it have fractional statistics in $2 + 1$ dimensions [Wil62]. Such composites, known as anyons [Wil62], are basically characterized by their peculiar statistics: under a half-rotation centered on a magnetic flux, leading to a two-particle permutation, the state of the system changes by a complex phase factor,

$$\psi(\mathbf{r}_2, \mathbf{r}_1) = e^{i\alpha}\psi(\mathbf{r}_1, \mathbf{r}_2), \quad (0 \le \alpha < 2\pi), \qquad (6.29)$$
$$\mathbf{r}_1 = (x_1, y_1), \quad \mathbf{r}_2 = (x_2, y_2),$$

rather than the standard bosons ($\alpha = 0$) or fermions ($\alpha = \pi$) statistics. Here, and throughout our discussion, $\mathbf{r}_\sigma$ designates the two-dimensional Cartesian coordinates of the first particle ($\sigma = 1$) and the second particle ($\sigma = 2$), respectively.

As has been noticed originally by Wilczek [Wil62], this consideration has a very clear mathematical explanation based on the fact that in two space dimensions the rotation group $\mathsf{SO}(2)$ is isomorphic with the Abelian unitary group $\mathsf{U}(1)$, the representations of which are labeled by real numbers. Due to the latter fact and the spin-statistics connection, the wave function of these particles may admit an arbitrary phase under rotations in $2 + 1$ spacetime dimensions. More precisely, circling a magnetic flux in the same direction by two half turns, the state of the system may not necessarily return to its original state, meaning that now the system can be described by a multivalued wave function.

Anyons have attracted much attention due to their own richness, both in theoretical treatment of fundamental concepts, as well as in their physical implications. Particles with fractional statistics were considered in the $\mathsf{O}(3)\,\sigma$ model due to the existence of solitons [WiZ83]. It also has been shown that anyons can be described as ordinary particles interacting with a Chern-Simons field [IeL90; IeL92]. A relativistic wave equation for anyons was formulated in [JaN91] and more recently in [HPV10]. Regarding the physical implications, anyons have a central role in the explanation of the quantum Hall effect [Lau88], in high-T_c superconductivity [FHL89], and more recently in topological quantum computation [DFN05; Col06; NSS08]. For a complete review underlying the fundamental theoretical descriptions and applications, we recommend [NSS08; IeL92; Wil90].

All this interest motivates us to study the subject. Particularly, we are interested in the enhanced classical theory of anyons in $2 + 1$ dimensions. In this respect it should be noted that classical theories for anyons have been considered before [Ply90; Gho95]. In these papers the authors follow a canonical quantization procedure [Dir58; GiT90] to derive the corresponding quantum theory for anyons. Here, we adopt a different construction to quantize the theory of particles with fractional statistics by the application of enhanced quantization and derive the corresponding enhanced classical theory. More precisely, we consider a rotationally invariant Hamiltonian operator with a quartic interaction as a basic example.

Maintaining the physical consideration behind anyons in $2 + 1$ dimensions, a coherent-state representation [Kl63a; Kl63b] of the quantum theory for anyons must exhibit the same property as (6.29). A direct consequence, therefore, is that the fiducial vectors must obey,

$$\langle \mathbf{r}_2, \mathbf{r}_1 | \eta \rangle = \eta(\mathbf{r}_2, \mathbf{r}_1) = e^{i\alpha} \eta(\mathbf{r}_1, \mathbf{r}_2) \,. \tag{6.30}$$

In choosing a coherent-state representation we first must provide a suitable fiducial vector consistent with the property (6.30). One possible way is to assume the fiducial vector has two separate parts $\eta(\mathbf{r}_1, \mathbf{r}_2) = A(\mathbf{r}_1, \mathbf{r}_2) S(\mathbf{r}_1, \mathbf{r}_2)$, where $S(\mathbf{r}_1, \mathbf{r}_2)$ is a symmetrical function and $A(\mathbf{r}_1, \mathbf{r}_2)$ has a nonsymmetrical form. In particular,

$$A(\mathbf{r}_2, \mathbf{r}_1) = (-1)^\gamma A(\mathbf{r}_1, \mathbf{r}_2) \,, \tag{6.31}$$

and the condition (6.30) is recovered for $\gamma = \alpha/\pi$. This construction leads to a proper description of nonrelativistic fermions for $\gamma = 1$, and it is generalized for anyons assuming $0 < \gamma < 2$ ($\gamma \neq 1$).

Wave functions for two or more anyon systems have been discussed before. One of the first proposals was a multivalued function with a complex nonsymmetrical part and a Gaussian-like symmetrical part [Hal84; Wu84]. Generalizations of wave functions of this form have been considered in several works (e.g. [Cho91; MLB91; IeL92; Pol91]) and in particular represent a bound state for a two-anyon system in a $(2 + 1)$-dimensional interaction with an external harmonic potential such as $\frac{1}{2}\Omega^2 r_i^2$ [Pol91]. Following these constructions, we study a family of fiducial vectors, parameterized by λ, suitable to discuss bosons, fermions, and anyons in the same framework, and given by

$$\eta_1(\mathbf{r}_1, \mathbf{r}_2) = N_\gamma (z_1 - z_2)^\gamma e^{-\frac{1}{2}\lambda(r_1^2 + r_2^2)}, \quad 0 \le \gamma < 2, \quad \gamma = \frac{\alpha}{\pi}, \qquad (6.32)$$

$$z_1 = x_1 + iy_1, \quad z_2 = x_2 + iy_2, \quad r_\sigma^2 = x_\sigma^2 + y_\sigma^2, \quad \lambda = \frac{\Omega}{\hbar}, \quad \Omega = \text{const.},$$

where z_1, z_2 are the positions of particles in a complex notation, Ω is an arbitrary positive constant, and N_γ are the corresponding normalization constants. It is clear that interchanging z_1 and z_2 is equivalent to interchanging the corresponding positions of the particles, and, as a result, one obtains the desired phase $\eta(\mathbf{r}_2, \mathbf{r}_1) = (-1)^\gamma \eta(\mathbf{r}_1, \mathbf{r}_2)$. We initially study bosons and fermions in the next subsection. Exact solutions and results for anyons are presented afterwards.

6.3.1　*General calculations for bosons and fermions*

In this subsection we examine, in detail, the choice of fiducial vectors (6.32) for the particular case of bosons ($\gamma = 0$) and fermions ($\gamma = 1$). More precisely, we are interested in obtaining the enhanced classical description for these particles, i.e., bosons and fermions, with a quartic interaction such as (6.25). To do so requires us to evaluate expectation values of the self-adjoint operators Q, P, Q^2, P^2, and Q^4, and the subsequent subsections are devoted to presenting those results. For details of these derivations, consult the original paper [AdK14].

Bosons ($\gamma = 0$)

From (6.32) the fiducial vectors for bosons are pure Gaussian functions,

$$\eta_1(\mathbf{r}_1, \mathbf{r}_2) = N_0 e^{-\frac{1}{2}\lambda(r_1^2 + r_2^2)}, \quad N_0 = \frac{\lambda}{\pi}. \qquad (6.33)$$

One of the useful properties of the usual Gaussian fiducial vectors is the vanishing of the expectation values of the position and momentum operators, which we refer to as physical centering. It is straightforward to see that $\langle Q_{x_i}^{2n+1} \rangle = 0$ for $n \in \mathbb{N}$, and since the Fourier transform of a Gaussian is another Gaussian, it follows that $\langle P_{x_i}^{2n+1} \rangle = 0$ as well. Intrinsic to the calculation of the enhanced Hamiltonian are the expectation values of $Q_{x_i}^2$,

$P_{x_i}^2$ and $Q_{x_i}^4$. For example, the expectation value of $Q_{x_1}^2$ is,

$$\langle Q_{x_1}^2 \rangle = N_0^2 \left(\frac{\pi}{\lambda}\right) \int_0^{2\pi} d\vartheta \cos^2 \vartheta \int_0^\infty dr_1 r_1^3 e^{-\lambda r_1^2} = \frac{1}{2\lambda} = \frac{\hbar}{2\Omega}, \qquad (6.34)$$

where ϑ is the polar angle. In addition, the expectation value for $P_{x_1}^2$ is

$$\langle P_{x_1}^2 \rangle = -\hbar^2 \lambda N_0^2 \int_{-\infty}^\infty d\mathbf{r}_1 \int_{-\infty}^\infty d\mathbf{r}_2 \left(\lambda x_1^2 - 1\right) e^{-\lambda\left(\mathbf{r}_1^2+\mathbf{r}_2^2\right)} = \frac{\hbar^2 \lambda}{2} = \frac{\hbar\Omega}{2}, \qquad (6.35)$$

and the remaining quadratic expectation values have the same result. [Remark: Here and in what follows we adopt the convention that one integration symbol for a single bold (= vector) integration variable means a double integration over the two coordinates; e.g., $\int_a^b d\mathbf{r} f(x,y) \equiv \int_a^b dx \int_a^b dy f(x,y)$.]

In order to discuss the enhanced classical theory for bosons with a quartic interaction one has to evaluate several additional expectation values such as $\langle Q_{x_\sigma}^4 \rangle$, $\langle Q_{x_\sigma}^2 Q_{x_{\sigma'}}^2 \rangle$, and $\langle Q_{x_\sigma}^2 Q_{y_{\sigma'}}^2 \rangle$. Fortunately, thanks to the structure of the Gaussian fiducial vectors (6.33), only $\langle Q_{x_1}^4 \rangle$ is independent,

$$\langle Q_{x_1}^4 \rangle = N_0^2 \left(\frac{\pi}{\lambda}\right) \int_0^{2\pi} d\vartheta \cos^4 \vartheta \int_0^\infty dr_1 r_1^5 e^{-\lambda r_1^2} = \frac{3}{4\lambda^2} = \frac{3\hbar^2}{4\Omega^2}. \qquad (6.36)$$

For the remaining terms, consider, for example, $\langle Q_{x_1}^2 Q_{x_2}^2 \rangle$ (which is equivalent to $\langle Q_{x_1}^2 Q_{y_1}^2 \rangle$ in the present case). From its form it follows that

$$\langle Q_{x_1}^2 Q_{x_2}^2 \rangle = N_0^2 \left(\frac{\pi}{\lambda}\right) \int_{-\infty}^\infty dx_1 x_1^2 e^{-\lambda x_1^2} \int_{-\infty}^\infty dx_2 x_2^2 e^{-\lambda x_2^2} = \frac{\langle Q_{x_1}^4 \rangle}{3}. \qquad (6.37)$$

Fermions ($\gamma = 1$)

For fermions the fiducial vectors (6.32) take the form

$$\eta_1(\mathbf{r}_1, \mathbf{r}_2) = N_1 (z_1 - z_2) e^{-\frac{1}{2}\lambda\left(\mathbf{r}_1^2 + \mathbf{r}_2^2\right)}, \qquad (6.38)$$

and the corresponding normalization constant is given by

$$N_1 = \left(\int_{-\infty}^\infty d\mathbf{r}_1 \int_{-\infty}^\infty d\mathbf{r}_2 |\mathbf{r}_1 - \mathbf{r}_2|^2 e^{-\lambda\left(\mathbf{r}_1^2+\mathbf{r}_2^2\right)}\right)^{-\frac{1}{2}} = \sqrt{\frac{\lambda^3}{2\pi^2}}. \qquad (6.39)$$

The expectation value of the coordinate operators Q_{x_1}, etc., are zero for these states. For example, $\langle Q_{x_1} \rangle$ has the form

$$\langle Q_{x_1} \rangle = N_1^2 \int_0^{2\pi} d\vartheta \cos \vartheta \int_0^\infty dr_1 r_1^2 e^{-\lambda r_1^2} \int_{-\infty}^\infty d\mathbf{r}_2 |\mathbf{r}_1 - \mathbf{r}_2|^2 e^{-\lambda \mathbf{r}_2^2} = 0. \qquad (6.40)$$

In addition, the expectation values for the momentum operator $\langle P_{x_1} \rangle$ reads,

$$\langle P_{x_1} \rangle = -i\hbar N_1^2 \int_{-\infty}^{\infty} d\mathbf{r}_1 \int_{-\infty}^{\infty} d\mathbf{r}_2 \{ (z_1^* - z_2^*) - \lambda\, x_1\, |\mathbf{r}_1 - \mathbf{r}_2|^2 \} e^{-\lambda(\mathbf{r}_1^2 + \mathbf{r}_2^2)} = 0,$$
(6.41)

and the remaining linear expectation values are also zero. The expectation value for $Q_{x_1}^2$ is

$$\langle Q_{x_1}^2 \rangle = N_1^2 \int_{-\infty}^{\infty} d\mathbf{r}_1 \int_{-\infty}^{\infty} d\mathbf{r}_2 \, (\mathbf{r}_1^2 + \mathbf{r}_2^2)\, x_1^2 e^{-\lambda(\mathbf{r}_1^2 + \mathbf{r}_2^2)} = \frac{3}{4\lambda} = \frac{3\hbar}{4\Omega}, \quad (6.42)$$

and for $P_{x_1}^2$,

$$\langle P_{x_1}^2 \rangle = -\hbar^2 \lambda N_1^2 \int_{-\infty}^{\infty} d\mathbf{r}_1 \int_{-\infty}^{\infty} d\mathbf{r}_2 \{ |\mathbf{r}_1 - \mathbf{r}_2|^2 \left(\lambda x_1^2 - 1 \right) - x_1^2 \} e^{-\lambda(\mathbf{r}_1^2 + \mathbf{r}_2^2)}$$

$$= \frac{3\hbar^2 \lambda}{4} = \frac{3\hbar\Omega}{4}. \quad (6.43)$$

In order to discuss rotationally invariant quartic interactions we have to evaluate, for example, the expectation values $\langle Q_{x_\sigma}^4 \rangle$, $\langle Q_{x_\sigma}^2 Q_{x_{\sigma'}}^2 \rangle$ and $\langle Q_{x_\sigma}^2 Q_{y_{\sigma'}}^2 \rangle$. The first one has the form

$$\langle Q_{x_1}^4 \rangle = N_1^2 \int_{-\infty}^{\infty} d\mathbf{r}_1 \int_{-\infty}^{\infty} d\mathbf{r}_2 \, |\mathbf{r}_1 - \mathbf{r}_2|^2 \, x_1^4 e^{-\lambda(\mathbf{r}_1^2 + \mathbf{r}_2^2)} = \frac{3}{2\lambda^2} = \frac{3\hbar^2}{2\Omega^2},$$
(6.44)

and the same result can be obtained for $\langle Q_{y_1}^4 \rangle$, etc. Next, it can easily be seen that expectation values like $\langle Q_{x_\sigma}^2 Q_{x_{\sigma'}}^2 \rangle$ and $\langle Q_{x_\sigma}^2 Q_{y_{\sigma'}}^2 \rangle$ enjoy the same property as (6.37). Finally, there exists one more interesting expectation value $\langle Q_{x_1} Q_{x_2} \rangle$, which vanishes for bosons, but is nonzero for fermions,

$$\langle Q_{x_1} Q_{x_2} \rangle = -2 N_1^2 \int_{-\infty}^{\infty} d\mathbf{r}_1 \int_{-\infty}^{\infty} d\mathbf{r}_2 x_1 x_2 \left(\mathbf{r}_1 \cdot \mathbf{r}_2 \right) e^{-\lambda(\mathbf{r}_1^2 + \mathbf{r}_2^2)}$$

$$= -\frac{1}{4\lambda} = -\frac{\hbar}{4\Omega}, \quad (6.45)$$

which again, due to the symmetry of the problem, is the same for $\langle Q_{y_1} Q_{y_2} \rangle$.

****

A parallel calculation evaluates similar moments for anyons in [AdK14] for a general value of γ, which as can be expected involves a number of special functions. The reader interested in those details is directed to that paper. Below we only summarize the results for evaluating the enhanced classical Hamiltonian for a special model.

6.4   Enhanced Classical Hamiltonian for a Quartic Model

In this section we present the enhanced classical Hamiltonian for the two-particle model with a quartic interaction given by

$$\mathcal{H}(\mathbf{P}, \mathbf{Q}) = \sum_{\sigma=1}^{2} \left[\frac{\mathbf{P}_\sigma^2}{2m} + \frac{m\varpi^2}{2} \mathbf{Q}_\sigma^2 \right] + gV, \quad V = \left(\mathbf{Q}_1^2 + \mathbf{Q}_2^2\right)^2, \quad (6.46)$$

$$\mathbf{P}_\sigma = (P_{x_\sigma}, P_{y_\sigma}), \quad \mathbf{Q}_\sigma = (Q_{x_\sigma}, Q_{y_\sigma}),$$

where m represent the mass of the particles, ϖ is the harmonic potential frequency, and g is the quartic interaction coupling constant.

The enhanced classical Hamiltonian is identified with the expectation value of the Hamiltonian operator (6.46) with respect to the coherent states,

$$H(\mathbf{p}, \mathbf{q}) = \langle \mathbf{p}, \mathbf{q} | \mathcal{H}(\mathbf{P}, \mathbf{Q}) | \mathbf{p}, \mathbf{q} \rangle \qquad (6.47)$$

$$= \sum_{\sigma=1}^{2} \left[\frac{\mathbf{p}_\sigma^2}{2m} + \frac{m\varpi^2}{2} \mathbf{q}_\sigma^2 + \frac{\langle \mathbf{P}_\sigma^2 \rangle}{2m} + \frac{m\varpi^2}{2} \langle \mathbf{Q}_\sigma^2 \rangle \right] + g \langle \mathbf{p}, \mathbf{q} | V | \mathbf{p}, \mathbf{q} \rangle,$$

where the coherent states are defined by

$$|\mathbf{p}, \mathbf{q}\rangle = \prod_{\sigma=1}^{2} U(\mathbf{q}_\sigma) V(\mathbf{p}_\sigma) |\eta\rangle, \qquad (6.48)$$

$$U(\mathbf{q}_\sigma) = e^{-i\mathbf{q}_\sigma \cdot \mathbf{P}_\sigma / \hbar}, \quad V(\mathbf{p}_\sigma) = e^{i\mathbf{p}_\sigma \cdot \mathbf{Q}_\sigma / \hbar}.$$

The enhanced classical Hamiltonian for bosons, fermions, or anyons is formally the same, except that the numerical values of the coefficients proportional to $\hbar$ have a different value for each type of particle considered. The expectation value of the potential V with respect to the coherent states (6.48) is

$$\langle \mathbf{p}, \mathbf{q} | V | \mathbf{p}, \mathbf{q} \rangle = \sum_{\sigma=1}^{2} \left[\left(\mathbf{q}_{x_\sigma}^2 + \mathbf{q}_{y_\sigma}^2 \right)^2 + 10 \, \mathbf{q}_\sigma^2 \langle Q_{x_1}^2 \rangle \right] \qquad (6.49)$$

$$+ 2 (\mathbf{q}_1 \cdot \mathbf{q}_2) \left(\langle Q_{x_1} Q_{x_2} \rangle + \langle Q_{y_1} Q_{y_2} \rangle \right) + \frac{16}{3} \langle Q_{x_1}^4 \rangle + 4 \langle Q_{x_1}^2 Q_{x_2}^2 \rangle.$$

Labeling $H_k(\mathbf{p}, \mathbf{q})$ as the enhanced Hamiltonian for bosons ($k = b$), for fermions ($k = f$), and for anyons ($k = \gamma$), and using the results of the previous subsections, we list below the enhanced classical Hamiltonians: for bosons

$$H_b(\mathbf{p}, \mathbf{q}) = \mathcal{H}(\mathbf{p}, \mathbf{q}) + \hbar \sum_{\sigma=1}^{2} \left(\frac{3g}{\Omega} \right) \mathbf{q}_\sigma^2 + \hbar \left(\frac{\Omega}{m} + \frac{m\varpi^2}{\Omega} \right) + \hbar^2 \left(\frac{3g}{\Omega^2} \right);$$

$$(6.50)$$

for fermions

$$H_f\left(\mathbf{p},\mathbf{q}\right) = \mathcal{H}\left(\mathbf{p},\mathbf{q}\right) + 6\hbar\sum_{\sigma=1}^{2}\left(\frac{3g}{\Omega}\right)\mathbf{q}_\sigma^2$$
$$+\hbar\left(\frac{3\Omega}{2m} + \frac{3m\varpi^2}{2\Omega}\right) + 2\hbar^2\left(\frac{3g}{\Omega^2}\right) ; \qquad (6.51)$$

and, after extensive calculations (see [AdK14]), for anyons

$$H_\gamma\left(\mathbf{p},\mathbf{q}\right) = \mathcal{H}\left(\mathbf{p},\mathbf{q}\right)$$
$$+\hbar\left[\frac{\Omega}{m}\left(1 + \gamma\frac{{}_2F_1\left(\frac{2-\gamma}{2},\frac{1-\gamma}{2};\frac{3}{2};1\right)}{{}_2F_1\left(\frac{1-\gamma}{2},-\frac{\gamma}{2};\frac{3}{2};1\right)} - \frac{\gamma}{2}\right) + \frac{m\varpi^2}{2\Omega}\left(2 + \gamma\right)\right]$$
$$+\frac{\hbar g\left(\gamma + 2\right)}{\Omega}\left[\sum_{\sigma=1}^{2}\frac{5\mathbf{q}_\sigma^2}{2} - \left(\mathbf{q}_1\cdot\mathbf{q}_2\right)\left(\frac{{}_2F_1\left(-\frac{1+\gamma}{2},-\frac{\gamma}{2};\frac{3}{2};1\right)}{{}_2F_1\left(\frac{1-\gamma}{2},-\frac{\gamma}{2};\frac{3}{2};1\right)} - 1\right)\right]$$
$$+\frac{\hbar^2\left(\gamma + 3\right)\left(\gamma + 2\right)}{6\Omega^2}\left({}_2F_1\left(\frac{1-\gamma}{2},-\frac{\gamma}{2};\frac{3}{2};1\right)\right)^{-1} \qquad (6.52)$$
$$\times\left\{5\ {}_2F_1\left(\frac{1-\gamma}{2},-\frac{\gamma}{2};\frac{5}{2};1\right) + \frac{\gamma\left(\gamma - 1\right)}{5}\ {}_2F_1\left(\frac{3-\gamma}{2},\frac{2-\gamma}{2};\frac{7}{2};1\right)\right\}.$$

where ${}_2F_1$ denotes a standard hypergeometric function, which is defined by

$$_2F_1(a,b;c;z) = \sum_{n=0}^{\infty}\frac{(a)_n\,(b)_n}{(c)_n}\frac{z^n}{n!} , \qquad (6.53)$$

where $(u)_0 = 1$ and for $n > 0$, $(u)_n = u(u+1)\cdots(u+n-1)$.

Finally, the *classical Hamiltonian* $H_c\left(\mathbf{p},\mathbf{q}\right)$, in which $\hbar \to 0$, has the same form for all three cases, namely

$$H_c\left(\mathbf{p},\mathbf{q}\right) = \lim_{\hbar\to0}H_k\left(\mathbf{p},\mathbf{q}\right)$$
$$= \tfrac{1}{2}\Sigma_{\sigma=1}^{2}[m^{-1}\mathbf{p}_\sigma^2 + m\varpi^2\mathbf{q}_\sigma^2] + g\left(\Sigma_{\sigma=1}^{2}\ \mathbf{q}_\sigma^2\right)^2 , \qquad (6.54)$$

as expected.

6.5　Comments

In this chapter we have focussed on a central question in enhanced quantization using canonical coherent states, namely, the choice of the fiducial vector and the issues that choice involves. Initially, it was argued that a good choice is largely dictated by the explicit form of the Hamiltonian operator under consideration, and, in many cases the choice of the unique ground state as the fiducial vector has several virtues. However, that choice can also be relaxed to consider other fiducial vectors, and we have illustrated that choice by focussing attention on a Hamiltonian operator with a quartic interaction. One reason behind this choice is basically due to the fact that it is, effectively, the simplest example in which $\mathcal{O}(\hbar; \mathbf{p}, \mathbf{q})$ coefficients of dynamical terms are involved that modify the classical description. Secondly, this choice can exhibit an example with symmetry, e.g., rotational invariance, and very likely similar properties can be extended to other models. In particular, fiducial vectors based on a Gaussian form are appropriate for bosons to deal with associated Hamiltonian operators and the associated $\hbar$-dynamical coefficients can be treated consistently with them.

Although commonly used, it is a fact that Gaussian fiducial vectors are not always suitable to consistently describe certain physical systems. In this respect, we have chosen two examples where this form can not be used: the enhanced quantization of fermions and anyons. In these cases the fiducial vectors can not be independent Gaussians, and, instead, they must involve cross correlations for fermions or anyons. This latter property has a necessary physical consequence, namely, ensuring that permutations of the variables of the coherent-state representation of Hilbert space vectors involve the required change of phase.

Earlier, non-Gaussian fiducial vectors were used in the coherent states constructed for fermions and anyons. We have calculated several expectation values for both systems and, in this regard, the exact calculations for anyons also include the corresponding results for fermions simply by choosing $\gamma = 1$. Using these results, we have calculated the enhanced classical Hamiltonian and after taking the limit in which $\hbar \to 0$, we have shown that bosons, fermions, and anyons all have the same classical Hamiltonian, despite the fundamental differences between their properties when $\hbar > 0$.

Chapter 7

Enhanced Quantization of Rotationally Invariant Models

WHAT TO LOOK FOR

Rotationally invariant models, either for vectors or square matrices, involve a high degree of symmetry that can help in finding solutions of such systems both classically and quantum mechanically. As the number of degrees of freedom approaches infinity, the quantum mechanical solutions based on canonical quantization are forced to become those of free systems even when the original classical systems were not free. This inevitable consequence of canonical quantization provides fairly simple examples of unnatural behavior of traditional quantization. On the other hand, enhanced quantization provides completely acceptable results for these models by appealing to the weak correspondence principle along with reducible representations for the basic kinematical operators. Thus, the models studied in this chapter offer relatively simple examples of systems that are satisfactorily solved by enhanced quantization procedures when conventional quantization procedures yield unnatural results.

7.1 Introduction and Background

In earlier work [Kla65; AKT66; Kla00] we have developed the quantum theory of real, classical vector models with Hamiltonians of the form

$$H_c(\overrightarrow{p}, \overrightarrow{q}) = \tfrac{1}{2}[\overrightarrow{p}^2 + m_0^2 \overrightarrow{q}^2] + \lambda(\overrightarrow{q}^2)^2 , \qquad (7.1)$$

where $\overrightarrow{p} = (p_1, p_2, \ldots, p_N)$, $\overrightarrow{q} = (q_1, q_2, \ldots, q_N)$, $\overrightarrow{p}^2 = \Sigma_{n=1}^N p_n^2$, $\overrightarrow{q}^2 = \Sigma_{n=1}^N q_n^2$; here $N \le \infty$, and when $N = \infty$, we require that $\overrightarrow{p}^2 + \overrightarrow{q}^2 < \infty$. In addition, $\overrightarrow{p}$ and $\overrightarrow{q}$ denote canonically conjugate sets of variables with a Poisson bracket such that $\{q_a, p_b\} = \delta_{ab}$. If $O \in \mathbf{SO}(N)$ denotes an $N \times N$ orthogonal rotation matrix, i.e., $OO^T = O^T O = \mathbb{1}$, the unit matrix, O^T is the transpose of O, and $\det(O) = 1$, then clearly $H_c(O\overrightarrow{p}, O\overrightarrow{q}) = H_c(\overrightarrow{p}, \overrightarrow{q})$. Also, the choice of a quartic interaction is not special since we are also able to analyze more general potentials of the form $V(\overrightarrow{q}^2)$.

In the next section, we briefly review the quantization process for this model but let us note here that if we assume, as is customary, that the solution for $N = \infty$ involves a ground state that is real, unique, and $\mathbf{SO}(\infty)$ invariant, then only trivial (= free) models exist. A proper solution is found outside standard procedures and involves reducible operator representations.

After that study, we take up the quantization of real, symmetric matrix models that have classical Hamiltonians of the form

$$H_c(p, q) = \tfrac{1}{2}[\mathrm{Tr}(p^2) + m_0^2 \mathrm{Tr}(q^2)] + \lambda \mathrm{Tr}(q^4) , \qquad (7.2)$$

where the variables $p \equiv \{p_{ab}\}_{a,b=1}^{N,N}$, and $p_{ba} = p_{ab}$, $q \equiv \{q_{ab}\}_{a,b=1}^{N,N}$, and $q_{ba} = q_{ab}$, $\mathrm{Tr}(p^2) = \Sigma_{a,b=1}^{N,N} p_{ab} p_{ba}$, $\mathrm{Tr}(q^2) = \Sigma_{a,b=1}^{N,N} q_{ab} q_{ba}$, and $\mathrm{Tr}(q^4) = \Sigma_{a,b,c,d=1}^{N,N,N,N} q_{ab} q_{bc} q_{cd} q_{da}$. The variables p and q are canonically conjugate and obey the Poisson bracket $\{q_{ab}, p_{cd}\} = \tfrac{1}{2}(\delta_{ac}\delta_{bd} + \delta_{ad}\delta_{bc})$. Clearly, this Hamiltonian is invariant under transformations where $p \to OpO^T$, and $q \to OqO^T$, when $O \in \mathbf{SO}(N)$. Other $\mathbf{SO}(N)$-invariant interactions may be considered as well. One example would be a classical Hamiltonian of the form

$$H_c(p, q) = \tfrac{1}{2}[\mathrm{Tr}(p^2) + m_0^2 \mathrm{Tr}(q^2)] + \lambda [\mathrm{Tr}(q^2)]^2 , \qquad (7.3)$$

which shares the same invariance properties, but is actually equivalent to the vector model considered above. Since that vector model requires special care in finding a solution, we may expect that some of the matrix models will also require an unconventional analysis.

We remark that the parameter N relates only to the size of the $N \times N$ real, symmetric matrices or in the form $N(N+1)/2$ to the number of independent elements in such matrices, and this limitation applies to the quantum story as well as the classical story. Thus our study of matrices is not related to the more common study that employs $1/N$-expansion techniques, etc.; see, e.g., [Wiki-i].

7.1.1 *Enhanced quantization and the weak correspondence principle*

Since this chapter will be the first time that we encounter reducible operator representations in any significant role, it is appropriate that we offer a brief review of how that possibility arises.

In conventional quantization, for a single degree of freedom, suitable canonical phase-space coordinates (p, q) are promoted to irreducible Hermitian operators, $p \to P$ and $q \to Q$, where $[Q, P] = i\hbar\mathbb{1}$, and the classical Hamiltonian $H_c(p, q)$ guides the choice of the Hamiltonian operator $\mathcal{H} = H_c(P, Q)$, modulo terms of order $\hbar$, which is then used in Schrödinger's equation. This is the standard classical/quantum connection.

Our procedures offer a different form of classical/quantum connection. Recall, again, that the quantum action functional is given by

$$A_Q = \int_0^T \langle \psi(t) | [i\hbar(\partial/\partial t) - \mathcal{H}] | \psi(t) \rangle \, dt , \qquad (7.4)$$

which, under general stationary variations, leads to Schrödinger's equation. However, if we limit ourselves to simple translations and simple uniform motion, rather than general variations—which, thanks to Galilean covariance, we can perform as macroscopic observers without disturbing the system—then the states in question are limited to $|\psi(t)\rangle \to |p(t), q(t)\rangle$. Here, $|p, q\rangle$ are canonical coherent states, given by

$$|p, q\rangle = e^{-iqP/\hbar} e^{ipQ/\hbar} |\eta\rangle , \qquad (7.5)$$

where P and Q need to be self adjoint to generate unitary transformations, and in the present case, the normalized fiducial vector $|\eta\rangle$ may be chosen as a general unit vector for which $\langle \eta | P | \eta \rangle = \langle \eta | Q | \eta \rangle = 0$. Thus, the restricted form of the quantum action functional becomes

$$A_{Q(R)} = \int_0^T \langle p(t), q(t) | [i\hbar(\partial/\partial t) - \mathcal{H}] | p(t), q(t) \rangle \, dt$$
$$= \int_0^T [p(t)\dot{q}(t) - H(p(t), q(t))] \, dt , \qquad (7.6)$$

namely the classical action functional enhanced by the fact that $\hbar > 0$ still. The Hamiltonian connection, called the *Weak Correspondence Principle* [Kla67], is given by

$$H(p, q) \equiv \langle p, q | \mathcal{H}(P, Q) | p, q \rangle$$
$$= \langle \eta | \mathcal{H}(P + p, Q + q) | \eta \rangle$$
$$= \mathcal{H}(p, q) + \mathcal{O}(\hbar; p, q) . \qquad (7.7)$$

At this point we have previously noted that in enhanced quantization the relation of the classical variables, (p, q), to the quantum variables, (P, Q), is entirely different from that of the usual procedure, and yet the new procedures have led to the same result as conventional canonical quantization, specifically, the same Hamiltonian operator, modulo terms of order $\hbar$, that is used in the standard approach. But, in (7.7) there is no requirement that P and Q are irreducible. If we admit other, nontrivial, operators in $\mathcal{H}(P, Q)$ which commute with both P and Q, the enhanced classical action functional still emerges. This relation between the quantum Hamiltonian and the enhanced classical Hamiltonian has been called the weak correspondence principle. In this chapter we will exploit this possibility, and in so doing, we will closely follow [Kl14a].

It is noteworthy that the models discussed in this chapter involve nonsingular potentials in the sense that the domain of the classical action functional for the free model is the same as the domain of the classical action functional for the interacting model. Classically, that means that the pseudofree model (the one that emerges as the coupling constant vanishes) is identical to the free model. That would suggest that conventional quantization techniques should do the job. However, we will learn that is not the case. Unlike some earlier examples, the resolution of our problems does not lie with affine quantization but instead in another direction allowed by enhanced quantization, namely, with reducible representations of the basic kinematical operators.

In an effort to show the power of enhanced quantization as compared to conventional quantization for the chosen problems, our presentation is basically short and to the point. Many additional details for the vector models are available in [Kla65; Kla00].

7.2   Rotationally Symmetric Vector Models

Let us consider the quantization of the vector model with a classical Hamiltonian given by [model #(4) in Chap. 1]

$$H_c(\overrightarrow{p}, \overrightarrow{q}) = \tfrac{1}{2}[\overrightarrow{p}^2 + m_0^2\,\overrightarrow{q}^2] + \lambda(\overrightarrow{q}^2)^2 \; ; \qquad (7.8)$$

a definition of the symbols appears below Eq. (7.1). As a consequence of rotational invariance, every classical solution is equivalent to a solution for $N = 1$ if $\overrightarrow{p} \,\|\, \overrightarrow{q}$ at time $t = 0$, or to a solution for $N = 2$ if $\overrightarrow{p} \,\|\!\!\!/\, \overrightarrow{q}$ at time $t = 0$. Moreover, solutions for $N = \infty$ may be derived from those for $N < \infty$ by the limit $N \to \infty$, provided we maintain $(\overrightarrow{p}^2 + \overrightarrow{q}^2) < \infty$. Indeed, there is the interesting perspective that the model for $N = \infty$

includes *all other models* since any other model with $N < \infty$ is obtained simply by setting $p_l = q_l = 0$ for all $l > N$.

A conventional canonical quantization begins with $\vec{p} \to \vec{P}$, $\vec{q} \to \vec{Q}$, which are irreducible operators that obey $[Q_l, P_n] = i\hbar\delta_{ln}\mathbb{1}$ as the only non-vanishing commutation relation. For a free model, with mass m and $\lambda = 0$, the quantum Hamiltonian $\mathcal{H}_0 = \frac{1}{2} : (\vec{P}^2 + m^2 \vec{Q}^2) :$, where $: (\cdot) :$ denotes normal ordering with respect to the ground state of the Hamiltonian. This expression has the feature that the Hamiltonian operator for $N = \infty$ is obtained as the limit of those for which $N < \infty$. Moreover, with the normalized ground state $|0\rangle$ of the Hamiltonian operator chosen as the fiducial vector for canonical coherent states, i.e.,

$$|\vec{p}, \vec{q}\rangle = \exp[-i\vec{q} \cdot \vec{P}/\hbar] \exp[i\vec{p} \cdot \vec{Q}/\hbar]|0\rangle, \qquad (7.9)$$

it follows that

$$\begin{aligned} H_0(\vec{p}, \vec{q}) &= \langle \vec{p}, \vec{q} | \tfrac{1}{2} : (\vec{P}^2 + m^2 \vec{Q}^2) : |\vec{p}, \vec{q}\rangle \\ &= \langle 0| \tfrac{1}{2} : [(\vec{P} + \vec{p})^2 + (\vec{Q} + \vec{q})^2] : |0\rangle \\ &= \tfrac{1}{2}(\vec{p}^2 + m^2 \vec{q}^2), \end{aligned} \qquad (7.10)$$

as desired, for all $N \le \infty$, provided that $(\vec{p}^2 + \vec{q}^2) < \infty$.

For comparison with later relations, we recall the characteristic function (i.e., the Fourier transform) of the ground-state distribution for the free vector model with mass m given by

$$\begin{aligned} C_0(\vec{f}) &= M_0 \int e^{i\vec{f} \cdot \vec{x}/\hbar} e^{-m\vec{x}^2/\hbar} \Pi_{n=1}^N dx_n \\ &= e^{-\vec{f}^2/4m\hbar}, \end{aligned} \qquad (7.11)$$

where M_0 is a normalization factor, and this result formally holds for $N \le \infty$.

However, canonical quantization of the interacting vector models with $\lambda > 0$ leads to trivial results for $N = \infty$. To show this, we assume that the Schrödinger representation of the ground state of an interacting model is real, unique, and rotationally invariant. As a consequence, the characteristic function of any ground-state distribution has the form (note:

$$\vec{f}^2 \equiv |f|^2 \equiv \Sigma_{n=1}^{N} f_n^2 \text{ and } r^2 \equiv \Sigma_{n=1}^{N} x_n^2)$$

$$C_N(\vec{f}) = \int e^{i\Sigma_{n=1}^{N} f_n x_n / \hbar} \, \Psi_0(r)^2 \, \Pi_{n=1}^{N} dx_n$$

$$= \int e^{i|f|r\cos(\theta)/\hbar} \, \Psi_0(r)^2 \, r^{N-1} \, dr \, \sin(\theta)^{N-2} \, d\theta \, d\Omega_{N-2}$$

$$\simeq M' \int e^{-\vec{f}^2 r^2 / 2(N-2)\hbar^2} \, \Psi_0(r)^2 \, r^{N-1} \, dr \, d\Omega_{N-2}$$

$$\to \int_0^{\infty} e^{-b\vec{f}^2/\hbar} \, w(b) \, db \,, \tag{7.12}$$

assuming convergence, where a steepest descent integral has been performed for θ, and in the last line we have taken the limit $N \to \infty$; additionally, $w(b) \geq 0$, and $\int_0^{\infty} w(b)\,db = 1$. This is the result based on symmetry [Sch38].

To examine the issue of uniqueness, let us assume that $w(b) = |\alpha|^2 \delta(b - 1/4m_1) + |\beta|^2 \delta(b - 1/4m_2)$, $m_1 \neq m_2$, where $|\alpha\beta| > 0$ and $|\alpha|^2 + |\beta|^2 = 1$. This implies that the Hamiltonian operator is reducible and that it has the form $\mathcal{H} = \mathcal{H}_1 \oplus \mathcal{H}_2$, and thus the ground state is given by $\alpha|0_1\rangle \oplus \beta|0_2\rangle$. However, for this Hamiltonian, the ground state is highly degenerate since *any* vector of the form $\gamma|0_1\rangle \oplus \delta|0_2\rangle$, with $|\gamma\delta| > 0$ and $|\gamma|^2 + |\delta|^2 = 1$, is also a ground state; such an argument applies for every choice of $w(b)$ not given by a single delta function. Uniqueness of the ground state then ensures that $w(b) = \delta(b - 1/4\tilde{m})$, for some $\tilde{m} > 0$, implying that the quantum theory is that of a free theory, i.e., *the quantum theory is trivial!* In addition, the classical limit of the resultant quantum theory is a free classical theory, which differs from the original, nonlinear classical theory.

This kind of result signals an *unnatural quantization*, and, moreover, a quantization that we interpret as an *unsatisfactory quantization*. Here is perhaps the simplest example where conventional quantization leads to an unnatural result; our goal is to use enhanced quantization to find an acceptable quantization of this model for $N = \infty$.

7.2.1   *Nontrivial quantization of vector models*

The way around this unsatisfactory result is to let the representations of $\vec{P}$ and $\vec{Q}$ be *reducible*. The weak correspondence principle, namely $H(\vec{p}, \vec{q}) \equiv \langle \vec{p}, \vec{q} | \mathcal{H} | \vec{p}, \vec{q} \rangle$, ensures that the enhanced classical Hamiltonian depends only on the proper variables, but, unlike conventional quantization procedures, there is no rule forbidding the Hamiltonian from being a function of other, non-trivial operators that commute with $\overrightarrow{P}$ and $\overrightarrow{Q}$, thus making the usual operators reducible. A detailed study [Kla65; Kla00] of the proper reducible representation—*still in accord with the argument above that limits the ground-state functional form to be a Gaussian*—leads to the following formulation. Let $\overrightarrow{R}$ and $\overrightarrow{S}$ represent a new set of canonical operators, independent of the operators $\overrightarrow{P}$ and $\overrightarrow{Q}$, and which obey the commutation relation $[S_l, R_n] = i\hbar\delta_{ln}\mathbb{1}$. We introduce two sets of compatible annihilation operators that annihilate the unit vector $|0, 0; \zeta\rangle$, namely

$$[m(\overrightarrow{Q} + \zeta\overrightarrow{S}) + i\overrightarrow{P}]|0, 0; \zeta\rangle = 0 \ ,$$
$$[m(\overrightarrow{S} + \zeta\overrightarrow{Q}) + i\overrightarrow{R}]|0, 0; \zeta\rangle = 0 \ . \qquad (7.13)$$

If we introduce eigenvectors $|\overrightarrow{x}, \overrightarrow{y}\rangle$ such that $\overrightarrow{Q}|\overrightarrow{x}, \overrightarrow{y}\rangle = \overrightarrow{x}|\overrightarrow{x}, \overrightarrow{y}\rangle$ and $\overrightarrow{S}|\overrightarrow{x}, \overrightarrow{y}\rangle = \overrightarrow{y}|\overrightarrow{x}, \overrightarrow{y}\rangle$, it follows that

$$\langle\overrightarrow{x}, \overrightarrow{y}|0, 0; \zeta\rangle = M \exp[-(\overrightarrow{x}^2 + 2\zeta\overrightarrow{x} \cdot \overrightarrow{y} + \overrightarrow{y}^2)/2\hbar] \ , \qquad (7.14)$$

where M is a normalization factor, and the condition $0 < \zeta < 1$ ensures normalizability. These two annihilation operators lead directly to two, related, free Hamiltonian operators,

$$\mathcal{H}_{0\,PQ} \equiv \tfrac{1}{2} : (\overrightarrow{P}^2 + m^2(\overrightarrow{Q} + \zeta\overrightarrow{S})^2) : \ ,$$
$$\mathcal{H}_{0\,RS} \equiv \tfrac{1}{2} : (\overrightarrow{R}^2 + m^2(\overrightarrow{S} + \zeta\overrightarrow{Q})^2) : \ , \qquad (7.15)$$

for which (7.14) is a common, unique, Gaussian ground state, which is also used to define normal ordering. Let new coherent states, which span the Hilbert space of interest, be defined with this ground state as the fiducial vector, as given by

$$|\overrightarrow{p}, \overrightarrow{q}\rangle \equiv \exp[-i\overrightarrow{q} \cdot \overrightarrow{P}/\hbar]\exp[i\overrightarrow{p} \cdot \overrightarrow{Q}/\hbar]|0, 0; \zeta\rangle \ , \qquad (7.16)$$

and it follows that

$$
\begin{aligned}
H(\overrightarrow{p}, \overrightarrow{q}) &= \langle\overrightarrow{p}, \overrightarrow{q}|\{\mathcal{H}_{0\,PQ} + \mathcal{H}_{0\,RS} + 4v : \mathcal{H}_{0\,RS}^2 :\}|\overrightarrow{p}, \overrightarrow{q}\rangle \\
&= \tfrac{1}{2}[\overrightarrow{p}^2 + m^2(1 + \zeta^2)\overrightarrow{q}^2] + v\zeta^4 m^4(\overrightarrow{q}^2)^2 \\
&\equiv \tfrac{1}{2}[\overrightarrow{p}^2 + m_0^2\overrightarrow{q}^2] + \lambda(\overrightarrow{q}^2)^2 \ , \qquad (7.17)
\end{aligned}
$$

as required; and this solution is valid for *all* N, $1 \leq N \leq \infty$, provided that $(\overrightarrow{p}^2 + \overrightarrow{q}^2) < \infty$. Observe in this case that $H(\overrightarrow{p}, \overrightarrow{q}) = H_c(\overrightarrow{p}, \overrightarrow{q})$ since there is no $\mathcal{O}(\hbar)$ term to distinguish them.

7.3 Rotationally Symmetric Matrix Models

As discussed in Sec. 7.1, we now turn our attention to the quantization of classical systems that have a classical Hamiltonian given by [model #(5) in Chap. 1]

$$H_c(p, q) = \tfrac{1}{2}[\text{Tr}(p^2) + m_0^2\text{Tr}(q^2)] + \lambda\,\text{Tr}(q^4) , \qquad (7.18)$$

where the variables p and q are $N \times N$ real, symmetric matrices, as explained after Eq. (7.2). The Hamiltonians for such models are invariant under matrix transformations $p \to OpO^T$ and $q \to OqO^T$, where $O \in \mathbf{SO}(N)$. The free model, with mass m and $\lambda = 0$, is readily quantized by promoting $p \to P$ and $q \to Q$, which are Hermitian, symmetric (in their indices) matrix operators with the property that the only nonvanishing commutator is

$$[Q_{ab}, P_{cd}] = i\hbar\tfrac{1}{2}(\delta_{ac}\delta_{bd} + \delta_{ad}\delta_{bc})\,\mathbb{1} . \qquad (7.19)$$

The free Hamiltonian operator is given by

$$\mathcal{H}_0 \equiv \tfrac{1}{2} : [\text{Tr}(P^2) + m^2\text{Tr}(Q^2)] : , \qquad (7.20)$$

and the normalized ground state $|0\rangle$ of this Hamiltonian is unique and rotationally invariant, is used to define normal ordering, and is given by

$$\langle x|0\rangle = M' \exp[-m\text{Tr}(x^2)/2\hbar] , \qquad (7.21)$$

where M' is a normalization factor, x is a real, symmetric $N \times N$ matrix, and $|x\rangle$ are eigenvectors of Q, namely, $Q|x\rangle = x|x\rangle$. Coherent states for this example are given by

$$|p, q\rangle = e^{-i\text{Tr}(qP)/\hbar}\, e^{i\text{Tr}(pQ)/\hbar}|0\rangle , \qquad (7.22)$$

and it follows that

$$\begin{aligned}
H_0(p, q) &= \langle p, q|\tfrac{1}{2} : [\text{Tr}(P^2) + m^2\text{Tr}(Q^2)] : |p, q\rangle \\
&= \langle 0|\tfrac{1}{2} : [\text{Tr}((P + p\mathbb{1})^2) + m^2\text{Tr}((Q + q\mathbb{1})^2)] : |0\rangle \\
&= \tfrac{1}{2}[\text{Tr}(p^2) + m^2\text{Tr}(q^2)] ,
\end{aligned} \qquad (7.23)$$

as desired.

Again, for comparison purposes, we compute the characteristic function for the ground-state distribution of the free model, which is given by

$$\begin{aligned}
C_0(f) &= M'^2 \int e^{i\text{Tr}(fx)/\hbar}\, e^{-m\text{Tr}(x^2)/\hbar}\, \Pi_{a\leq b=1}^{N,N}\, dx_{ab} \\
&= e^{-\text{Tr}(f^2)/4m\hbar} ,
\end{aligned} \qquad (7.24)$$

where now f denotes a real, symmetric $N \times N$ matrix, and this relation formally holds for all $N \leq \infty$.

When $\lambda > 0$, conventional quantization procedures would require a rescaling of the quartic interaction term by replacing λ by λ/N, as dictated by a perturbation analysis. However, under reasonable assumptions, just as in the vector case discussed in Sec. 7.2, we are able to show that when $\lambda > 0$ and $N = \infty$, the quantum theory for the matrix models is trivial ($=$ free) just as was the case for the vector models. To show this, we again appeal to the ground state which we assume to be real, unique, and rotationally invariant for the models under consideration. While in the vector models rotational invariance meant the ground state was a function of only one variable, namely $\overrightarrow{x}^2$, that is not the case for the matrix models. For example, the ground state could be a function of $\text{Tr}(x^2)$, $\text{Tr}(x^3)$, $\text{Tr}(x^4)$, $\det(x)$, etc. We recognize the fact of many invariant forms, but for simplicity we shall just display only two of them, namely, $\text{Tr}(x^2)$ and $\text{Tr}(x^4)$.

As before, we consider the characteristic function of the ground-state distribution in the Schrödinger representation given by

$$C_N(f) = \int e^{i\text{Tr}(fx)/\hbar} \, \Psi_0[\text{Tr}(x^2), \text{Tr}(x^4)]^2 \, \Pi_{a\leq b=1}^{N,N} dx_{ab} \, , \quad (7.25)$$

where, again, f is a real, symmetric $N \times N$ matrix. As defined, $C_N(f)$ is clearly invariant under rotations such as $f \to OfO^T$. As a real, symmetric matrix, we can imagine choosing $O \in \mathbf{SO}(N)$ so as to diagonalize f, namely the matrix f is now of the form where $f_{ab} = \delta_{ab}f_a$, with f_a being the diagonal elements. This means that the expression

$$\text{Tr}(fx) = \sum_{a=1}^{N} f_a x_{aa} \, , \quad (7.26)$$

and thus only the N *diagonal elements* of the real, symmetric $N\times N$ matrix x enter into the exponent of (7.25).

Let us introduce a suitable form of spherical coordinates in place of the $N^* \equiv N(N+1)/2$ integration variables x_{ab}, $a \leq b$. We choose the radius variable r so that $r^2 \equiv \text{Tr}(x^2) = \Sigma_{a,b=1}^{N,N} x_{ab}^2$, and the first N angles are identified with the diagonal elements of x leading to $x_{11} = r\cos(\theta_1)$, and for $2 \leq a \leq N$, we set $x_{aa} = r[\Pi_{l=1}^{a-1} \sin(\theta_l)]\cos(\theta_a)$; we also extend the latter notation to include x_{11}. The remaining $N(N-1)/2$ variables x_{ab} for $a < b$ are expressed in a similar fashion, but they each involve a factor $1/\sqrt{2}$ to account for their double counting in the definition of r. In fact, we need not specify the off-diagonal elements in detail as they do not appear in the Fourier exponent term. Expressed in these spherical variables, the characteristic function is given by

$$C_N(f) = K_N \int \exp\{ir\Sigma_{a=1}^N f_a \left[\Pi_{l=1}^{a-1}\sin(\theta_l)\right]\cos(\theta_a)/\hbar\} \tag{7.27}$$
$$\times \Psi_0[r^2, \mathrm{Tr}(x^4)]^2 \, r^{N^*-1} dr \, \Pi_{a=1}^N \left[\sin(\theta_a)^{N^*-(a+1)} d\theta_a\right] d\Omega_{N(N-1)/2} \,,$$

where K_N is the extra coefficient arising from the $2^{-N(N-1)/4}$ factor in the Jacobian. Observe that the multiple $\sin(\theta_a)$ factors in the Jacobian have huge powers with $N^* - (N+1)$ being the smallest of these. As $N \to \infty$, it follows that these factors all grow (since $N^* = N(N+1)/2$), and thus all such terms are approximately $N^2/2$ for $N \gg 1$. That fact forces each θ_a, $1 \le a \le N$, to be constrained to an interval of order $1/N$ around $\pi/2$. We can use that fact in a steepest descent evaluation of each of the θ_a integrals, $1 \le a \le N$. In those integrals the factors $\sin(\theta_a)$ in the Fourier exponent may be set equal to unity, and the terms $\cos(\theta_a)$ are effectively linear in their deviation from $\pi/2$. While the argument $\mathrm{Tr}(x^2) = r^2$ in the ground state is independent of any angles, that is not the case for the argument $\mathrm{Tr}(x^4)$ and any other rotationally invariant term that may be there. However, unlike the appearance of the factors $\sin(\theta_a)$ in the Jacobian, the sine or cosine functions of any angles appearing in the ground state most probably do *not* enter with enormous powers $O(N^2/2)$—e.g., even for $\det(x)$ the maximum power would be N— and therefore they should have very little influence on the stationary evaluation of the first N angle integrals. On the other hand, even though the terms $\cos(\theta_a)$ in the Fourier exponent are $O(1/N)$, the factors f_a are arbitrarily large and such terms can not be ignored.

As a consequence, we are led to an approximate evaluation of the integral (7.27) given by

$$C_N(f) \simeq K_N' \int \exp\{-r^2\Sigma_{a=1}^N f_a^2/2[N^* - (a+1)]\hbar^2\}$$
$$\times \Psi_0[r^2, \mathrm{Tr}(x^4)]^2 \, r^{N^*-1} dr \, d\Omega_{N(N-1)/2}$$
$$\to \int_0^\infty e^{-b\mathrm{Tr}(f^2)/\hbar} W(b) \, db \,, \tag{7.28}$$

where in the last line, assuming convergence, we have taken the limit $N \to \infty$; in addition, $W(b) \ge 0$, and $\int_0^\infty W(b)db = 1$. This is the result based on symmetry. If we insist on uniqueness of the ground state, then again it follows that $W(b) = \delta(b - 1/4\overline{m})$, which leads to a trivial (= free) theory for some mass, $\overline{m}$.

7.3.1 *Nontrivial quantization of matrix models*

In finding a nontrivial solution for matrix models, we are guided by the procedures used for vector models. For clarity and comparison, we start with the matrix model (7.3) and afterwards consider the matrix model given by (7.2). Initially, we introduce reducible representations of the basic variables P and Q, the $N{\times}N$ Hermitian, symmetric (in their indices) matrix operators, which satisfy Heisenberg's commutation relations. In addition, we introduce a second, and independent, set of similar matrix operators R and S. We choose a unit vector in Hilbert space, which we call $|0,0;\zeta\rangle$, $0 < \zeta < 1$, and require that

$$[m(Q + \zeta S) + i P]|0,0;\zeta\rangle = 0 \,,$$
$$[m(S + \zeta Q) + i R]|0,0;\zeta\rangle = 0 \,. \tag{7.29}$$

In terms of eigenvectors $|x,y\rangle$ for both Q and S, respectively, it follows that

$$\langle x,y|0,0;\zeta\rangle = M \, \exp\{-m\,\mathrm{Tr}(x^2 + 2\zeta xy + y^2)/2\hbar\} \,, \tag{7.30}$$

a Gaussian state in accord with the discussion following (7.28). There are two related, free-like, Hermitian Hamiltonian expressions of interest, given by

$$\mathcal{H}_{1\,PQ} = \tfrac{1}{2}\mathrm{Tr}\{[m(Q + \zeta S) - iP][m(Q + \zeta S) + iP]\}$$
$$= \tfrac{1}{2} : [\mathrm{Tr}(P^2) + m^2\mathrm{Tr}((Q + \zeta S)^2)] : \,, \tag{7.31}$$

and

$$\mathcal{H}_{1\,RS} = \tfrac{1}{2}\mathrm{Tr}\{[m(S + \zeta Q) - iR][m(S + \zeta Q) + iR]\}$$
$$= \tfrac{1}{2} : [\mathrm{Tr}(R^2) + m^2\mathrm{Tr}((S + \zeta Q)^2)] : \,, \tag{7.32}$$

and $|0,0;\zeta\rangle$ is the unique ground state for both of them. For coherent states, we choose

$$|p,q\rangle = \exp[-i\mathrm{Tr}(qP)/\hbar]\,\exp[i\mathrm{Tr}(pQ/\hbar)]\,|0,0;\zeta\rangle \,, \tag{7.33}$$

and, for the final, total Hamiltonian, it follows that

$$H(p,q) = \langle p,q|\{\mathcal{H}_{1\,PQ} + \mathcal{H}_{1\,RS} + 4v : \mathcal{H}_{1\,RS}^2 :\}|p,q\rangle$$
$$= \tfrac{1}{2}[\mathrm{Tr}(p^2) + m^2(1 + \zeta^2)\mathrm{Tr}(q^2)] + vm^4\zeta^4[\mathrm{Tr}(q^2)]^2$$
$$\equiv \tfrac{1}{2}[\mathrm{Tr}(p^2) + m_0^2\mathrm{Tr}(q^2)] + \lambda[\mathrm{Tr}(q^2)]^2 \,, \tag{7.34}$$

as desired. Once again, we observe that $H(p,q) = H_c(p,q)$.

The solution associated with Eq. (7.2) is different from that just presented, but, again, according to the analysis that led to (7.28), we are still obliged to look for a solution based on a Gaussian ground state. Once again we start with the canonical matrix operator pair P and Q, as well as the canonical matrix operator pair R and S that we used in the solution of the model (7.34). However, we now need yet another matrix operator pair, namely, the independent canonical matrix pair T and U. This time we start with a ground state denoted by the unit vector $|0,0,0;\xi\rangle$ which is annihilated by three operators, namely

$$[m(Q + \xi(S + U)) + iP]|0,0,0;\xi\rangle = 0 ,$$
$$[m(S + \xi Q) + iR]|0,0,0;\xi\rangle = 0 , \qquad (7.35)$$
$$[m(U + \xi Q) + iT]|0,0,0;\xi\rangle = 0 .$$

If we introduce eigenvectors $|x,y,z\rangle$ for the three operators $Q, S,$ and U, respectively, then it follows that in the Schrödinger representation, the ground state is given by

$$\langle x,y,z|0,0,0;\xi\rangle = M'' \exp\{-m[\text{Tr}(x^2 + y^2 + z^2 + 2\xi(xy + xz))]/2\hbar\} , \qquad (7.36)$$

and the condition $0 < \xi < 1/\sqrt{2}$ ensures normalizability.

There are now *four* Hamiltonian-like expressions of interest, the first three of which have the vector $|0,0,0;\xi\rangle$ as their common, unique ground state, namely,

$$\mathcal{H}_{2\,PQ} = \tfrac{1}{2}\text{Tr}\{[m(Q + \xi(S + U)) - iP][m(Q + \xi(S + U)) + iP]\}$$
$$= \tfrac{1}{2} : \{\text{Tr}(P^2) + m^2\text{Tr}[(Q + \xi(S + U))^2]\} : , \qquad (7.37)$$

$$H_{2\,RS} = \tfrac{1}{2}\text{Tr}\{[m(S + \xi Q) - iR][m(S + \xi Q) + iR]\}$$
$$= \tfrac{1}{2} : \{\text{Tr}(R^2) + m^2\text{Tr}[(S + \xi Q)^2]\} : , \qquad (7.38)$$

$$H_{2\,TU} = \tfrac{1}{2}\text{Tr}\{[m(U + \xi Q) - iT][m(U + \xi Q) + iT]\}$$
$$= \tfrac{1}{2} : \{\text{Tr}(T^2) + m^2\text{Tr}[(U + \xi Q)^2]\} : , \qquad (7.39)$$

and the fourth Hamiltonian expression, which also annihilates $|0,0,0;\xi\rangle$, is given by

$$\mathcal{H}_{2\,RSTU} = \text{Tr}\{[m(S + \xi Q) - iR][m(S + \xi Q) + iR]$$
$$\times [m(U + \xi Q) - iT][m(U + \xi Q) + iT]\} \qquad (7.40)$$
$$= \text{Tr}\{: [R^2 + m^2(S + \xi Q)^2] :: [T^2 + m^2(U + \xi Q)^2] :\} .$$

For coherent states, we choose

$$|p,q\rangle \equiv \exp[-i\text{Tr}(qP)/\hbar]\exp[i\text{Tr}(pQ)/\hbar]|0,0,0;\xi\rangle , \qquad (7.41)$$

and it follows, for the final, total Hamiltonian, that

$$
\begin{aligned}
H(p,q) &= \langle p,q| \{ H_{2\,PQ} + \mathcal{H}_{2\,RS} + H_{2\,TU} + v\mathcal{H}_{2\,RSTU} \} |p,q\rangle \\
&= \tfrac{1}{2}[\operatorname{Tr}(p^2) + m^2(1 + 2\xi^2)\operatorname{Tr}(q^2)] + vm^4\xi^4\operatorname{Tr}(q^4) \\
&\equiv \tfrac{1}{2}[\operatorname{Tr}(p^2) + m_0^2\operatorname{Tr}(q^2)] + \lambda\operatorname{Tr}(q^4) ,
\end{aligned}
\tag{7.42}
$$

as desired! This expression is valid for *all* $N \le \infty$. Again, we have $H(p,q) = H_c(p,q)$.

Critical commentary

It is worthwhile to examine a natural candidate that was *not* chosen for the last model, and to see why we did not choose that natural "solution". We start be asking why was it necessary to employ *three* sets of canonical pairs when it seems that *two* canonical pairs should be enough. Before focussing on just two operators, however, let us first assume that a new unit vector, $|0,0;\xi\rangle$—hoping for two pairs and not three—is annihilated by three operators, namely

$$
\begin{aligned}
A\,|0,0;\xi\rangle &\equiv (1/\sqrt{2m\hbar})[m(Q + \xi(S + U)) + iP]\,|0,0;\xi\rangle = 0 , \\
B\,|0,0;\xi\rangle &\equiv (1/\sqrt{2m\hbar})[m(S + \xi Q) + iR]\,|0,0;\xi\rangle = 0 , \\
C\,|0,0;\xi\rangle &\equiv (1/\sqrt{2m\hbar})[m(U + \xi Q) + iT]\,|0,0;\xi\rangle = 0 .
\end{aligned}
\tag{7.43}
$$

The matrix annihilation operators A, B, and C are associated with matrix creation operators, $A^\dagger$, $B^\dagger$, and $C^\dagger$, and

$$
[A_{ab}, A_{cd}^\dagger] = [B_{ab}, B_{cd}^\dagger] = [C_{ab}, C_{cd}^\dagger] = \tfrac{1}{2}(\delta_{ac}\delta_{db} + \delta_{ad}\delta_{cb})\mathbb{1} . \tag{7.44}
$$

For simplicity hereafter, we sometimes set $m = \hbar = 1$. Thus $\mathcal{H}_{2\,PQ} = \operatorname{Tr}(A^\dagger A)$, $\mathcal{H}_{2\,RS} = \operatorname{Tr}(B^\dagger B)$, and $\mathcal{H}_{2\,TU} = \operatorname{Tr}(C^\dagger C)$. For a fourth expression, let us introduce $\mathcal{F} \equiv \operatorname{Tr}(B^\dagger B^\dagger BB)$. With coherent states given (with $\hbar$) by

$$
|p,q\rangle = \exp[-i\operatorname{Tr}(qP)/\hbar]\exp[i\operatorname{Tr}(pQ)/\hbar]\,|0,0;\xi\rangle , \tag{7.45}
$$

it follows (restoring m part way through) that

$$
\begin{aligned}
W(p,q) &\equiv \langle p,q| \{ \operatorname{Tr}(A^\dagger A) + \operatorname{Tr}(B^\dagger B) + 4v\operatorname{Tr}(B^\dagger B^\dagger BB) \} |p,q\rangle \\
&= \tfrac{1}{2}[\operatorname{Tr}(p^2) + m^2(1 + \xi^2)\operatorname{Tr}(q^2)] + vm^4\xi^4\operatorname{Tr}(q^4) \\
&\equiv \tfrac{1}{2}[\operatorname{Tr}(p^2) + m_0^2\operatorname{Tr}(q^2)] + \lambda\operatorname{Tr}(q^4) ,
\end{aligned}
\tag{7.46}
$$

as desired—or so it would seem.

The evaluation carried out so far is insufficient to determine whether $\mathcal{F} \equiv \operatorname{Tr}(B^\dagger B^\dagger BB)$ is a genuine Hermitian *operator* or, instead, merely a *form* requiring restrictions on both kets *and* bras. For $\mathcal{F}$ to be an acceptable operator, it is necessary that

$$\langle \phi | \mathcal{F}^\dagger \mathcal{F} | \phi \rangle < \infty \tag{7.47}$$

for a dense set of vectors $|\phi\rangle$. If we are able to bring the expression $\mathcal{F}^\dagger \mathcal{F}$ into normal order, without potential divergences, then we can, for example, use coherent states to establish that $\mathcal{F}$ is an acceptable operator. After a lengthy calculation, it follows that

$$\begin{aligned}
\mathcal{F}^\dagger \mathcal{F} &\equiv \mathrm{Tr}(B^\dagger B^\dagger BB)\,\mathrm{Tr}(B^\dagger B^\dagger BB) \\
&= \; : \mathrm{Tr}(B^\dagger B^\dagger BB)\,\mathrm{Tr}(B^\dagger B^\dagger BB) : \\
&\quad + : \mathrm{Tr}(B^\dagger B^\dagger BBBB^\dagger) : + : \mathrm{Tr}(B^\dagger B^\dagger BBB^\dagger B) : \\
&\quad + : \mathrm{Tr}(B^\dagger B^\dagger BB^\dagger BB) : + \mathrm{Tr}(B^\dagger B^\dagger B^\dagger BBB) \\
&\quad + (\tfrac{5}{4} + \tfrac{1}{2}N)\,\mathrm{Tr}(B^\dagger B^\dagger BB) + \tfrac{1}{4}\mathrm{Tr}(B^\dagger B^\dagger)\mathrm{Tr}(BB) \,.
\end{aligned} \tag{7.48}$$

This calculation was performed for $N \times N$ matrices, and the last line of (7.48) has a coefficient N which means as $N \to \infty$ this term would diverge and cause suitable states with two or more excitations to have an infinite expectation value for the supposed operator $\mathcal{F} = \mathrm{Tr}(B^\dagger B^\dagger BB)$. Hence, $\mathcal{F}$ is a *form* rather than an acceptable operator, and this fact rules out this proposed quantum solution of the model represented by (7.2). A similar computation shows that $\mathrm{Tr}(B^\dagger BB^\dagger B)$ is also a form for similar reasons, and so is $\mathrm{Tr}(B^\dagger C^\dagger CB)$.

On the other hand, we now show that $\mathcal{H} \equiv \mathrm{Tr}(B^\dagger BC^\dagger C)\,[= \mathrm{Tr}(C^\dagger CB^\dagger B)]$ is a genuine operator for all $N \le \infty$! To do so, let us bring the expression $\mathcal{H}^\dagger \mathcal{H}$ into normal ordered form. It follows that

$$\begin{aligned}
\mathcal{H}^\dagger \mathcal{H} &\equiv \mathrm{Tr}(C^\dagger CB^\dagger B)\,\mathrm{Tr}(B^\dagger BC^\dagger C) \\
&= \; : \mathrm{Tr}(C^\dagger CB^\dagger B)\,\mathrm{Tr}(B^\dagger BC^\dagger C) : \\
&\quad + \tfrac{1}{2} : \mathrm{Tr}(C^\dagger CB^\dagger BC^\dagger C) : + \tfrac{1}{2} : \mathrm{Tr}(C^\dagger CB^\dagger CC^\dagger B) : \\
&\quad + \tfrac{1}{2} : \mathrm{Tr}(C^\dagger BB^\dagger CB^\dagger B) : + \tfrac{1}{2} : \mathrm{Tr}(C^\dagger CB^\dagger BB^\dagger B) : \\
&\quad + \tfrac{1}{4} : \mathrm{Tr}(C^\dagger CB^\dagger B) : + \tfrac{1}{4} : \mathrm{Tr}(C^\dagger BB^\dagger C) : \\
&\quad + \tfrac{1}{4} : \mathrm{Tr}(C^\dagger B)\,\mathrm{Tr}(B^\dagger C) : + \tfrac{1}{4} : \mathrm{Tr}(C^\dagger C)\,\mathrm{Tr}(B^\dagger B) : ,
\end{aligned} \tag{7.49}$$

with no factor of N. Thus, for all $N \le \infty$, the expression for $\mathcal{H}$ is an acceptable operator and not merely a form. And that is why we have chosen to use that operator to build our interaction in (7.42).

7.3.2 *Comments*

It should be observed that all the nonlinear models that were successfully treated have total Hamiltonians that are well defined and do not exhibit infinities even though they deal with an infinite number of degrees of freedom when $N = \infty$. This property arises because both the free portion of the Hamiltonian and the interaction portion are compatible, genuine operators. The spectrum of these total Hamiltonians can be computed, and that exercise has already been partially carried out for the vector model in [AKT66].

It should be noted that our procedures do not involve conventional $1/N$-expansion techniques. These methods also deal with matrix models, but, unlike our usage, they also use the parameter N for coefficients in the Hamiltonian operator and as coefficients of suitable Hilbert space vectors; see, e.g., [Hoo74; BIP78; Tho79; DMO07]. In particular, their results for $N = \infty$ are completely different than ours.

PART 4

Enhanced Quantization of Ultralocal Field Theories

Chapter 8

Enhanced Quantization of Ultralocal Models

WHAT TO LOOK FOR

Ultralocal models are field theories that have no spatial gradients, which leads to a large symmetry group. When canonically quantized, these models are perturbatively nonrenormalizable and trivial when studied nonperturbatively, which is an unsatisfactory outcome in either case. Enhanced quantization techniques, on the other hand, lead to acceptable, nontrivial quantization results for scalars (bosons), spinors (fermions), as well as the combination of scalars and spinors in a Yukawa-like interaction. These special models also provide guidance in how covariant scalar models can be usefully analyzed; that wisdom is used in the following chapter.

8.1 Introduction

Ultralocal models are idealized field theories that eliminate spatial derivatives. As such, they are generally nonphysical, but they offer considerable insight into other highly singular field theories. The interaction terms are invariably singular leading to pseudofree behavior when a nonlinear interaction is turned off. While operator methods are useful to study ultralocal scalar models, they can also be studied using affine quantization procedures. On the other hand, operator methods are presently used for the study of spinor models. The value of these models is based on their almost complete analytic solubility. In this chapter we are guided by [Kla70; Kl73a; Kla00]; however, the study of ultralocal models with both scalars and spinors together appears here for the first time.

It should be noted that M. Pilati relied on the techniques of ultralocal models for his study of strong-coupling quantum gravity in which all spatial derivatives are dropped; see [Pil82; Pil83; FrP85] as well as the few remarks in Sec. 10.14.2.

8.2   Canonical Quantization of Ultralocal Scalar Models

Consider the classical phase-space action functional

$$A_0 = \int \int_0^T \{\pi(t,x)\dot{\phi}(t,x) - \tfrac{1}{2}[\pi(t,x)^2 + m_0^2\phi(t,x)^2]\} \, dt \, d^s x \;, \qquad (8.1)$$

which describes the free ultralocal scalar field model where $x \in \mathbb{R}^s$, $s \geq 1$. This model differs from a relativistic free theory—which is discussed in the following chapter—by the absence of the term $[\vec{\nabla}\phi(t,x)]^2$, and therefore, for the ultralocal models, the temporal behavior of the field at one spatial point x is independent of the temporal behavior of the field at any spatial point $x' \neq x$. Traditional canonical quantization of the classical Hamiltonian, i.e., $\pi(x) \to \hat{\pi}(x)$ and $\phi(x) \to \hat{\phi}(x)$ where $H_c(\pi,\phi) \to \mathcal{H} = H_c(\hat{\pi},\hat{\phi})$, leads to an infinite ground state energy, so it is traditional to proceed differently. Formally speaking, the classical Hamiltonian density is first reexpressed as

$$\tfrac{1}{2}[\pi(x)^2 + m_0^2\phi(x)^2] = \tfrac{1}{2}[m_0\phi(x) - i\pi(x)][m_0\phi(x) + i\pi(x)] \;, \qquad (8.2)$$

and only then is it quantized directly. This procedure leads to the quantum Hamiltonian

$$\begin{aligned}
\mathcal{H}_0 &= \tfrac{1}{2}\int [m_0\hat{\phi}(x) - i\hat{\pi}(x)][m_0\hat{\phi}(x) + i\hat{\pi}(x)] \, d^s x \\
&= \tfrac{1}{2}\int [\hat{\pi}(x)^2 + m_0^2\hat{\phi}(x)^2 - \hbar m_0 \delta(0)] \, d^s x \\
&\equiv \tfrac{1}{2}\int \; : [\hat{\pi}(x)^2 + m_0^2\hat{\phi}(x)^2] : \; d^s x \;, \qquad (8.3)
\end{aligned}$$

a result that leads to conventional normal ordering symbolized, as we have done already, by $::$.

If we regularize this expression by a finite, s-dimensional, hypercubic, spatial lattice, and adopt a Schrödinger representation, then the Hamiltonian operator becomes

$$\mathcal{H}_0 = \tfrac{1}{2}\Sigma'_k [-\hbar^2 a^{-2s}\partial^2/\partial\phi_k^2 + m_0^2\phi_k^2 - \hbar m_0 a^{-s}] \, a^s \;. \qquad (8.4)$$

The primed sum Σ'_k denotes a sum over the spatial lattice, where $k = \{k_1,\ldots,k_s\}$, $k_j \in \mathbb{Z} \equiv \{0,\pm1,\pm2,\ldots\}$, sites on the spatial lattice. In this equation $a > 0$ is the lattice spacing, a^s is an elementary spatial cell volume, and on the lattice we have used $\hat{\pi}(x) \to -i\hbar a^{-s}\partial/\partial\phi_k$ and $\hat{\phi}(x) \to \phi_k$. The ground state of this Hamiltonian is just the product of a familiar Gaussian ground state,

$$\psi_k(\phi_k) = (m_0 a^s/\pi\hbar)^{1/4} e^{-m_0\phi_k^2 a^s/2\hbar}, \tag{8.5}$$

for a large number of independent, one-dimensional harmonic oscillators, and thus the characteristic functional (i.e., the Fourier transform) of the ground-state distribution is given by

$$
\begin{aligned}
C_0(f) &= \lim_{a\to 0} N_0 \int e^{(i/\hbar)\Sigma'_k f_k \phi_k a^s - (m_0/\hbar)\Sigma'_k \phi_k^2 a^s} \Pi'_k d\phi_k \\
&= \mathcal{N}_0 \int e^{(i/\hbar)\int f(x)\phi(x)\,d^sx - (m_0/\hbar)\int \phi(x)^2\,d^sx} \Pi'_x d\phi(x) \\
&= e^{-(1/4m_0\hbar)\int f(x)^2\,d^sx}.
\end{aligned} \tag{8.6}
$$

The primed product Π'_k runs over all N' sites on a spatial lattice. In the last two lines of this relation the continuum limit has been taken (with a formal version in the second line) in which a goes to zero, L, the number of sites on each edge, goes to infinity, but aL remains large and finite (at least initially).

Now, as just one example, suppose we introduce a quartic nonlinear interaction leading to the classical phase-space action functional,

$$A = \int\int_0^T \{\pi(t,x)\dot\phi(t,x) - \tfrac{1}{2}[\pi(t,x)^2 + m_0^2\phi(t,x)^2] - g_0\phi(t,x)^4\}\,dt\,d^sx. \tag{8.7}$$

Again, the temporal development of the field at one spatial point is independent of the temporal development at any other spatial point. Using the same lattice regularization as for the free theory and introducing normal ordering for the interaction, the characteristic functional of the ground-state distribution is necessarily of the form

$$
\begin{aligned}
C(f) &= \lim_{a\to 0} N \int e^{(i/\hbar)\Sigma'_k f_k \phi_k a^s - \Sigma'_k \widetilde{Y}(\phi_k, g_0, \hbar, a) a^s} \Pi'_k d\phi_k \\
&= \mathcal{N} \int e^{(i/\hbar)\int f(x)\phi(x)\,d^sx - \int Y(\phi(x), g_0, \hbar)\,d^sx} \Pi'_x d\phi(x) \\
&= e^{-(1/4\mathsf{m}\hbar)\int f(x)^2\,d^sx}.
\end{aligned} \tag{8.8}
$$

Here $\widetilde{Y}(\phi_k, g_0, \hbar, a)$ and $Y(\phi(x), g_0, \hbar)$ denote some non-quadratic functions that arise in the solution of the nonlinear Hamiltonian ground-state differential equation on the lattice and likewise in the formal continuum limit, respectively. Importantly, *the last line is a consequence of the Central Limit Theorem* (see below), yielding a *free theory* with a positive mass m, a factor that absorbs all trace of the quartic interaction. In short, a conventional canonical quantization effectively obliterates all trace of the quartic interaction. Summarizing, a conventional canonical quantization of this nonlinear field theory—which is a nonrenormalizable quantum field theory since all closed loops in a perturbation analysis diverge when integrated (if the spatial volume is infinite) or summed (if the spatial volume is finite) over spatial momentum—has rigorously led to a trivial (= free) theory, even though the original classical theory was nontrivial.

A brief review of the Central Limit Theorem may be helpful. Consider the characteristic functional (with a symmetric distribution and $\hbar = 1$)

$$C(f) = \lim_{a \to 0} \widetilde{N} \int e^{i\Sigma'_k f_k \phi_k a^s} - \Sigma'_k \widetilde{Y}(\phi_k, a) a^s \, \Pi'_k d\phi_k$$

$$= \lim_{a \to 0} \Pi'_k \widetilde{N}' \int e^{i f_k \lambda} - \widetilde{Y}(\lambda a^{-s}, a) a^s \, d\lambda$$

$$= \lim_{a \to 0} \Pi'_k \{ 1 - \tfrac{1}{2} f_k^2 \langle \lambda^2 \rangle + \tfrac{1}{24} f_k^4 \langle \lambda^4 \rangle - \cdots \} , \qquad (8.9)$$

where $\langle \lambda^{2p} \rangle \equiv \widetilde{N}' \int \lambda^{2p} \exp[-\widetilde{Y}(\lambda a^{-s}, a) a^s] \, d\lambda$. To achieve an acceptable continuum limit, it is essential that $\langle \lambda^2 \rangle = A_2 a^s$, for very small a^s and fixed $A_2 > 0$. If, the resultant distribution is *narrow*—very roughly like a Gaussian—it follows, for small a^s, that $\langle \lambda^{2p} \rangle = \mathcal{O}(\langle \lambda^2 \rangle^p) = \mathcal{O}(a^{sp})$ for $p \in \mathbb{N}^+$; as a consequence, the continuum limit is $\exp[-\tfrac{1}{2} A_2 \int f(x)^2 \, d^s x]$. A different result obtains if the distribution has a dominant, narrow central portion 'resting' on a *wide* pedestal the integral of the latter being $\mathcal{O}(a^s)$ with a width that, for small a^s, no longer depends on a. If such is the case, then, thanks to the pedestal, $\langle \lambda^{2p} \rangle = \mathcal{O}(a^s)$ for *all* moments, which, in the continuum limit, leads to a compound or generalized Poisson distribution (and possibly a Gaussian term as well). A Gaussian term and/or a Poisson term exhaust the possibilities for such formal integrals. In this chapter, we will make heavy use of Poisson distributions. For further details regarding the Central Limit Theorem, see, e.g., [Luk70; Wiki-j].

An examination of free-field ultralocal operator properties confirms their unsuitability for nonlinear interactions. The field operator $\hat{\phi}(x)$ for the free ultralocal scalar model involves the sum of the annihilation and creation operators, $A(x)$ and $A(x)^\dagger$, which satisfy the canonical commutation relations (CCR) $[A(x), A(y)] = 0$ and $[A(x), A(y)^\dagger] = \delta(x - y)\mathbf{1}$, along with $A(x)|0\rangle = 0$, for the 'no-particle' state $|0\rangle$ and all $x \in \mathbb{R}^s$. While the bilinear combination $A(x)^\dagger A(x)$, when smeared, is a proper operator, other combinations such as $A(x)^\dagger A(x)^\dagger$ and $A(x) A(x)$ are not proper operators; for a demonstration of such assertions, see a similar study for fermion operators below in Sec. 8.6.2. As a consequence, the basic operators $A(x)$ and $A(x)^\dagger$ can not be used for nonlinear interacting ultralocal scalar models.

The question naturally arises: Can we change quantization procedures to yield a nontrivial result? We first study that problem by an affine quantization, followed later by a 'secret recipe' that rapidly leads to the same result.

8.3 Quantization of Ultralocal Scalar Models

We start by modifying the classical Hamiltonian before quantization [model #(6) in Chap. 1]. This modification is completely different from the one used for the conventional quantization discussed above. For both $g_0 = 0$ and $g_0 > 0$, let us consider

$$H_c(\pi, \phi) = \int \{ \tfrac{1}{2}[\pi(x)^2 + m_0^2 \phi(x)^2] + g_0 \phi(x)^4 \} \, d^s x \qquad (8.10)$$
$$= \int \{ \tfrac{1}{2}[\pi(x)\phi(x)\phi(x)^{-2}\phi(x)\pi(x) + m_0^2 \phi(x)^2] + g_0 \phi(x)^4 \} \, d^s x \,.$$

If the classical Poisson bracket for the field $\phi(x)$ and the momentum $\pi(x')$ is given by $\{\phi(x), \pi(x')\} = \delta(x - x')$, then the Poisson bracket between the field $\phi(x)$ and the dilation field $\kappa(x') \equiv \pi(x')\phi(x')$ is given by $\{\phi(x), \kappa(x')\} = \delta(x - x')\phi(x)$. On quantization, we will treat the classical product $\pi(x)\phi(x)$ as the dilation field $\kappa(x)$, and we will invoke affine quantum commutation relations for which $\phi(x) \to \hat\phi(x)$ and $\kappa(x) \to \hat\kappa(x)$ such that $[\hat\phi(x), \hat\kappa(x')] = i\hbar\delta(x - x')\hat\phi(x)$. Note well: If we choose affine commutation relations for which $\hat\phi(x)$ and $\hat\kappa(x)$, when smeared, are self-adjoint operators, then it follows that the canonical momentum operator $\hat\pi(x)$, when smeared, is only a *form* and *not* an operator due to the local operator product involved; recall that a form requires restrictions on kets *and* bras. In short, when quantizing fields, one can choose either affine field variables or canonical field variables, since both systems generally can not exist simultaneously because of the local operator product involved.

As compared to Chap. 4, where affine variables were first introduced, it is noteworthy that the affine field variables $\hat\phi(x)$ and $\hat\kappa(x)$ we presently deal with comprise a reducible representation of operators since the spectrum of $\hat\phi(x)$, when smeared with a positive test function, runs over the whole real line, except for $\hat\phi(x) = 0$.

To see what affine quantization leads to, let us work formally and focus on twice the kinetic energy density. Thus, classically,

$$\pi(x)^2 = \pi(x)\phi(x)\phi(x)^{-2}\phi(x)\pi(x)$$
$$= \kappa(x)\phi(x)^{-2}\,\kappa(x)\,, \tag{8.11}$$

which on quantization becomes

$$\hat{\kappa}(x)\hat{\phi}(x)^{-2}\hat{\kappa}(x) = \tfrac{1}{4}[\hat{\pi}(x)\hat{\phi}(x) + \hat{\phi}(x)\hat{\pi}(x)]\hat{\phi}(x)^{-2}[\hat{\pi}(x)\hat{\phi}(x) + \hat{\phi}(x)\hat{\pi}(x)]$$
$$= \tfrac{1}{4}[2\hat{\pi}(x)\hat{\phi}(x) + i\hbar\delta(0)]\hat{\phi}(x)^{-2}[2\hat{\phi}(x)\hat{\pi}(x) - i\hbar\delta(0)]$$
$$= \hat{\pi}(x)^2 + i\tfrac{1}{2}\hbar\delta(0)[\hat{\phi}(x)^{-1}\hat{\pi}(x) - \hat{\pi}(x)\hat{\phi}(x)^{-1}]$$
$$+\tfrac{1}{4}\hbar^2\delta(0)^2\hat{\phi}(x)^{-2}$$
$$= \hat{\pi}(x)^2 + \tfrac{3}{4}\hbar^2\delta(0)^2\hat{\phi}(x)^{-2}\,. \tag{8.12}$$

The factor $\tfrac{3}{4}$ will have an essential role to play.

Guided by this calculation we introduce the same lattice regularization to again quantize the classical free theory (i.e., $g_0 = 0$), but this time focussing on an affine quantization, namely

$$\mathcal{H}'_0 = \tfrac{1}{2}\Sigma'_k[-\hbar^2 a^{-2s}\partial^2/\partial\phi_k^2 + m_0^2\phi_k^2 + F\hbar^2 a^{-2s}\phi_k^{-2} - 2E'_0]\,a^s\,, \tag{8.13}$$

where $F \equiv (\tfrac{1}{2} - ba^s)(\tfrac{3}{2} - ba^s)$; here $b > 0$ is a constant factor with dimensions $(\text{length})^{-s}$, and E'_0 is explained below. Note that F is a regularized form of $\tfrac{3}{4}$, and it becomes that number in the continuum limit. Again the ground state of this Hamiltonian is a product over one-dimensional ground states for each independent degree of freedom, and for a small dimensionless cell volume, $ba^s \ll 1$, each ground state wave function has the form

$$\psi_k(\phi_k) = (ba^s)^{1/2}\,e^{-m_0\phi_k^2\,a^s/2\hbar}\,|\phi_k|^{-(1-2ba^s)/2}\,. \tag{8.14}$$

This new form of ground-state wave function corresponds to a pseudofree model analogous to those introduced in Chap. 3. We now see that it also arises upon quantization of the free classical model by exploiting an unconventional factor-ordering ambiguity and insisting on securing affine kinematical variables, $\hat{\phi}(x)$ and $\hat{\kappa}(x)$.

In effect, as the continuum limit is approached, the new factor in (8.14) serves to "mash the measure" (see below for a fuller explanation of this phrase) in the sense that under a change of m_0 the usual *free* ground-state distributions are *mutually singular* (with disjoint support) while the *pseudofree* ground-state distributions are *equivalent* (with equal support). This very fact can be seen as the reason for the divergence-free character of the affine quantization for these models.

The characteristic functional for the pseudofree ground-state distribution is given by

$$C_{pf}(f) = \lim_{a\to 0} \int \Pi'_k \{(ba^s)\, e^{if_k\phi_k a^s/\hbar} - m_0\phi_k^2 a^s/\hbar\, |\phi_k|^{-(1-2ba^s)}\} \Pi'_k d\phi_k$$

$$= \lim_{a\to 0} \Pi'_k \{1 - (ba^s)\int [1 - \cos(f_k\phi a^s/\hbar)]$$

$$\times e^{-m_0\phi^2 a^s/\hbar}\, d\phi/|\phi|^{(1-2ba^s)}\}$$

$$= \exp\{-b\int d^s x \int [1 - \cos(f(x)\lambda/\hbar)]\, e^{-bm\lambda^2/\hbar}\, d\lambda/|\lambda|\}, \qquad (8.15)$$

where $m_0 \equiv ba^s m$ and $\phi a^s \equiv \lambda$; note the multiplicative renormalization involved in $m_0 = ba^s m$. The final distribution described by (8.15) is a generalized Poisson distribution [Luk70].

The reader is encouraged to become convinced that only the number $\frac{3}{4}$ could have led to the desirable result in (8.15). Note that the ground-state energy $E'_0 = \frac{1}{2}\hbar mb^2 a^s$, and observe that the ground-state energy on the lattice is finite and it vanishes in the continuum limit. Indeed, it is remarkable that the pseudofree Hamiltonian that annihilates the pseudofree ground state wave function has a discrete spectrum with *uniform spacing* as well as a *vanishing* zero-point energy [Kla00]; these are two very good reasons to embrace this version of the ground state.

Finally, we consider the affine quantization of an ultralocal model with a quartic interaction as described by the classical action in (8.11). In that case the lattice form of the quantum Hamiltonian acquires the additional term $g_0\Sigma'_k\phi_k^4 a^s$ [with no normal ordering needed, but rather $g_0 = (ba^s)^3 g$, with $g \geq 0$] along with a suitable change of E'_0. Now the lattice regularized ground-state wave function at each site has, for small values of ba^s, the form

$$\psi_k = (ba^s)^{1/2}\, e^{-\frac{1}{2}\tilde{y}(\phi_k, \hbar, a)\, a^s}\, |\phi_k|^{-(1-2ba^s)/2}, \qquad (8.16)$$

for some appropriate nonquadratic function $\tilde{y}(\phi_k, \hbar, a)$. In turn, the characteristic functional of the interacting ground-state distribution becomes, in the continuum limit,

$$C(f) = \exp\{-b\int d^s x \int [1 - \cos(f(x)\lambda/\hbar)]\, e^{-b^{-1}y(b\lambda, \hbar)}\, d\lambda/|\lambda|\}, \qquad (8.17)$$

where $\tilde{y}(\phi, \hbar, a)$ is related to $b^{-1}y(b\lambda, \hbar)$ by $\lambda = \phi a^s$ and multiplicative renormalization of suitable coefficients. For ultralocal models with symmetric potentials, general arguments from the theory of infinite divisibility [Luk70] show that *only* the free (Gaussian) or the nonfree (generalized Poisson) distributions, with characteristic functions for ground-state distributions illustrated above, are allowed by ultralocal symmetry and uniqueness of the ground state.

Generalized Poisson distributions also include other characteristic functionals such as

$$C(f) = e^{-b' \int d^s x \int [1 - \cos(f(x)\omega)/\hbar] e^{-Z(\omega,\hbar)} d\omega/|\omega|^{2\gamma}} , \qquad (8.18)$$

for $\frac{1}{2} < \gamma < \frac{3}{2}$, as well as our featured expressions for which $\gamma = \frac{1}{2}$. All such models were analyzed in [Kla00], but it has now been recognized that models for which $\gamma > \frac{1}{2}$ are effectively renormalized functions of the models for which $\gamma = \frac{1}{2}$. This is made clear if we make a transformation of integration variables, e.g, for both positive variables, such as, $\omega^{2\gamma-1} \propto [-\ln(\lambda)]^{-1}$ valid for $0 < \omega + \lambda \ll 1$, and then each variable growing monotonically until they both reach ∞ together.

Note that following the route of affine quantization for the ultralocal model has resulted in overcoming the triviality of quantization that conventional canonical procedures invariably lead to for such models. We will see below that appropriate affine coherent states are central to establishing the classical/quantum connection by giving rise to a suitable enhanced classical action functional.

8.3.1 *Basic operators for ultralocal scalar models*

Observe that the affine commutation relation $[\hat{\phi}(x), \hat{\kappa}(y)] = i\hbar\delta(x-y)\hat{\phi}(x)$ defines a (formal) Lie algebra [Coh61] that has the appearance of a current algebra [ItZ80]. As such, there are *bilinear* realizations of the affine commutation relations involving conventional annihilation and creation operators, unlike the conventional canonical commutation relations, which, instead, involve *linear* relations in terms of annihilation and creation operators. Specifically, let us introduce conventional annihilation operators $A(x,\lambda)$ and creation operators $A(x,\lambda)^\dagger$, $x \in \mathbb{R}^s$, $\lambda \in \mathbb{R}$, which satisfy the conventional commutation relations $[A(x,\lambda), A(x',\lambda')] = 0$ and $[A(x,\lambda), A(x',\lambda')^\dagger] = \delta(x - x')\delta(\lambda - \lambda')\mathbb{1}$. In turn, these operators act on a separable Hilbert space in which there is a normalized, unique (up to a factor) 'no-particle' vector $|0\rangle$ such that $A(x,\lambda)|0\rangle = 0$ for all (x,λ). This means that the formal operators $A(x,\lambda)$ and $A(x,\lambda)^\dagger$ also determine a conventional Fock representation, with a new 'component', λ, at each point x. Next, we introduce a real, nonvanishing, c-number function $c(\lambda)$ that satisfies the conditions that $\int [\lambda^2/(1 + \lambda^2)] c(\lambda)^2 d\lambda < \infty$ and $\int c(\lambda)^2 d\lambda = \infty$, when integrated over the whole real line. The important function $c(\lambda)$ is called the *model function* for reasons that will become clear.

Because of their frequent occurrence, we introduce the operator

$B(x, \lambda) \equiv A(x, \lambda) + c(\lambda)\mathbb{1}$ and its adjoint $B(x, \lambda)^\dagger = A(x, \lambda)^\dagger + c(\lambda)\mathbb{1}$. Evidently, $[B(x, \lambda), B(x', \lambda')] = 0$ and $[B(x, \lambda), B(x', \lambda')^\dagger] = \delta(x - x')\delta(\lambda - \lambda')\mathbb{1}$, but because $c \notin L^2(\mathbb{R})$, the operators $B(x, \lambda)$ and $B(x, \lambda)^\dagger$ do *not* form a Fock representation. Armed with this 'substructure' of familiar operators, we are now ready to construct the affine field operators that characterize the ultralocal scalar models.

8.3.2 *Solution of scalar ultralocal models*

The ultralocal field operators $\hat{\phi}(x)$ and $\hat{\kappa}(x)$, say at time $t = 0$, are defined by

$$\hat{\phi}(x) = \int B(x, \lambda)^\dagger \lambda B(x, \lambda)\, d\lambda , \qquad (8.19)$$

$$\hat{\kappa}(x) = -\tfrac{1}{2}i\hbar \int B(x, \lambda)^\dagger \left[\lambda(\partial/\partial\lambda) + (\partial/\partial\lambda)\lambda\right] B(x, \lambda)\, d\lambda , \qquad (8.20)$$

and it is straightforward to show that these operators satisfy the affine commutation relation. Local products of the field operators follow the rules of an operator product expansion. Consider

$$\hat{\phi}(x)\hat{\phi}(y) = \int B(x, \lambda)^\dagger \lambda B(x, \lambda)\, d\lambda \cdot \int B(y, \lambda')^\dagger \lambda' B(y, \lambda')\, d\lambda'$$
$$= \delta(x - y) \int B(x, \lambda)^\dagger \lambda^2 B(x, \lambda)\, d\lambda + !\,\hat{\phi}(x)\hat{\phi}(y)\,! , \qquad (8.21)$$

where we have introduced $!\,(\cdot)\,!$ as normal ordering for the operators B and $B^\dagger$. As $y \to x$, it is clear that the first term dominates. We select the first term by formally dividing by $\delta(0)$, but that expression has dimensions we should not ignore. To preserve dimensions we introduce the parameter $b > 0$ with dimensions $(\text{length})^{-s}$, and define the local product (R for 'renormalized')

$$\hat{\phi}_R^2(x) \equiv b \int B(x, \lambda)^\dagger \lambda^2 B(x, \lambda)\, d\lambda . \qquad (8.22)$$

A proper evaluation using smearing functions and their limiting behavior leads to the same result. Higher powers follow the same rule, and are given by

$$\hat{\phi}_R^p(x) \equiv b^{p-1} \int B(x, \lambda)^\dagger \lambda^p B(x, \lambda)\, d\lambda , \qquad (8.23)$$

etc. *Note: ALL local powers are bilinear in the basic operators $B^\dagger$ and B!*

Observe that a sum of such terms could represent the potential density in forming a Hamiltonian operator. Desired properties of the Hamiltonian operator $\mathcal{H} = \int \mathcal{H}(x)\, d^s x$ are that $\mathcal{H} \geq 0$ as well as having a unique (up to a factor) ground state, $\mathcal{H}|0\rangle = 0$. In order to arrange that property, we need $\mathcal{H}(x)|0\rangle = 0$. A general analysis of operator possibilities [Kla70; Kla00] leads to a generic form for $\mathcal{H}(x)$ given by

$$\mathcal{H}(x) = \int B(x, \lambda)^\dagger \, \mathsf{h}(\partial/\partial\lambda, \lambda) \, B(x, \lambda) \, d\lambda \, , \tag{8.24}$$

and it is easy to arrange that $\mathcal{H}(x)|0\rangle = 0$ if we insist that

$$\mathsf{h}(\partial/\partial\lambda, \lambda) \, c(\lambda) = 0 \tag{8.25}$$

since, in this case, we also find that

$$\mathcal{H}(x) = \int A(x, \lambda)^\dagger \, \mathsf{h}(\partial/\partial\lambda, \lambda) \, A(x, \lambda) \, d\lambda \, . \tag{8.26}$$

Consequently, we choose

$$\mathsf{h}(\partial/\partial\lambda, \lambda) \equiv -\tfrac{1}{2}\hbar^2(\partial^2/\partial\lambda^2) + \tfrac{1}{2}\hbar^2 c(\lambda)^{-1} \partial^2 \, c(\lambda)/\partial\lambda^2 \, . \tag{8.27}$$

Observe that the Hamiltonian h is *completely independent of the overall scale of the model function* $c(\lambda)$; in other words, $c'(\lambda) = k_c \, c(\lambda)$, $k_c > 0$, leads to the identical Hamiltonian and thus the very same dynamical spectrum and eigenfunctions regardless of the value of the scale factor k_c. However, we will find a reason to fix k_c later in the analysis. It is also noteworthy that $\hat{\phi}(x)$ and $\mathcal{H}$ determine $\hat{\kappa}(x)$ since

$$[\hat{\phi}_R^2(x), \mathcal{H}] = 2i\hbar\hat{\kappa}(x) \, . \tag{8.28}$$

These important equations link the operator representation determined by $c(\lambda)$ to the dynamics of the model. Note that $\mathcal{H} \geq 0$ requires that $\mathsf{h} \geq 0$; however to ensure a nondegenerate ground state requires that $\mathsf{h} > 0$. Consequently, to ensure a unique ground state, we need to require that $c \notin L^2(\mathbb{R})$, which provides the justification for that criterion. Local field powers require that $|\int \lambda^p c(\lambda)^2 \, d\lambda| < \infty$, which, noting that $c(-\lambda) = c(\lambda)$ for a symmetric potential, supports the requirement that $\int [\lambda^2/(1 + \lambda^2)]c(\lambda)^2 \, d\lambda < \infty$. [Remark: Although we do not discuss this case at length, for a general potential without even symmetry, it follows that the model function has no special even symmetry in λ.]

The foregoing discussion contains enough information to evaluate the truncated (T) vacuum expectation values in the sense that they are reduced to quadrature. In particular, for an even potential and for N even, it follows that

$$\langle 0|\,\hat{\phi}(t_N, x_N)\hat{\phi}(t_{N-1}, x_{N-1}) \cdots \hat{\phi}(t_1, x_1)|0\rangle^T$$
$$= \delta(x_N - x_{N-1})\delta(x_{N-1} - x_{N-2}) \cdots \delta(x_2 - x_1) \tag{8.29}$$
$$\times \int c(\lambda)\,\lambda\, e^{-i(t_N - t_{N-1})\mathsf{h}}\lambda \cdots \lambda\, e^{-i(t_2 - t_1)\mathsf{h}}\lambda\, c(\lambda)\, d\lambda \, ;$$

if N is odd, this expectation value vanishes since $c(-\lambda) = c(\lambda)$.

There is also another set of expectation values that provides a useful way to characterize each model. These expressions may be obtained first by setting all time variables $t_j = 0$ in (8.29) followed by smearing every field with the same smooth function $f(x)$. For N even (or odd), the result is given by

$$\langle 0| [\int f(x) \hat{\phi}(x) d^s x]^N |0\rangle^T = \int f(x)^N d^s x \int \lambda^N c(\lambda)^2 d\lambda , \qquad (8.30)$$

or stated otherwise,

$$\langle 0| [e^{(i/\hbar) \int f(x) \hat{\phi}(x) d^s x} - \mathbb{1}] |0\rangle^T = \int d^s x \int [e^{i f(x) \lambda/\hbar} - 1] c(\lambda)^2 d\lambda . \qquad (8.31)$$

Finally, (8.31) leads to the important relation

$$C(f) \equiv \langle 0| e^{(i/\hbar) \int f(x) \hat{\phi}(x) d^s x} |0\rangle$$
$$= \exp\{-\int d^s x \int [1 - e^{i f(x) \lambda/\hbar}] c(\lambda)^2 d\lambda\} , \qquad (8.32)$$

which relates the characteristic function of the ground-state distribution for a given model to the model function itself, as we have already determined.

Free and pseudofree models for ultralocal models

The singular behavior outlined for anharmonic oscillators in Chap. 2 can also be studied for ultralocal models. The classical interacting ultralocal model depicted in (8.7) represents a free classical theory when $g_0 = 0$, but it is important to realize that the limit of the interacting theories when $g_0 \to 0$ is *not* the free theory, and the reason for that is the *domain* of the resultant functional is not the *domain* of the genuine free theory. In particular, the domain of the free model involves the requirement that $\int \int [\dot{\phi}(t, x)^2 + \phi(t, x)^2] \, dt \, d^s x < \infty$, while the domain of the pseudofree action functional, i.e., the functional derived from the interacting action functionals as $g_0 \to 0$, also involves the subsidiary condition that $\int \int \phi(t, x)^4 \, dt \, d^s x < \infty$. Since there are many functions $\phi(t, x)$ that satisfy the first requirement but fail the second requirement, e.g., $\phi_{sing}(t, x) = |x|^{-s/3} \exp(-t^2 - x^2)$, that means that the pseudofree theory differs from the free theory. In simple terms, this means that the classical ultralocal interacting models are *not* continuously connected to their own classical ultralocal free theory as the coupling constant goes to zero.

This distinction between free and pseudofree classical ultralocal models arises again for the quantum ultralocal models. Ultralocal scalar models have typically been defined (setting $\hbar = 1$) so that the model function

$c(\lambda) = (b/|\lambda|)^{1/2} \exp[-\frac{1}{2}y(\lambda)]$, where typically $y(0) = 0$; note: in fact, the scale factor k_c for this model function has been omitted from this expression, which makes little difference since that factor does not change the definition of $h(\partial/\partial\lambda, \lambda)$. An interacting model—although *not* the example related to (8.7)—is given by $y_g(\lambda) = b\, m\lambda^2 + b^3\, g\, \lambda^4$; note: the variables $m_0 [= (ba^s)m]$ and m, $g_0 [= (ba^s)^3 g]$ and g, etc., are, respectively, multiplicatively related to one another. It follows, in this case, that, as $g \to 0$, the pseudofree theory is characterized by $y_0(\lambda) = bm\lambda^2$; the result, however, is *not* a Gaussian distribution, i.e., *not* a free theory, which means that the interacting theories of this example are not continuously connected to their own free theory. We could also examine the case when $y_g(\lambda) = bm\lambda^2 + b^{-1}gw(b\lambda)$, for a wide class of functions w, each of which leads to the same pseudofree theory as before when $g \to 0$. On the other hand, for sufficiently different interactions, there can generally be many different pseudofree theories for the same free theory, similar to what has already been discussed for the oscillator cases previously.

To study the issue of possibly different ultralocal pseudofree theories, it is helpful to study the situation when more than one scalar field is involved. In particular, let us consider three different classical action functionals that involve two ultralocal scalar fields, specifically

$$A_1 = \iint_0^T \{ \tfrac{1}{2}[\dot{\phi}_1^2 + \dot{\phi}_2^2 - m_0^2\phi_1^2 - m_0^2\phi_2^2] - g_{01}\phi_1^4 - g_{02}\phi_2^4 \} \, dt\, d^s x \, , \quad (8.33)$$

$$A_2 = \iint_0^T \{ \tfrac{1}{2}[\dot{\phi}_1^2 + \dot{\phi}_2^2 - m_0^2\phi_1^2 - m_0^2\phi_2^2] - g_2\phi_1^2\phi_2^2 \} \, dt\, d^s x \, , \quad (8.34)$$

$$A_3 = \iint_0^T \{ \tfrac{1}{2}[\dot{\phi}_1^2 + \dot{\phi}_2^2 - m_0^2\phi_1^2 - m_0^2\phi_2^2] - g_3[\phi_1^2 + \phi_2^2]^2 \} \, dt\, d^s x \, . \quad (8.35)$$

For simplicity we have made the two masses the same. The first case (A_1) involves two independent fields, and the natural characteristic functional is the product of two separate characteristic functionals such as we have studied above. The second case (A_2) involves a situation that is invariant under interchanging the two fields, i.e., $\phi_1 \leftrightarrow \phi_2$, and most importantly, the interaction term involves both fields. In this case, we would expect the model function to be of the form $c_2(\lambda_1, \lambda_2) \propto e^{-z_{sym}(\lambda_1,\lambda_2)/2}$ for some symmetric function (which includes some form of singularity at the origin). The third case (A_3) has a higher degree of symmetry in that it involves $\mathbf{SO}(2)$ rotational invariance of the two fields. In that case, we would expect the model function to have the form $c_3(\lambda_1, \lambda_2) \propto e^{-y(\lambda_1^2+\lambda_2^2)/2}/(\lambda_1^2 + \lambda_2^2)^{1/2}$. The three cases each have potentially different behavior at the origin $\lambda_1 = \lambda_2 = 0$. However, the third case involves interactions that are part of the first two cases, so this is a good reason to accept the third form of the singularity for all three cases. This proposal may seem surprising for A_1, but if one of the variables is absent altogether, then the proposed singularity integrates to the former case. For example, if we consider

$$\int [1 - e^{if(x)\lambda_1}] e^{-bm\lambda_1^2 - gb^5\lambda_1^6} (\lambda_1^2 + \lambda_2^2)^{-1} d\lambda_1 d\lambda_2 , \qquad (8.36)$$

it turns out that this is proportional to

$$\int [1 - e^{if(x)\lambda_1}] e^{-bm\lambda_1^2 - gb^5\lambda_1^6} |\lambda_1|^{-1} d\lambda_1 \qquad (8.37)$$

taking note of the fact that

$$\int (\lambda_1^2 + \lambda_2^2)^{-1} d\lambda_2 = \pi |\lambda_1|^{-1} . \qquad (8.38)$$

Thus, the result of the integral in (8.36) is equivalent since the factor π is just another example of a scale factor k_c that has no effect on the Hamiltonian h.

Note carefully the difference in the interaction terms. The classical free models (for $g_{01} = g_{02} = g_{012} = 0$) are identical for A_1 and A_2, but both action functionals have pseudofree models different from the free model, and, indeed, *the two pseudofree models are different from each other*. The quantum theory for A_1 would correspond to a characteristic functional that was the product of two separate characteristic functionals, one for field ϕ_1, and one for field ϕ_2. The quantum theory for A_2, however, would lead to a characteristic functional that involves both fields in a nontrivial way. To illustrate the situation we have in mind for A_1, there would be two separate model functions $c_1(\lambda)$ and $c_2(\lambda)$, each of which is much like the model function for a single ultralocal field. On the other hand, for A_2, the model function c_{12} would have to involve *both* field variables, i.e., in the latter case, $c_{12}(\lambda_1, \lambda_2)$, and the pseudofree theories for this case would also be quite different. For this latter case, for example, the pseudofree model function, for possibly different masses, would be of the form

$$c_{12}(\lambda_1, \lambda_2) = b^{1/2} (\lambda_1^2 + \lambda_2^2)^{-1/2} e^{-\frac{1}{2} b [m_1 \lambda_1^2 + m_2 \lambda_2^2]} . \qquad (8.39)$$

8.3.3 *Weak correspondence principle*

The weak correspondence principle connects the Hamiltonian operator and the enhanced classical Hamiltonian. The Hamiltonian operator involves a model function $c(\lambda)$ that satisfies $\int c(\lambda)^2 d\lambda = \infty$. However, for an even model function, i.e., $c(-\lambda) = c(\lambda)$, we now impose the condition $b \int \lambda^2 c(\lambda)^2 d\lambda \equiv 1$; in addition, we impose the requirement that $b^{2p-1} \int \lambda^{2p} c(\lambda)^2 d\lambda = 1 + \mathcal{O}(\hbar)$, $p \in \mathbb{N}^+$. The former condition is arranged by choosing the scaling parameter k_c that still remains free; the latter condition is arranged by having deviations from unity involve $\hbar$. [Remark: If $c(-\lambda) \neq c(\lambda)$—as is the case for a nonsymmetric potential—there is a separate k_c for $\lambda > 0$ and $\lambda < 0$.]

Following [Kla00], we choose affine coherent states constructed, for suitable $u(x)$ and $v(x) \equiv \ln|\phi(x)|$ functions, as

$$|u, v\rangle \equiv e^{i\int u(x)\hat{\phi}_R^2(x)\,d^sx/\hbar}\, e^{-i\int v(x)\hat{\kappa}(x)\,d^sx/\hbar}|0\rangle\,, \qquad (8.40)$$

where the basic operators obey the affine commutation relation $[\hat{\phi}_R^2(x), \hat{\kappa}(y)] = 2i\hbar\delta(x-y)\hat{\phi}_R^2(x)$. In a sense, however, these affine coherent states are actually 'canonical coherent states' since (for $\hbar = b = 1$)

$$\begin{aligned}
|u, v\rangle &\equiv e^{i\int d^sx\, u(x)\int B(x,\lambda)^\dagger \lambda^2 B(x,\lambda)\,d\lambda} \\
&\quad \times e^{-i\int d^sx\, v(x)\int B(x,\lambda)^\dagger \sigma B(x,\lambda)\,d\lambda}|0\rangle \\
&= {}_!e^{\int d^sx\int B(x,\lambda)^\dagger[e^{iu(x)\lambda^2}e^{-iv(x)\sigma}-1]B(x,\lambda)\,d\lambda}{}_!|0\rangle \\
&= e^{\int d^sx\int B(x,\lambda)^\dagger[e^{iu(x)\lambda^2}e^{-iv(x)\sigma}-1]c(\lambda)\,d\lambda}|0\rangle \\
&= N\,e^{\int d^sx\int A(x,\lambda)^\dagger[e^{iu(x)\lambda^2}e^{-iv(x)\sigma}-1]c(\lambda)\,d\lambda}|0\rangle\,, \qquad (8.41)
\end{aligned}$$

where $\sigma \equiv -\frac{1}{2}i\hbar[\lambda(\partial/\partial\lambda) + (\partial/\partial\lambda)\lambda]$ and N is a normalization factor. Consequently, $A(x,\lambda)|u,v\rangle = [e^{iu(x)\lambda^2}e^{-iv(x)\sigma} - 1]c(\lambda)|u,v\rangle$.

Thus, with $\epsilon(\phi(x)) \equiv sign[\phi(x)] = \pm 1$, and using $\phi(x) \equiv \epsilon(\phi(x))e^{v(x)} = \pm e^{v(x)}$ and $\pi(x) \equiv 2\epsilon(\phi(x))u(x)e^{v(x)} = \pm 2u(x)e^{v(x)}$, we find that

$$\begin{aligned}
H(\pi,\phi) &= \int d^sx\int \langle u,v|A(x,\lambda)^\dagger\, h\, A(x,\lambda)|u,v\rangle\,d\lambda \\
&= \int d^sx\int c(\lambda)[e^{iv(x)\sigma}e^{-iu(x)\lambda^2}-1]\,h\,[e^{iu(x)\lambda^2}e^{-iv(x)\sigma}-1]c(\lambda)\,d\lambda \\
&= \int d^sx\int c(\lambda)[h(|\phi(x)|^{-1}\partial/\partial\lambda + \pi(x)\epsilon(\phi(x))\lambda, |\phi(x)|\lambda)]c(\lambda)\,d\lambda \\
&= \int\{\tfrac{1}{2}[\pi(x)^2 + m^2\phi(x)^2] + g\phi(x)^4\}\,d^sx + \mathcal{O}(\hbar;\pi,\phi)\,, \qquad (8.42)
\end{aligned}$$

and the quantum correction term $\mathcal{O}(\hbar;\pi,\phi)$ is given by

$$\begin{aligned}
\mathcal{O}(\hbar;\pi,\phi) = \int d^sx\{&g\phi(x)^4[\langle\lambda^4\rangle - 1] \qquad\qquad\qquad (8.43) \\
&- \tfrac{1}{2}\hbar^2\phi(x)^{-2}\langle|\lambda|^{1/2}(\partial/\partial\lambda)|\lambda|^{-1}(\partial/\partial\lambda)|\lambda|^{1/2}\rangle\}\,,
\end{aligned}$$

where $\langle(\cdot)\rangle \equiv \int c(\lambda)(\cdot)c(\lambda)\,d\lambda$.

Remark: In [ZhK94], canonical coherent states were used for this analysis which was not appropriate. In [Kla00], instead of using the 'natural' choice of affine fields, namely, $\hat{\phi}(x)$ and $\hat{\kappa}(x)$, different affine coherent states based on $\hat{\phi}_R^2(x)$ and $\hat{\kappa}(x)$ were used. This is because the formal commutator $[\hat{\phi}(x), \mathcal{H}] = i\hbar\hat{\pi}(x)$, and $\hat{\pi}(x)$, even when smeared, does not include such affine coherent states in its domain. However, affine coherent states generated by $\hat{\phi}_R^2(x)$ and $\hat{\kappa}(x)$ do not have that problem. To consider the enhanced classical action functional for this model, we first note that the weak correspondence principle, just discussed, yields the enhanced classical Hamiltonian portion correctly. For the rest of the enhanced classical action functional, we note (again with $\hbar = b = 1$) that

$$
\begin{aligned}
i\langle u,v|(\partial/\partial t)|u,v\rangle &= \int d^s x\,\langle 0|\,[-\dot{u}(t,x)\,e^{2v(t,x)}\,\hat{\phi}_R^2(x) + \dot{v}(t,x)\,\hat{\kappa}(x)]\,|0\rangle \\
&= \int d^s x \int c(\lambda)\,[-\dot{u}(t,x)\,e^{2v(t,x)}\,\lambda^2 + \dot{v}(t,x)\,\sigma]\,c(\lambda)\,d\lambda \\
&= -\tfrac{1}{2}\int d^s x\,[(\partial/\partial t)\{\pi(t,x)/\phi(t,x)\}\,\phi(t,x)^2] \\
&= \tfrac{1}{2}\int d^s x\,[\pi(t,x)\,\dot{\phi}(t,x) - \phi(t,x)\,\dot{\pi}(t,x)]\,,
\end{aligned}
\tag{8.44}
$$

which is a perfectly acceptable result. This expression confirms the consistency of the present calculation with the principles of enhanced quantization.

8.3.4 *A rapid quantization of ultralocal scalar fields*

Armed with the foregoing discussion, there is a simple yet useful story that leads to the quantization of ultralocal scalar models very quickly. We start with the ultralocal free theory on a lattice—defined by: spacing $a > 0$, sites on an axis $L < \infty$, and total spatial lattice sites $N' \equiv L^s$—which has a ground state given by the product of many Gaussian factors and thus a characteristic function for the ground-state distribution given (for $\hbar = 1$) by

$$
C_{latt}(f) = M' \int e^{i\Sigma_k' f_k \phi_k\,a^s\, -\, m_0 \Sigma_k' \phi_k^2\,a^s}\,\Pi_k'\,d\phi_k\,,
\tag{8.45}
$$

which, in the continuum limit—defined by: $a \to 0$, $L \to \infty$, and aL is large and fixed—leads to suitable measures that are *mutually singular* (i.e., disjoint support) for different values of m_0. Mutual singularity leads to divergences when perturbations of the mass or higher-order interaction terms are considered. The source of these divergences can be exposed if we change integration variables from $\{\phi_k\}$ to 'hyperspherical variables' $\{\kappa, \eta_k\}$ defined by $\phi_k \equiv \kappa\eta_k$, $\kappa^2 \equiv \Sigma_k'\phi_k^2$, $1 \equiv \Sigma_k'\eta_k^2$, $0 \le \kappa < \infty$, and $-1 \le \eta_k \le 1$, which leads to

$$
\begin{aligned}
C_{latt}(f) = M' \int & e^{i\kappa\Sigma_k' f_k \eta_k\,a^s\, -\, m_0\kappa^2\Sigma_k'\eta_k^2\,a^s}\,\kappa^{N'-1}\,d\kappa \\
& \times 2\delta(1 - \Sigma_k'\eta_k^2)\,\Pi_k'\,d\eta_k\,;
\end{aligned}
\tag{8.46}
$$

do not confuse the hyperspherical radius κ with the dilation function $\kappa(x) = \pi(x)\phi(x)$.

In the continuum limit, when $N' \to \infty$, the overwhelming effect of the term N' in the hyperspherical radius factor $\kappa^{N'-1}$ leads to mutually singular measures if m_0 changes, as a steepest-descent evaluation would make clear. However, mutually singular measures can be replaced by *equivalent* measures (i.e., equal support) if we choose a pseudofree ground-state distribution that has the lattice form $[\Pi'_k |\phi_k|^{-(1-R/N')}] \exp[-m_0 \Sigma'_k \phi_k^2 \, a^s]$, where $R > 0$ is fixed and finite. The effect of that change (with $\hbar = 1$ still) results in

$$C'_{latt}(f) = M'' \int e^{i\Sigma'_k f_k \phi_k \, a^s - m_0 \Sigma'_k \phi_k^2 \, a^s} \, \Pi'_k \{ |\phi_k|^{-(1-R/N')} \, d\phi_k \} \, ,$$

(8.47)

which leads, in the continuum limit, to

$$\begin{aligned}
C'(f) &= M'' \lim_{a \to 0} \Pi'_k \int e^{i f_k \phi \, a^s - m_0 \phi^2 \, a^s} \, |\phi|^{-(1-2ba^s)} \, d\phi \\
&= M''' \lim_{a \to 0} \Pi'_k \int e^{i f_k \lambda - bm\lambda^2} \, |\lambda|^{-(1-2ba^s)} \, d\lambda \\
&= \lim_{a \to 0} \Pi'_k \{ 1 - (ba^s) \int [1 - \cos(f_k \lambda)] \, e^{-bm\lambda^2} \, |\lambda|^{-(1-2ba^s)} \, d\lambda \} \\
&= e^{-b \int d^s x \int [1 - \cos(f(x)\lambda)] e^{-bm\lambda^2} \, d\lambda / |\lambda|} \, ,
\end{aligned}$$

(8.48)

where we have set $R = 2ba^s N'$, $m_0 = (ba^s)m$, and $\phi a^s = \lambda$. This functional implicitly determines the ground state, thus the Hamiltonian operator, and finally the Euclidean action functional with the required counterterm.

A similar story can be carried out for interacting ultralocal models as well. The modification introduced by the factor $\kappa^{-(N'-R)}$ is referred to as *measure mashing*; effectively, it 'tricks' the system into appearing—from the aspect of this variable only—as a *quantum mechanical problem in R-dimensions with R < ∞*. This is the 'secret recipe' to turning a term-by-term *infinite* perturbation series (about the *free* model) into a term-by-term *finite* perturbation series (about the *pseudofree* model). Measure mashing is the 'royal route' to obtaining a rapid quantization of ultralocal models, and possibly other models as well.

8.4 Ultralocal and Rotational Vector Models Combined

In the previous section we have dealt with ultralocal models with classical Hamiltonians such as

$$H_c(\pi, \phi) = \int \{ \tfrac{1}{2} [\pi(x)^2 + m_0^2 \phi(x)^2] + g_0 \phi(x)^4 \} d^s x , \qquad (8.49)$$

and in Chap. 7 we dealt with rotationally invariant vector models with classical Hamiltonians such as

$$H_c(\pi, \phi) = \int \tfrac{1}{2} [\pi(x)^2 + m_0^2 \phi(x)^2] \, d^s x + g_0' \{ \int \phi(x)^2 \, d^s x \}^2 . \qquad (8.50)$$

This latter expression assumes the more familiar form studied in Chap. 7 given by

$$H_c(\vec{p}, \vec{q}) = \tfrac{1}{2} [\vec{p}^2 + m_0^2 \vec{q}^2] + g_0' (\vec{q}^2)^2 \qquad (8.51)$$

when we introduce $\pi(x) = \Sigma_{n=1}^{\infty} p_n h_n(x)$ and $\phi(x) = \Sigma_{n=1}^{\infty} q_n h_n(x)$, where $\{ h_n(x) \}_{n=1}^{\infty}$ form a complete, orthonormal set of functions. In this brief section we consider a model for which the classical Hamiltonian is given by [model #(7) in Chap. 1]

$$H_c(\pi, \phi) = \int \{ \tfrac{1}{2} [\pi(x)^2 + m_0^2 \phi(x)^2] + g_0 \phi(x)^4 \} \, d^s x + g_0' \{ \int \phi(x)^2 \, d^s x \}^2 ; \qquad (8.52)$$

namely, it *combines an ultralocal and a rotationally invariant interaction together.*

On the surface, these two forms of interaction seem incompatible because ultralocal models use affine field operators while rotationally invariant models use canonical operators. We treat this problem initially as an ultralocal problem with affine field operators and introduce a reducible representation that permits us to include the latter interaction with coupling constant g_0'. Based on the foregoing ultralocal discussion, we can present the result as follows: The field operators are given (for $t = 0$) by

$$\hat{\phi}(x) = \int B(x, \lambda, \zeta)^\dagger \lambda B(x, \lambda, \zeta) \, d\lambda \, d\zeta , \qquad (8.53)$$

$$\hat{\kappa}(x) = -\tfrac{1}{2} i \hbar \int B(x, \lambda, \zeta)^\dagger [(\partial/\partial\lambda)\lambda + \lambda(\partial/\partial\lambda)] B(x, \lambda, \zeta) \, d\lambda \, d\zeta , \qquad (8.54)$$

where as before $B(x, \lambda, \zeta) = A(x, \lambda, \zeta) + c(\lambda, \zeta)$. Here, $c(\lambda, \zeta)$ is a new model function chosen so that $\int c(\lambda, \zeta)^2 d\lambda d\zeta = \infty$ and, e.g., $\int \theta(\pm\lambda) \lambda c(\lambda, \zeta)^2 d\lambda d\zeta \equiv \pm 1$ as well as $\int \theta(\pm\lambda) \lambda^p c(\lambda, \zeta)^2 d\lambda d\zeta = \pm 1 + \mathcal{O}(\hbar)$, $p \in \mathbb{N}^+$, where, again, $\theta(u) = 1$ for $u > 0$, otherwise, $\theta(u) = 0$; as before, these values for integrals of λ and λ^p are first arranged by rescaling the model function and second by arranging $\hbar$ deviations from unity. As usual,

$|0\rangle$ is a 'no-particle' state for which $A(x, \lambda, \zeta)|0\rangle = 0$ for all arguments. The Hamiltonian operator is given by

$$\mathcal{H} = \int d^s x \int A(x, \lambda, \zeta)^\dagger \, \mathsf{h}(\partial/\partial\lambda, \lambda, \zeta) \, A(x, \lambda, \zeta) \, d\lambda \, d\zeta$$
$$+ \int d^s x \int A(x, \lambda, \zeta)^\dagger \, \widetilde{\mathsf{h}}(\partial/\partial\zeta, \lambda, \zeta) \, A(x, \lambda, \zeta) \, d\lambda \, d\zeta \qquad (8.55)$$
$$+ g' : \{ \int d^s x \int A(x, \lambda, \zeta)^\dagger \, \widetilde{\mathsf{h}}(\partial/\partial\zeta, \lambda, \zeta) A(x, \lambda, \zeta) \, d\lambda \, d\zeta \}^2 : \, ,$$

where the operator $A(x, \lambda, \zeta)$ (and its adjoint) can be replaced by $B(x, \lambda, \zeta)$ (and its adjoint) because we require that

$$\mathsf{h}(\partial/\partial\lambda, \lambda, \zeta) c(\lambda, \zeta) = 0 \qquad \text{and} \qquad \widetilde{\mathsf{h}}(\partial/\partial\zeta, \lambda, \zeta) c(\lambda, \zeta) = 0 \quad (8.56)$$

as well. Specifically, with $0 < \xi < 1$, we choose

$$\mathsf{h}(\partial/\partial\lambda, \lambda, \zeta) = -\tfrac{1}{2}\hbar^2 \partial^2/\partial\lambda^2 + \tfrac{1}{2}\hbar^2 (\lambda^2 + \zeta^2)^{-1}$$
$$+ \tfrac{1}{2} m^2 (\lambda + \xi\zeta)^2 + g\lambda^4 - \mathsf{e}_0 \, , \qquad (8.57)$$

and

$$\widetilde{\mathsf{h}}(\partial/\partial\zeta, \lambda, \zeta) = -\tfrac{1}{2}\hbar^2 \partial/\partial\zeta^2 + \tfrac{1}{2}\hbar^2 (\lambda^2 + \zeta^2)^{-1} + \tfrac{1}{2} m^2 (\zeta + \xi\lambda)^2 - \widetilde{\mathsf{e}}_0 \, .$$
$$(8.58)$$

The form of the model function $c(\lambda, \zeta)$ is not known exactly, but we can offer a class of model functions that includes a pseudofree model ($y = 0$) and interacting model functions ($y \neq 0$)

$$c(\lambda, \zeta) = k_c(b)^{1/2} (\lambda^2 + \zeta^2)^{-1/2}$$
$$\times e^{-bm(\lambda^2 + 2\xi\lambda\zeta + \zeta^2)/2\hbar - y(b\lambda, b\zeta, \hbar)/b} \, , \quad (8.59)$$

for suitable functions $y(b\lambda, b\zeta, \hbar)$ and where k_c is chosen to ensure the requirement, e.g., that $\int \theta(\pm\lambda)\lambda c(\lambda, \zeta)^2 \, d\lambda \, d\zeta = \pm 1$.

The weak correspondence principle connects the Hamiltonian operator and the enhanced classical Hamiltonian. Again, we choose affine coherent states constructed, for suitable $\phi(x) = \pm e^{v(x)}$ and $\pi(x) = 2u(x)e^{v(x)}$ functions, as

$$|u, v\rangle = e^{i \int \pi(x)\hat{\phi}_R^2(x) \, d^s x/\hbar} \, e^{-i \int v(x)\hat{\kappa}(x) \, d^s x/\hbar} |0\rangle \, . \qquad (8.60)$$

As noted previously, such coherent states are actually canonical coherent states since (with $b = 1$)

$$|u, v\rangle \equiv N \, e^{\int d^s x \int A(x, \lambda, \zeta)^\dagger [e^{iu(x)\lambda^2/\hbar} e^{-iv(x)\sigma/\hbar} - 1] c(\lambda, \zeta) \, d\lambda \, d\zeta} |0\rangle \, ,$$
$$(8.61)$$

where again $\sigma \equiv -\frac{1}{2} i\hbar [\lambda(\partial/\partial\lambda) + (\partial/\partial\lambda)\lambda]$ and N is a normalization factor. Thus, we find that

$$
\begin{aligned}
H(\pi, \phi) &= \langle u, v | \mathcal{H} | u, v \rangle \\
&= \int d^s x \int \{ c(\lambda, \zeta) [\mathsf{h}(\phi(x)^{-1}\partial/\partial\lambda + \pi(x)\lambda, \phi(x)\lambda, \zeta) \\
&\quad + \widetilde{\mathsf{h}}(\partial/\partial\zeta, \phi(x)\lambda, \zeta)] c(\lambda, \zeta) \, d\lambda \, d\zeta \} \\
&\quad + g' \{ \int d^s x \int c(\lambda, \zeta) [\widetilde{\mathsf{h}}(\partial/\partial\zeta, \phi(x)\lambda, \zeta)] c(\lambda, \zeta) \, d\lambda \, d\zeta \}^2 \\
&= \int \{ \tfrac{1}{2} [\pi(x)^2 + m^2(1 + \xi^2)\phi(x)^2] + g_0 \phi(x)^4 \} d^s x \\
&\quad + \tfrac{1}{4} m^4 \xi^4 g' \{ \int \phi(x)^2 \, d^s x \}^2 + \mathcal{O}(\hbar; \pi, \phi) , \qquad (8.62)
\end{aligned}
$$

which shows, in these parameters, that $m_0^2 = (1 + \xi^2)m^2$, $g_0 = g$, and $g_0' = \frac{1}{4} m^4 \xi^4 g'$, rather like the rotationally invariant vector model. We leave it as an exercise for the reader to determine what is the integral representation for $\mathcal{O}(\hbar; \pi, \phi)$ in the present case.

Unlike ultralocal models alone, observe that an additional variable, ζ, is needed just as was the case for rotationally invariant models. This concludes our discission of combined ultralocal and rotationally invariant models.

8.5 Multi-Dimensional Scalar Ultralocal Models

The models in this section have a classical action functional given by [model #(8) in Chap. 1]

$$
A_C = \int \int_0^T \{ \tfrac{1}{2} [\overrightarrow{\dot{\phi}}(t, x)^2 - m_0^2 \overrightarrow{\phi}(t, x)^2] - g_0 [\overrightarrow{\phi}(t, x)^2]^2 \} \, dt \, d^s x , \qquad (8.63)
$$

where $\overrightarrow{\phi}(t, x) = \{ \phi_1(t, x), \phi_2(t, x), \dots, \phi_N(t, x) \}$ for $1 \le N \le \infty$. It may be noted that this model involves a rotational symmetry like the vector models of Chap. 7, but now does so for every point $x \in \mathbb{R}^s$. Viewed conventionally, like all ultralocal models, these vector ultralocal models are nonrenormalizable when treated by perturbation theory, due to the absence of spatial gradients, and trivial (= free) when viewed nonperturbatively, thanks again to the Central Limit Theorem. Such results classify the conventional quantization of these models as unnatural, and we will show how to obtain natural results. The way to approach these models is first as an $N < \infty$ vector ultralocal model, and then adjust parameters so that a satisfactory result is obtained as $N \to \infty$. In the latter stage, we will draw upon lessons from the rotationally symmetric vector models treated in Chap. 7. This analysis also draws on certain parts of [ZhK94].

Much of the necessary framework can be taken over from a previous section dealing with one-component scalar ultralocal models. This means that for $N < \infty$, we can assume, at time $t = 0$, the basic field operators are given by

$$\hat{\phi}_n(x) = \int B(x, \vec{\lambda})^\dagger \lambda_n B(x, \vec{\lambda}) \, d\vec{\lambda} \, , \qquad (8.64)$$

where $1 \le n \le N < \infty$, $\vec{\lambda} = \{\lambda_1, \lambda_2, \dots, \lambda_N\}$, and $d\vec{\lambda} = \Pi_{n=1}^N d\lambda_n$. [Remark: The symbol n is not to be confused with the use of n for the dimension of spacetime.] Just as in the previous section, $B(x, \vec{\lambda}) = A(x, \vec{\lambda}) + c(\vec{\lambda})$, where $c(\vec{\lambda})$ is a new model function (with the usual integration constraints), and $|0\rangle$ is a 'no-particle' state for which $A(x, \vec{\lambda})|0\rangle = 0$ for all arguments. The Hamiltonian operator is given by

$$\mathcal{H} = \int d^s x \int A(x, \vec{\lambda})^\dagger \mathsf{h}(\nabla_\lambda, \vec{\lambda}) A(x, \vec{\lambda}) \, d\vec{\lambda} \, , \qquad (8.65)$$

where

$$\mathsf{h}(\nabla_\lambda, \vec{\lambda}) = -\tfrac{1}{2} \hbar^2 [\nabla_\lambda^2 - c(\vec{\lambda})^{-1} \nabla_\lambda^2 c(\vec{\lambda})] \, , \qquad (8.66)$$

and $\nabla_\lambda^2 \equiv \Sigma_{n=1}^N \partial^2 / \partial \lambda_n^2$. It follows that the time-dependent field operator is given by

$$\hat{\phi}_n(t, x) = e^{it\mathcal{H}/\hbar} \hat{\phi}(x) e^{-it\mathcal{H}/\hbar} = \int B(x, \vec{\lambda})^\dagger e^{it\mathsf{h}/\hbar} \lambda_n e^{-it\mathsf{h}/\hbar} B(x, \vec{\lambda}) \, d\vec{\lambda} \qquad (8.67)$$

with the convention that $e^{-it\mathsf{h}/\hbar} c(\vec{\lambda}) = c(\vec{\lambda})$. Thus the truncated (T) vacuum expectation values are given (for $\hbar = 1$) by

$$\langle 0 | \hat{\phi}_{n_R}(t_R, x_R) \hat{\phi}_{n_{R-1}}(t_{R-1}, x_{R-1}) \cdots \hat{\phi}_{n_1}(t_1, x_1) | 0 \rangle^T$$
$$= \delta(x_R - x_{R-1}) \cdots \delta(x_2 - x_1) \int c(\vec{\lambda}) \lambda_{n_R} e^{-i(t_R - t_{R-1})\mathsf{h}}$$
$$\times \cdots e^{-i(t_3 - t_2)\mathsf{h}} \lambda_{n_2} e^{-i(t_2 - t_1)\mathsf{h}} \lambda_{n_1} c(\vec{\lambda}) \, d\vec{\lambda} \, . \qquad (8.68)$$

At equal times and smeared in space with the same functions $\{f_n(x)\}_{n=1}^N$, this relation, with $\hat{\phi}_n(f_n) \equiv \int f_n(x) \hat{\phi}_n(x) \, d^s x$ (no summation), becomes

$$\langle 0 | [\Sigma_n \hat{\phi}_n(f_n)]^R | 0 \rangle^T = \int \int [\Sigma_n f_n(x) \lambda_n]^R c(\vec{\lambda})^2 \, d\vec{\lambda} \, d^s x \, , \qquad (8.69)$$

which leads to the characteristic function

$$\langle 0 | e^{i\Sigma_n \hat{\phi}_n(f_n)} | 0 \rangle = e^{-\int d^s x \int [1 - e^{i\Sigma_n f_n(x)\lambda_n}] c(\vec{\lambda})^2 \, d\vec{\lambda}} \, . \qquad (8.70)$$

At this point we choose the model function appropriate to a pseudofree model, specifically, as

$$c_{pf}(\vec{\lambda})^2 = k_{cN}^2 \, b \, |\vec{\lambda}|^{-N} e^{-bm_N \vec{\lambda}^2} \, , \qquad (8.71)$$

where $\vec{\lambda}^2 = \Sigma_{n=1}^N \lambda_n^2$, $|\vec{\lambda}| = [\vec{\lambda}^2]^{1/2}$, and k_{cN} is the free parameter in the overall scale of the model function. The subscript N on two of the parameters will be chosen so that a proper result emerges in the limit $N \to \infty$.

Observe that the model function is invariant under orthogonal rotations of the vector $\vec{\lambda}$, just as the classical model requires. Much as was done in the study of the rotationally invariant vector models in Chap. 7, the exponent of the characteristic function (8.70) can be reexpressed in spherical coordinates, which, with $|\vec{f}(x)|^2 \equiv \Sigma_{n=1}^N f_n(x)^2 \geq 0$ and $\tau \equiv |\vec{\lambda}| \geq 0$, leads to

$$\int [1 - e^{i \Sigma_n f_n(x) \lambda_n}] c(\vec{\lambda})^2 d\vec{\lambda} \tag{8.72}$$
$$= k_{cN}^2 b \int [1 - e^{i |\vec{f}(x)| \tau \cos(\theta)}] e^{-b m_N \tau^2} \tau^{-1} d\tau \, \sin(\theta)^{N-2} \, d\theta \, d\Omega_{N-2} \, .$$

The next step is to evaluate the θ integral by a steepest descent analysis valid for large N. This will change the first exponent to become $-|\vec{f}(x)|^2 \tau^2 / 2(N-2)$ as well as introduce an overall multiplicative factor $[\pi/2(N-2)]^{1/2}$ for the remaining integrals. A rescaling of the variable τ can change the first exponent to read $-|\vec{f}(x)|^2 \tau^2$ with a cost of introducing a related term in the factor $\exp[-b m_N \tau^2] \to \exp[-b m_N \tau^2 2(N-2)]$. The factor m_N is chosen to fix that term to be $\exp[-b\widetilde{m}\tau^2]$ for all N with some mass $\widetilde{m} > 0$. Finally, we insist that $k_{cN}^2 [\pi/2(N-2)]^{1/2} \int d\Omega_{N-2} = 1$ for all N. After these changes we arrive at an *N-independent characteristic functional* for the pseudofree model suitable to use for *all* N including $N = \infty$ given by

$$C_{pf}(\vec{f}) = \exp\{-b \int d^s x \int_0^\infty [1 - e^{-|\vec{f}(x)|^2 \tau^2}] e^{-b\widetilde{m}\tau^2} \, d\tau/\tau\} \, . \tag{8.73}$$

The result in (8.73) has some similarities to Eq. (7.12) which was the result based on symmetry when $N = \infty$ for the rotationally invariant vector models. In that case, it was necessary to choose the weight function as a delta function for a single mass to ensure that the ground state was nondegenerate. However, in the present situation, the uniqueness of the ground state is ensured by the form of the Hamiltonian (8.65) and the fact that $h > 0$. Thus, various interacting models are based on this pseudofree model function with the addition of higher-order terms in the exponent.

As it stands, the expression (8.73) is not in the canonical form of a characteristic functional which involves an explicit Fourier transformation. To remedy this situation, we first note that the integration variable τ can be extended to the whole real line leading to

$$C_{pf}(\vec{f})$$
$$= \exp\{-\tfrac{1}{2} b \int d^s x \int_{-\infty}^\infty [1 - e^{-|\vec{f}(x)|^2 \tau^2}] e^{-b\widetilde{m}\tau^2} \, d\tau/|\tau|\} \, . \tag{8.74}$$
$$= \exp\{-\tfrac{1}{2} b \int d^s x \int_{-\infty}^\infty \int_{-\infty}^\infty [1 - e^{i 2\sqrt{\pi} |\vec{f}(x)| \tau \zeta}] e^{-b\widetilde{m}\tau^2 - \pi \zeta^2} \, d\tau \, d\zeta/|\tau|\}$$
$$= \exp\{-\tfrac{1}{2} b \int d^s x \int_{-\infty}^\infty \int_{-\infty}^\infty [1 - e^{i |\vec{f}(x)| \tau}] e^{-b\widetilde{m}\tau^2/(4\pi\zeta^2) - \pi \zeta^2} \, d\tau \, d\zeta/|\tau|\}.$$

In this expression we see that the model function has *two* variables, τ and ζ, thus implying, in this form, that we deal with reducible affine variables, which is not unexpected since the rotationally invariant models required reducible operator representations as well.

This concludes our ultralocal scalar discussions. Next, we turn our attention to ultralocal spinors, and later to ultralocal scalars and ultralocal spinors combined.

8.6　Quantization of Ultralocal Spinor Models

8.6.1　*Preliminary considerations: Free models*

The discussion in this section is principally based on [Kl73a]; it also includes some updating of that earlier work [model #(9) in Chap. 1].

To begin with, the ultralocal *free* spinor model follows the line of most free models. We introduce spinor fields composed of two types of creation and annihilation operators that satisfy the *canonical anticommutation relations* (CAR): $\hat{\psi}_{m\,0}(x) = U_0(m)\,a_m(x) + L_0(m)\,b_m(x)^{\dagger}$ and its adjoint $\hat{\psi}_{m\,0}(x)^{\dagger} = U_0(m)\,a_m(x)^{\dagger} + L_0(m)\,b_m(x)$, where $\{a_m(x), a_n(y)\} = 0$ while $\{a_m(x), a_n(y)^{\dagger}\} = \delta_{m,n}\delta(x-y)\,\mathbb{1}$ (and the same for the independent operators b and $b^{\dagger}$); in addition, $a_m(x)|0\rangle = b_m(x)|0\rangle = 0$ for the 'no particle' state $|0\rangle$. Here, $U_0(m) = 1$ for $1 \leq m \leq M_0$ and zero otherwise, while $L_0(m) = 1$ for $M_0 + 1 \leq m \leq 2M_0$ and zero otherwise, leading to 'upper' (U_0) and 'lower' (L_0) components for the free model. The ultralocal free field model Hamiltonian operator is given by

$$\mathcal{H}_0 = \mu \int [U_0(m)\,a_m(x)^{\dagger} a_m(x) + L_0(m)\,b_m(x)^{\dagger} b_m(x)]\,d^s x \,, \quad (8.75)$$

with summation on m understood—here and later—and this expression represents a form of 'second quantization'. The quantum action functional for this example is given by

$$A_Q = \int \langle \psi(t)|\,[i\hbar(\partial/\partial t) - \mathcal{H}_0]\,|\psi(t)\rangle \, dt \,, \quad (8.76)$$

which, with

$$|\psi(t)\rangle = \int \psi_m(x,t)\,[U_0(m)\,a_m(x)^{\dagger} + L_0(m)\,b_m(x)^{\dagger}]\,|0\rangle \, d^s x \,, \quad (8.77)$$

leads to

$$A_{Q(R)} = \int_0^T \int [i\hbar\psi_m(x,t)^* \dot{\psi}_m(x,t) - \mu\psi_m(x,t)^* \psi_m(x,t)]\,d^s x \, dt \,. \quad (8.78)$$

In turn, this action functional represents a form of 'first quantization'.

Observe that this latter expression is *not* the usual first-quantized 'co-variant' free model with the gradients removed that has a first-quantized Hamiltonian density $\mu[U_0(m)\psi_m(x)^*\psi_m(x) - L_0(m)\psi_m(x)^*\psi_m(x)]$, which implies the Hamiltonian spectrum is unbounded below. A lower bound is restored exploiting the famous 'Dirac sea', which assumes the negative energy levels are fully occupied so that, after ignoring that level, the Hamiltonian is effectively bounded below by zero. [Remark: In one approach (which we do not choose to follow), the negative sign is reversed if the 'functions' involved are treated as independent "Grassmann variables" (a.k.a., anti-commuting c-numbers) [Wiki-k].] In our analysis of interacting ultralocal spinor models, we, too, will encounter a Hamiltonian operator that is non-negative.

8.6.2 *Preliminary considerations: Interacting models*

If we were to deal with nonlinear interactions as perturbations of free models we would encounter terms such as: (i) $a_m(x)^\dagger b_n(x)^\dagger$, (ii) $a_m(x)b_n(x)$, etc. The first example (i) is 'not an operator', i.e., it has only the zero vector in its domain as is easily seen by evaluating $\|\int d^s x f(x) a_m(x)^\dagger b_n(x)^\dagger |\phi\rangle\|$ for a smooth smearing function $f(x)$. The second example (ii) is 'not closable', i.e., let $|\psi_K\rangle \to 0$ as $K \to \infty$ in the sense that $\| |\psi_K\rangle \| \to 0$, while, on the other hand, let $\mathcal{A}$ be an operator for which $\mathcal{A}|\psi_K\rangle \to |\phi_{\mathcal{A}}\rangle \in \mathfrak{H}$ and $|\phi_{\mathcal{A}}\rangle \neq 0$, as $K \to \infty$, then $\mathcal{A}$ is is not densely defined and so its adjoint cannot be defined, and $\mathcal{A}$ is called non-closable. As an example, suppose $|\psi_K\rangle = \int\int f_K(x,y) a(x)^\dagger b(y)^\dagger d^s x\, d^s y\, |0\rangle$, where $f_K(x,y) = \exp[-x^2 - y^2 - K(x-y)^2]$ and $\mathcal{A} = \int g(z) a(z) b(z)\, d^s z$ for a smooth function $g(z)$. Thus such local operator products are not usable. Therefore, ultralocal spinor fields that obey the CAR apply only to free models with no interaction. In order to deal with nonlinear interactions, we will need another anticommutation rule. Remark: A similar analysis applies for two bosons, or one boson and one fermion, for the appropriate annihilation and creation operators.

We motivate the new anticommutation rule by first discussing *deuterons*, which are composed of two fermion particles: one proton and one neutron (ignoring quarks). While the separate particles behave as fermions, the composite particle acts like a boson meaning that arbitrarily many deuterons can enjoy the same quantum state simultaneously. This is true up to a point, namely, until the fermions have enough separate energy to break the bond that holds them together and the deuteron disappears in favor of its two constituents, the proton and neutron. In that sense the deuteron is not a 'genuine' boson. In dealing with the mathematics of ultralocal fermions, we need to allow for idealized combinations within which such bosons are *genuine* bosons stable at all energies. We cannot derive the needed new anticommutation rules from the CAR (as we could for the CCR), but instead we must deduce them indirectly.

To begin with, we introduce two sets of conventional fermion annihilation and creation operators, $a(x, \vec{\xi})$ and $b(x, \vec{\xi})$, $\vec{\xi} \equiv \{\xi_1, \dots, \xi_{2M}\} \in \mathbb{R}^{2M}$, $M \geq 1$, and their adjoints. The usual anticommutation rules apply: namely, $\{a(x, \vec{\xi}), a(x', \vec{\xi'})^\dagger\} = \{b(x, \vec{\xi}), b(x', \vec{\xi'})^\dagger\} = \delta(x-x')\delta(\vec{\xi}-\vec{\xi'})\,\mathbb{1}$, with all other anticommutation relations vanishing. In addition, we introduce a normalized, 'no-particle' vector $|0\rangle$ such that $a(x, \vec{\xi})|0\rangle = b(x, \vec{\xi})|0\rangle = 0$, which also makes the representation of the operators a and b a Fock representation— except for one point. We create the fermionic component of the Hilbert space by repeatedly applying fermionic creation operators $a^\dagger$ and $b^\dagger$ smeared with smooth functions, such as $f(x, \vec{\xi})$, which are restricted to be *antisymmetric* in $\vec{\xi}$-space, i.e., $f(x, -\vec{\xi}) = -f(x, \vec{\xi})$.

For the rest of the Hilbert space, we introduce a set of conventional *boson* (sic) annihilation and creation operators, $A(x, \vec{\xi})$ and $A(x, \vec{\xi})^\dagger$, which obey the only nonvanishing, standard commutation rule $[A(x, \vec{\xi}), A(y, \vec{\xi'})^\dagger] = \delta(x - y)\delta(\vec{\xi} - \vec{\xi'})\,\mathbb{1}$. Once again we insist that $A(x, \vec{\xi})|0\rangle = 0$, where $|0\rangle$ is the same 'no-particle' vector used above for the fermions. The Hilbert space is completed by repeatedly applying the creation operator $A(x, \vec{\xi})^\dagger$ smeared with smooth functions, such as $F(x, \vec{\xi})$, which are *symmetric* in $\vec{\xi}$-space, i.e,, $F(x, -\vec{\xi}) = F(x, \vec{\xi})$. We also introduce a new model function $c(\vec{\xi})\,[= c(-\vec{\xi})]$, much as before, which is nonvanishing, real, and has similar integration properties, namely, $\int [\vec{\xi}^2/(1+\vec{\xi}^2)]c(\vec{\xi})^2\, d\vec{\xi} < \infty$ and $\int c(\vec{\xi})^2\, d\vec{\xi} = \infty$, where $\vec{\xi}^2 \equiv \Sigma_{m=1}^{2M}\xi_m^2$, and $d\vec{\xi} \equiv \Pi_{m=1}^{2M}d\xi_m$. Again we introduce the operators $B(x, \vec{\xi}) \equiv A(x, \vec{\xi}) + c(\vec{\xi})\,\mathbb{1}$ and $B(x, \vec{\xi})^\dagger \equiv A(x, \vec{\xi})^\dagger + c(\vec{\xi})\,\mathbb{1}$, which satisfy the commutation relation $[B(x, \vec{\xi}), B(y, \vec{\xi'})^\dagger] = \delta(x - y)\delta(\vec{\xi} - \vec{\xi'})\,\mathbb{1}$.

The foregoing discussion has led to a new 'substructure' of familiar operators with which we can build the basic field operators for the spinor ultralocal models.

Solution of spinor ultralocal models

The basic spinor field operators are defined, at $t = 0$ and for $1 \leq m \leq M$, by

$$\hat{\psi}_m(x) = \int B(x, \vec{\xi})^\dagger \xi_m a(x, \vec{\xi})\, d\vec{\xi}, \qquad (8.79)$$

and for $M + 1 \leq m \leq 2M$ by

$$\hat{\psi}_m(x) = \int b(x, \vec{\xi})^\dagger \xi_m B(x, \vec{\xi}) \, d\vec{\xi} \, . \qquad (8.80)$$

In turn, the adjoints to these operators, for $1 \leq m \leq M$, read as

$$\hat{\psi}_m(x)^\dagger = \int a(x, \vec{\xi})^\dagger \xi_m B(x, \vec{\xi}) \, d\vec{\xi} \, , \qquad (8.81)$$

and for $M + 1 \leq m \leq 2M$ by

$$\hat{\psi}_m(x)^\dagger = \int B(x, \vec{\xi})^\dagger \xi_m b(x, \vec{\xi}) \, d\vec{\xi} \, . \qquad (8.82)$$

As was the case for the free ultralocal spinor models, we use $U(r) = 1$ for $1 \leq r \leq M$, and zero otherwise, for 'upper' components and $L(r) = 1$ for $M + 1 \leq r \leq 2M$, and zero otherwise, for 'lower' components, which leads to

$$\hat{\psi}_r(x) = \int [U(r) \, B(x, \vec{\xi})^\dagger \xi_r \, a(x, \vec{\xi}) + L(r) \, b(x, \vec{\xi})^\dagger \xi_r \, B(x, \vec{\xi})] \, d\vec{\xi} \, , \qquad (8.83)$$

and

$$\hat{\psi}_s(x)^\dagger = \int [U(s) \, a(x, \vec{\xi})^\dagger \xi_s \, B(x, \vec{\xi}) + L(s) \, B(x, \vec{\xi})^\dagger \xi_s \, b(x, \vec{\xi})] \, d\vec{\xi} \, . \qquad (8.84)$$

It is also useful to introduce 'upper' and 'lower' vectors defined as

$$\hat{\psi}_\uparrow(x) = \int B(x, \vec{\xi})^\dagger \vec{\xi} \, a(x, \vec{\xi}) \, d\vec{\xi} \, , \qquad (8.85)$$

$$\hat{\psi}_\downarrow(x) = \int b(x, \vec{\xi})^\dagger \vec{\xi} \, B(x, \vec{\xi}) \, d\vec{\xi} \, , \qquad (8.86)$$

and their adjoints. When dealing with either $\hat{\psi}_\uparrow(x)$ or $\hat{\psi}_\downarrow(x)$ it is generally sufficient to let m satisfy $1 \leq m \leq M$ in both cases.

The first observation is to note that these operators satisfy certain anticommutation properties, namely

$$\{\hat{\psi}_r(x), \hat{\psi}_s(y)\} = 0 \, , \qquad (8.87)$$

and its adjoint, when $x \neq y$. On the other hand, we observe that

$$\{\hat{\psi}_r(x), \hat{\psi}_s(y)^\dagger\} = \delta(x - y) \, R_{rs}(x) \, , \qquad (8.88)$$

where

$$R_{rs}(x) \equiv \int [U(r)U(s) + L(r)L(s)] \, B(x, \vec{\xi})^\dagger \xi_r \xi_s \, B(x, \vec{\xi}) \qquad (8.89)$$
$$+ U(r)U(s) \, a(x, \vec{\xi})^\dagger \xi_r \xi_s \, a(x, \vec{\xi}) + L(r)L(s) \, b(x, \vec{\xi})^\dagger \xi_r \xi_s \, b(x, \vec{\xi})] \, d\vec{\xi} \, .$$

This complicated expression obviously satisfies

$$[R_{rs}(x), R_{tu}(y)] = 0 \qquad (8.90)$$

for all x and y, just as boson operators do. Indeed, this expression contains a *genuine* boson operator $\int B^\dagger \xi_r \xi_s B \, d\vec{\xi}$ which is *exactly* like a renormalized local field product of two ultralocal scalar fields. This term realizes our 'perfect deuteron', namely, a *genuine boson* operator arising from the local product of two *fermion*-like operators.

Moreover, there are several local operator products of interest, which are defined just like the local products for the scalar case—again multiplying by the positive factor b [*not* to be confused with the fermion operator $b(x, \vec{\xi})$] to preserve dimensions. Next we list a few local products omitting obviously implicit conditions such as $L(r)U(s)$ on the first term, etc.:

$$(\hat{\psi}_{\downarrow r}\hat{\psi}_{\uparrow s})_R(x) = b \int b(x, \vec{\xi})^{\dagger} \xi_r \xi_s \, a(x, \vec{\xi}) \, d\vec{\xi} \,, \tag{8.91}$$

$$(\hat{\psi}_{\uparrow r}^{\dagger}\hat{\psi}_{\uparrow s})_R(x) = b \int a(x, \vec{\xi})^{\dagger} \xi_r \xi_s \, a(x, \vec{\xi}) \, d\vec{\xi} \,, \tag{8.92}$$

$$(\hat{\psi}_{\uparrow r}\hat{\psi}_{\uparrow s}^{\dagger})_R(x) = b \int B(x, \vec{\xi})^{\dagger} \xi_r \xi_s \, B(x, \vec{\xi}) \, d\vec{\xi} \,, \tag{8.93}$$

$$(\hat{\psi}_{\uparrow r}\hat{\psi}_{\uparrow s}^{\dagger}\hat{\psi}_{\downarrow t})_R(x) = b^2 \int B(x, \vec{\xi})^{\dagger} \xi_r \xi_s \xi_t \, a(x, \vec{\xi}) \, d\vec{\xi} \,, \tag{8.94}$$

etc. Note that it follows that

$$\{\hat{\psi}_r(x), \hat{\psi}_s(y)\} = \delta(x - y) \, S_{rs}(x) \,, \tag{8.95}$$

where

$$S_{rs}(x) = L(r)U(s) \int b(x, \vec{\xi})^{\dagger} \xi_r \xi_s \, a(x, \vec{\xi}) \, d\vec{\xi} \,. \tag{8.96}$$

Clearly, the spinor field operators $\hat{\psi}(x)$ and $\hat{\psi}(x)^{\dagger}$ are *not* conventional fermion operators. However, they are constructs of genuine fermion and genuine boson operators that underlie them; that is all that is important.

Assuming $a \leftrightarrow b$ symmetry, the Hamiltonian $\mathcal{H} = \int \mathcal{H}(x) \, d^s x$ for such fields is reduced to the generic form where

$$\begin{aligned}
\mathcal{H}(x) = &\int A(x, \vec{\xi})^{\dagger} \mathsf{h}(\nabla_{\xi}, \vec{\xi}) \, A(x, \vec{\xi}) \, d\vec{\xi} \\
&+ \int a(x, \vec{\xi})^{\dagger} \tilde{\mathsf{h}}(\nabla_{\xi}, \vec{\xi}) \, a(x, \vec{\xi}) \, d\vec{\xi} \\
&+ \int b(x, \vec{\xi})^{\dagger} \tilde{\mathsf{h}}(\nabla_{\xi}, \vec{\xi}) \, b(x, \vec{\xi}) \, d\vec{\xi} \,, \tag{8.97}
\end{aligned}$$

where $B^{\dagger}, B$ have been replaced by $A^{\dagger}, A$ because we choose

$$\begin{aligned}
\mathsf{h}(\nabla_{\xi}, \vec{\xi}) &\equiv -\tfrac{1}{2}\hbar^2 \, \nabla_{\xi}^2 + \tfrac{1}{2}\hbar^2 \, c(\vec{\xi})^{-1}\nabla_{\xi}^2 \, c(\vec{\xi}) \\
\tilde{\mathsf{h}}(\nabla_{\xi}, \vec{\xi}) &\equiv -\tfrac{1}{2}\hbar^2 \, \nabla_{\xi}^2 + v(\vec{\xi}) \,, \tag{8.98}
\end{aligned}$$

where $\nabla_{\xi}^2 \equiv \Sigma_{m=1}^{2M} \, \partial^2/\partial\xi_m^2$. If the potential term is composed solely of the operator $R(x)$ and its local powers, it follows [Kl73a] that the spectrum of h and $\tilde{\mathsf{h}}$ are identical. Instead, if the potential $v(\vec{\xi})$ has a variety of terms, then the spectrum of h and $\tilde{\mathsf{h}}$ are different. In any case, to ensure that the ground state is unique and that $\mathcal{H} \geq 0$, it is necessary that $\mathsf{h} > 0$ and $\tilde{\mathsf{h}} > 0$.

Once again, and for N even, there is enough information to determine the truncated (T) vacuum expectation values as given, for example, by

$$\langle 0|\hat{\psi}_{r_N}(t_N, x_N)\hat{\psi}^\dagger_{r_{N-1}}(t_{N-1}, x_{N-1})\cdots\hat{\psi}_{r_2}(t_2, x_2)\hat{\psi}^\dagger_{r_1}(t_1, x_1)|0\rangle^T$$

$$= \delta(x_N - x_{N-1})\delta(x_{N-1} - x_{N-2})\cdots\delta(x_2 - x_1) \quad\quad (8.99)$$

$$\times \int c(\vec{\xi})\xi_{r_N} e^{-i(t_N - t_{N-1})\bar{h}}\xi_{r_{N-1}}\cdots\xi_{r_3} e^{-i(t_3 - t_2)h}\xi_{r_2} e^{-i(t_2 - t_1)\bar{h}}\xi_{r_1} c(\vec{\xi})\, d\vec{\xi}.$$

In this expression, all r values are required to be upper r, i.e., $1 \le r \le M$, otherwise the expression vanishes. If N is odd, such expressions also vanish for any choice of r values.

Some possible examples of model functions are given (for $\hbar = 1$) by

$$c_{pf}(\vec{\xi}) = b^{1/2}[\vec{\xi}^2]^{-M/2}e^{-\frac{1}{2}bm\vec{\xi}^2}, \quad\quad (8.100)$$

for a typical pseudofree model, where, again, $\vec{\xi}^2 = \Sigma^{2M}_{m=1}\xi^2_m$, and

$$c_g(\vec{\xi}) = c_{pf}(\vec{\xi})\, e^{-\frac{1}{2}gb^{-1}y(b\vec{\xi})}, \qu\quad (8.101)$$

for a general interacting model. As $g \to 0$ such models pass continuously to the pseudofree model $c_{pf}(\vec{\xi})$. A perturbation analysis of the interacting model, about the pseudofree model, is obtained by an expansion of $c_g(\vec{\xi})$ in powers of g, which, as claimed, leads to a series that is term-by-term finite.

This model does not lend itself to a straightforward enhanced quantization viewpoint. For example, consider the rendering of the formal Hamiltonian "$\mathcal{H} = \int d^s x \{\mu_0 \hat{\psi}(x)^\dagger \hat{\psi}(x) + g_0 [\hat{\psi}(x)^\dagger \mathcal{M}\hat{\psi}(x)]^2\}$" as follows:

$$\mathcal{H} = \int d^s x \{\mu(\hat{\psi}^\dagger_r\hat{\psi}_r)_R(x) + g(\hat{\psi}^\dagger_r\hat{\psi}_s\hat{\psi}^\dagger_t\hat{\psi}_u)_R(x)\mathcal{M}_{rs}\mathcal{M}_{tu}\} \qu\quad (8.102)$$

$$= \int d^s x \int \{B(x, \vec{\xi})^\dagger h B(x, \vec{\xi}) + a(x, \vec{\xi})^\dagger \tilde{h} a(x, \vec{\xi}) + b(x, \vec{\xi})^\dagger \tilde{h} b(x, \vec{\xi})\}\, d\vec{\xi}.$$

The expression in quotation marks represents the formal Hamiltonian expression obtained by replacing c-numbers by q-numbers; observe that if conventional fermion operators were involved, the nonlinear interaction would vanish. The next line translates the intent of the formal line into renormalized products of our fermion-like operators, with summation on indices understood, a procedure that retains the nonlinear interaction. The last line represents the Hamiltonian expressed in bilinear terms of the fundamental boson and fermion operators. An attempt to reduce this operator form to strictly 'first' quantization spinor terms does not seem possible, and consequently we do not try to do so in this case.

On the other hand, a brief discussion of simpler yet successful enhanced treatments of 'first/'second quantization' examples is given in [Kl12d].

This concludes out discussion of ultralocal spinor models.

8.7   Ultralocal Scalar-Spinor Models

In this section we offer a brief overview of an ultralocal model that combines spinor and scalar fields [model #(10) in Chap. 1]. The motivating problem is given by the 'first quantized' action functional

$$A = \int\int_0^T \{\tfrac{1}{2}[\dot\phi(t,x)^2 - m_0^2\phi(t,x)^2] + i\hbar\psi(t,x)^*\dot\psi(t,x) - \mu_0\psi(t,x)^*\beta\psi(t,x)$$
$$-g_0\psi(t,x)^*\phi(t,x)\psi(t,x) - g_0'\phi(t,x)^4\} \, dt \, d^sx \; . \qquad (8.103)$$

Of course, additional potentials, such as $[\psi(t,x)^*\mathcal{M}\psi(t,x)]^2$, may also be considered. The classical equations of motion that arise from this action functional are

$$\ddot\phi(t,x) + m_0^2\phi(t,x) = -g_0[\psi(t,x)^*\psi(t,x)] - 4g_0'\phi(t,x)^3 \; , \qquad (8.104)$$

and

$$i\hbar\dot\psi(t,x) - \mu_0\beta\psi(t,x) = g_0\phi(t,x)\psi(t,x) \qquad (8.105)$$

and its adjoint. When $g_0 > 0$, these equations imply that the field $\phi(t,x)$ contains a 'component' of $\psi(t,x)^*$ and $\psi(t,x)$—and vice versa; this would especially be clear if the spinor fields were Grassmann variables.

Guided by this classical requirement, and based on our earlier discussions, we adopt the basic time-zero fields

$$\hat\phi(x) = \int B(x,\vec\xi,\lambda)^\dagger \lambda B(x,\vec\xi,\lambda) \, d\vec\xi \, d\lambda \qquad (8.106)$$
$$+ \int [a(x,\vec\xi,\lambda)^\dagger \lambda a(x,\vec\xi,\lambda) + b(x,\vec\xi,\lambda)^\dagger \lambda b(x,\vec\xi,\lambda)] \, d\vec\xi \, d\lambda \; ,$$

where $B(x,\vec\xi,\lambda) = A(x,\vec\xi,\lambda) + c(\vec\xi,\lambda)$, etc. In addition, we have

$$\hat\psi_r(x) = \int [U(r)B(x,\vec\xi,\lambda)^\dagger \xi_r a(x,\vec\xi,\lambda)$$
$$+ L(r)b(x,\vec\xi,\lambda)^\dagger \xi_r B(x,\vec\xi,\lambda)] \, d\vec\xi \, d\lambda \; , \qquad (8.107)$$

and its adjoint.

The Hamiltonian operator $\mathcal{H}$ is given by

$$\mathcal{H} = \int d^sx \int [A(x,\vec\xi,\lambda)^\dagger \mathsf{h}(\nabla_\xi,\nabla_\lambda,\vec\xi,\lambda) A(x,\vec\xi,\lambda)$$
$$+ a(x,\vec\xi,\lambda)^\dagger \tilde{\mathsf{h}}(\nabla_\xi,\nabla_\lambda,\vec\xi,\lambda) a(x,\vec\xi,\lambda)$$
$$+ b(x,\vec\xi,\lambda)^\dagger \tilde{\mathsf{h}}(\nabla_\xi,\nabla_\lambda,\vec\xi,\lambda) b(x,\vec\xi,\lambda)] \, d\vec\xi \, d\lambda \; , \qquad (8.108)$$

where, as before $B(x,\vec\xi,\lambda)$ has been replaced by $A(x,\vec\xi,\lambda)$ because $c(\vec\xi,\lambda)$ has been chosen so that

$$\mathsf{h}(\nabla_\xi,\nabla_\lambda,\vec\xi,\lambda) = -\tfrac{1}{2}\hbar^2\{[\nabla_\xi^2 + \nabla_\lambda^2]$$
$$- c(\vec\xi,\lambda)^{-1}[\nabla_\xi^2 + \nabla_\lambda^2]c(\vec\xi,\lambda)\} \; , \qquad (8.109)$$

as well as

$$\tilde{h}(\nabla_\xi, \nabla_\lambda, \vec{\xi}, \lambda) = -\tfrac{1}{2}\hbar^2[\nabla_\xi^2 + \nabla_\lambda^2] + v(\vec{\xi}, \lambda) \ . \tag{8.110}$$

Here $\nabla_\xi^2 \equiv \Sigma_{m=1}^{2M} \partial^2/\partial\xi_m^2$ and $\nabla_\lambda^2 \equiv \partial^2/\partial\lambda^2$. The potential v and the model function c both may depend on $\hbar$.

The time-dependent scalar field operator is given by

$$\begin{aligned}
\hat{\phi}(t, x) &= e^{i\mathcal{H}t/\hbar} \hat{\phi}(x) e^{-i\mathcal{H}t/\hbar} \\
&= \int \{ B(x, \vec{\xi}, \lambda)^\dagger e^{iht/\hbar} \lambda e^{-iht/\hbar} B(x, \vec{\xi}, \lambda) \\
&\quad + a(x, \vec{\xi}, \lambda)^\dagger e^{i\tilde{h}t/\hbar} \lambda e^{-i\tilde{h}t/\hbar} a(x, \vec{\xi}, \lambda) \\
&\quad + b(x, \vec{\xi}, \lambda)^\dagger e^{i\tilde{h}t/\hbar} \lambda e^{-i\tilde{h}t/\hbar} b(x, \vec{\xi}, \lambda) \} \, d\vec{\xi} \, d\lambda \ . \tag{8.111}
\end{aligned}$$

The time-dependent spinor field operator is likewise given by

$$\begin{aligned}
\psi_r(t, x) &= e^{i\mathcal{H}t/\hbar} \psi_r(x) e^{-i\mathcal{H}t/\hbar} \\
&= \int \{ U(r) B(x, \vec{\xi}, \lambda)^\dagger e^{iht/\hbar} \xi_r e^{-i\tilde{h}t/\hbar} a(x, \vec{\xi}, \lambda) \tag{8.112} \\
&\quad + L(r) b(x, \vec{\xi}, \lambda)^\dagger e^{i\tilde{h}t/\hbar} \xi_r e^{-iht/\hbar} B(x, \vec{\xi}, \lambda) \} \, d\vec{\xi} \, d\lambda \ ,
\end{aligned}$$

and its adjoint.

A reasonable example of the Hamiltonian expressions for h and $\tilde{h}$ is given by

$$\mathsf{h} = -\tfrac{1}{2}\hbar^2 \nabla_\xi^2 - \tfrac{1}{2}\hbar^2 \partial^2/\partial\lambda^2 + \tfrac{1}{2}m^2\lambda^2 + g\vec{\xi}^2\lambda + g'\lambda^4 - \mathsf{e}_0 \ , \tag{8.113}$$

$$\tilde{\mathsf{h}} = -\tfrac{1}{2}\hbar^2 \nabla_\xi^2 - \tfrac{1}{2}\hbar^2 \partial^2/\partial\lambda^2 + \mu\vec{\xi}^2 + g\vec{\xi}^2\lambda + g'\lambda^4 - \tilde{\mathsf{e}}_0 \ . \tag{8.114}$$

These expressions are sufficient to construct various truncated vacuum expectation values.

This concludes our discussion of the present model.

PART 5

Enhanced Quantization of
Covariant Scalars and Gravity

Chapter 9

Enhanced Quantization of Covariant Scalar Field Models

WHAT TO LOOK FOR

Conventional quantization of covariant scalar field models φ_n^4 for spacetime dimensions $n \geq 5$ is trivial; hence it is unnatural. However, an alternative $\mathcal{O}(\hbar)$ counterterm, different from the usual ones, leads to nontrivial results for all $n \geq 5$, and provides a different quantization for $n \leq 4$ as well. In this chapter we initially determine the counterterm that provides these desirable properties as simply and directly as possible. Further properties are analyzed in later sections. The same counterterm also resolves models such as φ_n^p for all even p and all $n \geq 2$, including all cases that are traditionally regarded as nonrenormalizable. In so doing, we can also treat 'mixed models' such as $g\varphi_3^4 + g'\varphi_3^{44}$, etc.

9.1 Introduction

The basic classical models discussed in this chapter are those of quartic self-interacting covariant scalar fields with a classical action functional of the form [model #(11) in Chap. 1]

$$A = \int_0^T \int \{ \tfrac{1}{2} [\dot{\phi}(x,t)^2 - (\vec{\nabla}\phi)(x,t)^2 - m_0^2 \phi(x,t)^2] - g_0\,\phi(x,t)^4 \}\, d^s x\, dt \,,$$

$$(9.1)$$

where $s \geq 1$ is the number of spatial dimensions and $x \in \mathbb{R}^s$; note that these models differ from ultralocal models of the previous chapter by the addition of spatial derivatives. The given model is typically denoted by φ_n^4, where $n = s + 1$ is the number of spacetime dimensions. Some remarks will also be made about interaction powers greater than four. With sufficiently smooth initial data, the classical solutions for such models are nontrivial and well defined for all $n \geq 2$ [Ree76].

The standard, textbook procedures for the quantization of covariant scalar quantum fields, such as φ_n^4, lead to self-consistent results for spacetime dimensions $n = 2, 3$, but for spacetime dimensions $n \geq 4$, the same procedures lead to triviality (that is, to a free or generalized free theory) either rigorously for $n \geq 5$ [Aiz81; Frö82; FFS92], or on the basis of renormalization group [Cal88] and Monte Carlo [FSW82] studies for $n = 4$. A trivial result implies that the (free) classical model obtained as the classical limit of the proposed quantum theory does not reproduce the (nonfree) classical model with which one started, and, in this case, it is fair to say that such a quantization is *unnatural* or *unfaithful*, since its classical limit differs from the original classical theory. Effectively, this result follows from the natural presumption that the counterterms needed for renormalization of the quantization process are those suggested by traditional perturbation theory, an approach that has its own limitations since the models φ_n^p, for $p > 2n/(n-2)$, are deemed to be *nonrenormalizable* because they are ill-defined by a conventional perturbation analysis due to the need for infinitely many distinct counterterms [ItZ80]. However, the conventional quantization process is inherently ambiguous, and it allows for a wide variety of nonclassical terms (proportional to $\hbar$ so that the right classical limit is formally obtained); for example, an harmonic oscillator normally has a positive zero-point energy except when it is one of the degrees of freedom of a free field. Such an ambiguity in quantization may well allow for other, nontraditional counterterms, also proportional to $\hbar$, which lead instead to a nontrivial quantization and may indeed be *faithful* in the sense that the classical limit of the alternative quantum theory leads to the very same nonlinear classical theory with which one started. Such an alternative quantization procedure may have a better claim to be a proper quantization than one that leads to an unfaithful result.

Some issues need to be addressed immediately: For some problems, a trivial result for a quartic interacting, quantum scalar field is appropriate. For example, if one is studying a second-order phase transition of some crystalline substance on a cubic lattice of *fixed lattice spacing*, then the quantum formulation of the field at each lattice site is unambiguous, and in that case the long correlation length makes a self-interacting continuum field a suitable approximating model. Alternatively, when faced with triviality, some authors argue that unfaithful quantizations can serve as "effective theories", valid in their low-order perturbation series, for low-energy questions, assuming that some future theory will resolve high-energy issues in a suitable fashion. While this view is of course possible, it seems more likely that an unfaithful quantization is simply the result of an inappropriate quantization procedure, as our discussion will illustrate.

Stated succinctly, our goal is nothing less than finding an alternative counterterm that leads to a genuine, faithful quantization of self-interacting scalar fields for all spacetime dimensions. Auxiliary fields are not wanted, and as we shall see they are not needed. No claim is made that, if successful, these alternative quantization procedures are "right" or even "to be preferred"; but they would seem to offer a valid, self-consistent, alternative quantization procedure that possibly may have some advantages in certain cases.

Building on the lessons learned from the study of ultralocal scalar models, we shall begin our presentation of covariant scalar models with an approach involving another example of the 'secret recipe' that quickly leads to the solution of certain models, for instance, a nontrivial quantization of φ_n^4 for all $n \geq 2$, including an alternative quantization for $n = 2, 3$, drawing on [Kl14b]. Following that discussion, we expand our study to offer an alternative approach to the quantization of such models, based, effectively, on an enhanced affine quantization approach.

9.2 A Rapid Quantization of Covariant Scalar Models

Examples of the family of models we consider have classical (but imaginary time) action functionals given (now for $x \in \mathbb{R}^n$) by

$$I(\phi) = \int \{ \tfrac{1}{2}[(\nabla \phi)(x)^2 + m_0^2 \phi(x)^2] + g_0 \phi(x)^4 \} \, d^n x . \tag{9.2}$$

In turn, the quantization of such an example may be addressed by the formal functional integral

$$S(h) \equiv \mathcal{M} \int e^{(1/\hbar) \int h(x) \phi(x) \, d^n x - (1/\hbar)[I(\phi) + F(\phi, \hbar)/2]} \, \Pi_x \, d\phi(x) , \tag{9.3}$$

where $\mathcal{M}$ is chosen so that $S(0) = 1$, $h(x)$ is a smooth source function, and $F(\phi, \hbar)$ represents the as-yet-unspecified counterterm to control divergences, which should formally vanish as $\hbar \to 0$ so that the proper classical limit formally emerges. Traditionally, a perturbation expansion of the quartic interaction term guides the choice of the counterterm(s) in order to cancel any divergences that are encountered. Here, on the other hand, we seek to find a different counterterm that permits a perturbation expansion without any divergences. Superficially, it appears like we are seeking a 'needle in a haystack'; happily, in our case, we are able to find the 'needle'!

Although the formal functional integral (9.3) is essentially undefined, it can be given meaning by first introducing a lattice regularization in which the spacetime continuum is replaced with a periodic, hypercubic lattice with lattice spacing $a > 0$ and with $L < \infty$ sites along each axis. The sites themselves are labeled by multi-integers $k = \{k_0, k_1, \ldots, k_s\} \in \mathbb{Z}^n$, $n = s + 1$, and h_k and ϕ_k denote field values at the kth lattice site; in particular, k_0 is designated as the Euclidean time variable. This regularization results in the L^n-dimensional integral

$$
\begin{aligned}
S_{latt}(h) &\equiv M \int e^{(1/\hbar)Z^{-1/2}\Sigma_k h_k \phi_k\, a^n} \\
&\quad \times e^{-(1/2\hbar)[\Sigma_{k,k^*}(\phi_{k^*} - \phi_k)^2\, a^{n-2} + m_0^2 \Sigma_k \phi_k^2\, a^n]} \\
&\quad \times e^{-(1/\hbar)g_0 \Sigma_k \phi_k^4\, a^n - (1/2\hbar)\Sigma_k F_k(\phi,\hbar)\, a^n}\, \Pi_k\, d\phi_k \\
&\equiv \langle e^{(1/\hbar)Z^{-1/2}\Sigma_k h_k \phi_k\, a^n} \rangle .
\end{aligned}
\tag{9.4}
$$

Here we have introduced the field-strength renormalization constant Z. Each of the factors Z, m_0^2, and g_0 are treated as bare parameters implicitly dependent on the lattice spacing a, k^* denotes one of the n nearest neighbors in the positive direction from the site k, and M is chosen so that $S_{latt}(0) = 1$. The counterterm $F_k(\phi,\hbar)$ also implicitly depends on a, and the notation $F_k(\phi,\hbar)$ means that the formal, locally generated counterterm $F(\phi,\hbar)$ may, when lattice regularized, depend on finitely-many field values located within a small, finite region of the lattice around the site k. Noting that a primed sum Σ'_k, below and elsewhere, denotes a sum over a single spatial slice all sites of which have the same Euclidean time k_0, we are led to choose the counterterm

$$
\tfrac{1}{2}\Sigma_k F_k(\phi,\hbar)\, a^n \equiv \tfrac{1}{2}\Sigma_{k_0} a\, [\hbar^2 \Sigma'_k \mathcal{F}_k(\phi)\, a^s]
\tag{9.5}
$$

with

$$
\begin{aligned}
\mathcal{F}_k(\phi) &\equiv \tfrac{1}{4}(1 - 2ba^s)^2 a^{-2s} \left(\Sigma'_t \frac{J_{t,k}\phi_k}{[\Sigma'_m J_{t,m}\phi_m^2]} \right)^2 \\
&\quad - \tfrac{1}{2}(1 - 2ba^s)a^{-2s} \Sigma'_t \frac{J_{t,k}}{[\Sigma'_m J_{t,m}\phi_m^2]} \\
&\quad + (1 - 2ba^s)a^{-2s} \Sigma'_t \frac{J_{t,k}^2 \phi_k^2}{[\Sigma'_m J_{t,m}\phi_m^2]^2} .
\end{aligned}
\tag{9.6}
$$

Here b is a fixed, positive parameter with dimensions (length)$^{-s}$ and $J_{k,l} = 1/(2s+1)$ for $l = k$ and for l one of the $2s$ spatially nearest neighbors to k; otherwise, $J_{k,l} \equiv 0$. Although $\mathcal{F}_k(\phi)$ does not depend only on ϕ_k, it nevertheless becomes a local potential in the formal continuum limit. [Remark: If $J_{k,l}$ is taken as $\delta_{k,l}$, the resultant counterterm is that appropriate to the ultralocal models.]

At first sight, it seems impossible that such a counterterm could be successful since it is strong for *small* field values while the usual divergence difficulties are due to *large* field values. It has even been suggested that such a term would vanish when the renormalization group is invoked. If the reader entertains similar thoughts, he or she is encouraged to read further.

While much of our discussion is straightforward, our discovery of the special counterterm will be indirect. To begin with, we observe that if $\langle [\Sigma_k g_0 \phi_k^4 a^n]^p \rangle$ denotes a *spacetime* lattice expectation value and $\langle [\Sigma_k' g_0 \phi_k^4 a^s]^p \rangle$ denotes a *spatial* lattice expectation value at a fixed Euclidean time k_0, then

$$| \langle [\Sigma_k g_0 \phi_k^4 a^n]^p \rangle |$$
$$= \Sigma_{k_{0_{(1)}}} \cdots \Sigma_{k_{0_{(p)}}} a^p \, | \langle \{ [\Sigma_{k_{(1)}}' g_0 \phi_{k_{(1)}}^4 a^s] \cdots [\Sigma_{k_{(p)}}' g_0 \phi_{(p)}^4 a^s] \} \rangle | \qquad (9.7)$$
$$\le \Sigma_{k_{0_{(1)}}} \cdots \Sigma_{k_{0_{(p)}}} a^p \, | \langle [\Sigma_{k_{(1)}}' g_0 \phi_{k_{(1)}}^4 a^s]^p \rangle \cdots \langle [\Sigma_{k_{(p)}}' g_0 \phi_{k_{(p)}}^4 a^s]^p \rangle |^{1/p} ,$$

based on Hölder inequalities. It follows that if the *spatial* expectation value is finite then the *spacetime* expectation value is finite; clearly, if the spatial term diverges, then the spacetime term also diverges. Next, we recall that the ground state of a system determines the Hamiltonian operator (up to an additive constant), which in turn determines the lattice action including the required counterterm. Thus, in order to further the search for a new counterterm, our initial focus will be to examine general properties of the ground-state wave function.

9.2.1 *Determining the proper counterterm*

We now return to the lattice-regularized functional integral (9.4). In order for this mathematical expression to be physically relevant following a Wick rotation to real time, we impose the requirement of *reflection positivity* [GlJ87], which is assured if the counterterm satisfies

$$\Sigma_k F_k(\phi, \hbar) \equiv \Sigma_{k_0} \{ \Sigma_k' F_k(\phi, \hbar) \} , \qquad (9.8)$$

where the sum over the Euclidean-time lattice is outside the braces. Thus each element $\{ \Sigma_k' F_k(\phi, \hbar) \}$ of the temporal sum involves fields all of which have the same temporal value k_0, but they may involve several other fields at nearby sites to k in spatial directions only.

The lattice formulation (9.4) is a functional representation of the operator expression

$$S_{latt}(h) = N \operatorname{Tr}\{\mathbb{T}\, e^{(1/\hbar)\Sigma_{k_0}[Z^{-1/2}\Sigma'_k h_k \hat{\phi}_k\, a^n - \mathcal{H}\, a]}\}\,, \qquad (9.9)$$

where $N = [\operatorname{Tr}\{e^{-(La)\mathcal{H}/\hbar}\}]^{-1}$ is a normalization factor, $\mathbb{T}$ signifies time ordering, $\hat{\phi}_k$ is the field operator, and $\mathcal{H}$ denotes the Hamiltonian operator. When the function h is limited to a single spatial slice, e.g., $h_k \equiv a^{-1}\delta_{k_0,0} f_k$, this expression reduces to

$$S'_{latt}(f) = N \operatorname{Tr}\{e^{-(1/\hbar)(L-1)\mathcal{H}\, a} \cdot e^{(1/\hbar)[Z^{-1/2}\Sigma'_k f_k \hat{\phi}_k\, a^s - \mathcal{H}\, a]}\}$$
$$\qquad (9.10)$$

which, for large La, is realized by

$$S'_{latt}(f) = \int e^{(1/\hbar)Z^{-1/2}\Sigma'_k f_k \phi_k\, a^s}\, \Psi_0(\phi)^2\, \Pi'_k d\phi_k\,, \qquad (9.11)$$

and where Π'_k denotes a product over a single spatial slice at fixed k_0, $\Psi_0(\phi)$ denotes the unique, normalized ground-state wave function of the Hamiltonian operator $\mathcal{H}$ for the problem, expressed in the Schrödinger representation, with the property that $\mathcal{H}\Psi_0(\phi) = 0$. After a change of integration variables from the $N' \equiv L^s$ variables $\{\phi_k\}$ to 'hyperspherical coordinates' $\{\kappa, \eta_k\}$, defined by $\phi_k \equiv \kappa \eta_k$, $\kappa^2 \equiv \Sigma'_k \phi_k^2$, $1 \equiv \Sigma'_k \eta_k^2$, where $0 \leq \kappa < \infty$, and $-1 \leq \eta_k \leq 1$, Eq. (9.11) becomes

$$S'_{latt}(h) = \int e^{(1/\hbar)Z^{-1/2}\kappa\Sigma'_k h_k \eta_k\, a^s}\, \Psi_0(\kappa\eta)^2\, \kappa^{N'-1} d\kappa$$
$$\times 2\delta(1 - \Sigma'_k \eta_k^2)\, \Pi'_k d\eta_k. \qquad (9.12)$$

We define the continuum limit as $a \to 0$, $L \to \infty$, with La large, fixed, and finite. In the continuum limit $N' = L^s \to \infty$ and divergences will generally arise in a perturbation expansion, because, when parameters like m_0 or g_0, are changed, the measures in (9.12) become, in the continuum limit, *mutually singular* due to the overwhelming influence of N' in the measure factor $\kappa^{N'-1}$. However, we can avoid that conclusion provided the ground-state distribution $\Psi_0(\phi)^2$ contains a factor that serves to 'mash the measure', i.e., effectively change $\kappa^{N'-1}$ to κ^{R-1}, where $R > 0$ is fixed and finite. Specifically, we want the ground-state wave function to have the form

$$\Psi_0(\phi) = \text{``} M'\, e^{-U(\phi, \hbar, a)/2}\, \kappa^{-(N'-R)/2}\text{''}$$
$$= M'\, e^{-U(\kappa\eta, \hbar, a)/2}\, \kappa^{-(N'-R)/2}\, \Pi'_k [\Sigma'_l J_{k,l}\, \eta_l^2]^{-(1-R/N')/4}$$
$$= M'\, e^{-U(\phi, \hbar, a)/2}\, \Pi'_k [\Sigma'_l J_{k,l}\, \phi_l^2]^{-(1-R/N')/4}. \qquad (9.13)$$

The first line (in quotes) indicates the qualitative κ-behavior that will effectively mash the measure, while the second and third lines illustrate a specific functional dependence on field variables that leads to the desired factor. Here we choose the constant coefficients $J_{k,l} \equiv 1/(2s+1)$ for $l = k$ and for l equal to each of the $2s$ spatially nearest neighbors to the site k; otherwise, $J_{k,l} \equiv 0$. Thus, $\Sigma'_l J_{k,l} \phi_l^2$ provides a *spatial average of field-squared values at and nearby the site k*.

As part of the ground-state distribution, the denominator factor is dominant for *small-field* values, and its form is no less fundamental than the rest of the ground-state distribution that is determined by the gradient, mass, and interaction terms that fix the *large-field* behavior. The factor R/N' appears in the local expression of the small-field factor, and on physical grounds that quotient should not depend on the number of lattice sites in a spatial slice nor on the specific parameters mentioned above that define the model. Therefore, we can assume that $R \propto N'$, and so we set

$$R \equiv 2ba^s N' , \qquad (9.14)$$

where $b > 0$ is a fixed factor with dimensions $(\text{length})^{-s}$ to make R dimensionless. Even though the ground-state distribution diverges when certain of the η_k-factors are simultaneously zero, these are all integrable singularities since when any subset of the $\{\eta_k\}$ variables are zero, there are always fewer zero factors arising from the singularities thanks to the local averaging procedure; indeed, this very fact has motivated the averaging procedure. Thus, the benefits of measure mashing are: If a perturbation expansion of a certain term in the exponent $U(\phi, \hbar, a)$ is made after $\kappa^{N'-1}$ is changed to κ^{R-1}, then the various measures are *equivalent* and the terms in the perturbation expansion are all finite. Since the spatial perturbation-expansion terms are finite, then, according to the inequality (9.7), so too are the corresponding spacetime perturbation-expansion terms.

To obtain the required functional form of the ground-state wave function for small-field values, we choose our counterterm $F_k(\phi, \hbar)$ to build that feature into the Hamiltonian. In particular, the counterterm is a specific potential term of the form $\frac{1}{2}\Sigma'_k F_k(\phi, \hbar) a^s \equiv \frac{1}{2}\hbar^2 \Sigma'_k \mathcal{F}_k(\phi) a^s$ where, with $T(\phi) \equiv \Pi'_r [\Sigma'_l J_{r,l} \phi_l^2]^{-(1-2ba^s)/4}$,

$$\mathcal{F}_k(\phi) \equiv \frac{a^{-2s}}{T(\phi)} \frac{\partial^2 T(\phi)}{\partial \phi_k^2} , \qquad (9.15)$$

which leads to the expression in (9.6).

More generally, the full, nonlinear Hamiltonian operator including the desired counterterm is defined as

$$\mathcal{H} = -\tfrac{1}{2}\hbar^2 a^{-2s}\sum_k' \frac{\partial^2}{\partial\phi_k^2} a^s + \tfrac{1}{2}\sum_{k,k*}' (\phi_{k*} - \phi_k)^2 a^{s-2} \tag{9.16}$$

$$+ \tfrac{1}{2}m_0^2\sum_k'\phi_k^2 a^s + g_0\sum_k'\phi_k^4 a^s + \tfrac{1}{2}\hbar^2\sum_k'\mathcal{F}_k(\phi) a^s - E_0 .$$

Indeed, this latter equation implicitly determines the ground-state wave function $\Psi_0(\phi)$ via $\mathcal{H}\Psi_0(\phi) = 0$!

The author has used several different arguments previously [Kla11] to suggest this alternative model for nontrivial scalar field quantization. The present discussion argues, simply, that an unconventional $\mathcal{O}(\hbar)$ counterterm added to the naive Hamiltonian is all that is needed to achieve viable results. After all, conventional quantization assumes—even when using the favored Cartesian coordinates—that the proper Hamiltonian operator is open to various quantum corrections that only experiment, or mathematical consistency, can resolve. Based on that argument alone, there seems to be enough justification for our proposal.

It is important to observe that no normal ordering applies to the terms in the Hamiltonian. Instead, local field operator products are determined by an operator product expansion [ItZ80]. In addition, note that the counterterm $\hbar^2\Sigma_k'\mathcal{F}_k(\phi)$ does *not* depend on any parameters of the model and specifically not on g_0. This is because the counterterm is really a counterterm for the *kinetic energy*. This fact follows because not only is $\mathcal{H}\Psi_0(\phi) = 0$, but then $\mathcal{H}^q\Psi_0(\phi) = 0$ for all integers $q \geq 2$. While $[\Sigma_k'\partial^2/\partial\phi_k^2]\Psi_0(\phi)$ may be a square-integrable function, the expression $[\Sigma_k'\partial^2/\partial\phi_k^2]^q\Psi_0(\phi)$ will surely not be square integrable for suitably large q. To ensure that $\Psi_0(\phi)$ is in the domain of $\mathcal{H}^q$, for all q, the derivative term and the counterterm must be considered together to satisfy domain requirements, hence our claim that the counterterm should be considered as a 'renormalization' of the kinetic energy.

We recall the concept of a classical pseudofree theory from Chap. 2, which is well illustrated by an anharmonic oscillator model with the classical action functional $A_g \equiv A_0 + g\,A_I = \int_0^T \{\tfrac{1}{2}[\dot{x}(t)^2 - x(t)^2] - g\,x(t)^{-4}\}\,dt$. It is clear that the domain $D(A_g)$ of allowed paths, $\{x(t)\}_0^T$, satisfies $D(A_g) = D(A_0) \cap D(A_I)$, and it follows that $\lim_{g\to 0} D(A_g) \equiv D(A_0') \subset D(A_0)$. In brief, the interacting model is *not* continuously connected to its own free theory (A_0), but to a pseudofree theory (A_0'), which has the same formal action functional as the free theory but a strictly smaller domain. As described in Chap. 3, this behavior also exists in the quantum theory since, when $g \to 0$, the propagator for the interacting model does not pass continuously to the free quantum propagator but to a pseudofree quantum propagator with a different set of eigenfunctions and eigenvalues. We now argue that a similar behavior may be seen in our scalar field models.

Since the counterterm (9.6) does not depend on the coupling constant, it follows that the counterterm remains even when $g_0 \to 0$, which means that the interacting quantum field theory does *not* pass to the usual free quantum field theory as $g_0 \to 0$, but instead it passes to a pseudofree quantum field theory. To help justify this property, let us first consider the classical (Euclidean) action functional (9.2). Regarding the separate components of that expression, and assuming $m_0 > 0$, a multiplicative inequality [LSU68; Kla00] states that

$$\{\int \phi(x)^4 \, d^n x\}^{1/2} \leq \tilde{C} \int [(\nabla \phi)(x)^2 + m_0^2 \phi(x)^2] \, d^n x , \qquad (9.17)$$

where $\tilde{C} = (4/3)[m_0^{(n-4)/2}]$ when $n \leq 4$ (the renormalizable cases), and $\tilde{C} = \infty$ when $n \geq 5$ (the nonrenormalizable cases), which means, in the latter case, there are fields $\phi(x)$ for which the integral on the left side of the inequality diverges while the integral on the right side is finite; for example, $\phi_{sing}(x) = |x|^{-\alpha} e^{-x^2}$, for $n/4 \leq \alpha < (n-2)/2$. In other words, for φ_n^4 models in particular, there are different pseudofree and free classical theories when $n \geq 5$. Thus, for $n \geq 5$, it is reasonable to assume that the pseudofree quantum field theory is different from the free quantum field theory.

Moreover, it is significant that a similar inequality

$$|\int \phi(x)^p \, d^n x|^{2/p} \leq \tilde{C}' \int [(\nabla \phi)(x)^2 + m_0^2 \phi(x)^2] \, d^n x , \qquad (9.18)$$

where $\tilde{C}' = (4/3)[m_0^{[n(1-2/p)-2]}]$ holds when $p \leq 2n/(n-2)$ (the renormalizable cases), and $\tilde{C}' = \infty$ when $p > 2n/(n-2)$ (the nonrenormalizable cases) [Kla00]. Since our choice of counterterm did not rely on the original power (i.e., four) of the interaction term, it follows that the *very same counterterm* applies to any interaction of the form $\int \phi(x)^p \, d^n x$, as well as a weighted sum of such terms for different p values, provided that the potential has an appropriate lower bound.

Consequently, the quantum models developed above with the unconventional counterterm provide viable candidates for those quantum theories normally classified as nonrenormalizable, and they do so in such a manner that in a perturbation analysis—about the pseudofree theory, *not* about the free theory—divergences do not arise because all the underlying measures are equivalent and not mutually singular. Later, we discuss these divergence-free properties from a perturbation point of view, following [Kla11], in an analysis that also determines the dependence of Z, m_0^2, and g_0 on the parameters a and N'.

Having found an unconventional counterterm that conveys good properties to the nonrenormalizable models, it is natural—for reasons discussed in the following section—to extend such good behavior to the traditionally renormalizable models by using the unconventional counterterm for them as well. *Thus, we are led to adopt the lattice regularized Hamiltonian $\mathcal{H}$ in (9.16), including the counterterm, for all spacetime dimensions $n \geq 2$.* Of course, the conventional quantizations for φ_n^4, $n = 2, 3$, are still valid for suitable physical applications of their own.

Additionally, the lattice Hamiltonian (9.16) also determines the lattice Euclidean action including the unconventional counterterm, which is then given by

$$I = \tfrac{1}{2}\Sigma_{k,k^*}(\phi_{k^*} - \phi_k)^2\, a^{n-2} + \tfrac{1}{2}m_0^2\Sigma_k\phi_k^2\, a^n$$
$$+ g_0\Sigma_k\phi_k^4\, a^n + \tfrac{1}{2}\Sigma_k\hbar^2\mathcal{F}_k(\phi)\, a^n \,, \qquad (9.19)$$

where, in this expression, k^* refers to all n nearest neighbors of the site k in the positive direction. Although the lattice form of the counterterm involves averages over field-squared values in nearby spatial regions of the central site, it follows that the continuum limit of the counterterm is local in nature, as remarked previously. It is noteworthy that preliminary Monte Carlo studies based on the lattice action (9.19) support a nontrivial behavior of the φ_4^4 model exhibiting a positive renormalized coupling constant [Sta10].

9.2.2   *Mixed models*

There already exist well-defined, conventional results for φ_2^4 and φ_3^4, and, yes, for different applications, we are proposing alternative quantizations for these models. The unusual counterterm (9.6) depends on the number of spatial dimensions, but it does not depend on the power p of the interaction φ_n^p. Thus the counterterm applies to '*mixed models*' like $g_0\varphi_3^4 + g_0'\varphi_3^8$, $g_0''\varphi_5^4 + g_0'''\varphi_5^8 + g_0''''\varphi_5^{138}$, etc., each of which exhibits reasonable properties when the coupling constants are turned on and off, in different orders, again and again. Unlike the conventional approach, there are no divergences in perturbative expansions about the pseudofree model in the new formulation after operator product expansions are introduced for the local operator products. Although all of the terms in a perturbation expansion are finite, they cannot be computed by using Feynman diagrams because the validity of such diagrams relies on the functional integral for the free model having a Gaussian distribution; however, this is not the case for the pseudofree model. For the important example of φ_4^4, as already noted, traditional methods lead to a renormalizable theory, yet nonperturbative methods support a trivial (= free) behavior. On the other hand, our procedures for φ_4^4 are expected to be divergence free and nontrivial.

In summary, the argument that has led to the new counterterm, and thereby the divergence-free perturbation theory about the pseudofree model, has been based on the freedom to choose $\mathcal{O}(\hbar)$ corrections to a naive form of the quantum Hamiltonian. We, too, ask that the extra terms are chosen in a manner guided by experiment, and at the very least, experimental confirmation requires that the classical limit of the chosen quantum Hamiltonian agree with the nonlinear classical theory one started with. Of course, there may still be other forms of counterterms that have genuine quantum theories and agree with the classical limit. While our choice is 'minimal', e.g., involving no new dimensional parameters, experiment may yet favor another choice of counterterm.

9.3   Additional Details Regarding Covariant Models

9.3.1   *Preliminaries*

The previous section was devoted to a rapid quantization of the classical (Euclidean-time) model in (9.2), while this section and those that follow are devoted to further examination of some details that have led to the new formulation. We focus on φ_n^4, $n \geq 2$, models that are formulated as Euclidean functional integrals; certain other models may be treated by analogous procedures.

Just as before, we suppose Euclidean spacetime is replaced by a periodic, hypercubic lattice with L sites on an edge, $L < \infty$, and a uniform lattice spacing of a, $a > 0$. Let the sites be labeled by $k = \{k_0, k_1, k_2, \ldots, k_s\}$, where $k_j \in \mathbb{Z}$, for all j, k_0 denotes the future time direction under a Wick rotation, and $s = n - 1$. We denote lattice sums (and products) over all lattice points by Σ_k (and Π_k), and, importantly, lattice sums (and products) over just a spatial slice at a fixed value of k_0 by Σ'_k (and Π'_k). The total number of lattice sites is $N = L^n$, while the number of lattice sites in a spatial slice is $N' = L^s$.

In eventually taking the continuum limit we shall do so in two steps. First, we let $L \to \infty$ and $a \to 0$ together so that the full spacetime volume

$V = (La)^n$ remains large but finite; so too for the spatial volume $V' = (La)^s$. Second, when needed, we take the limit that both V and V' diverge. In this fashion, we can discuss continuum properties for finite spatial and spacetime volumes, which would not have been possible if we had let $L \to \infty$ before taking the limit $a \to 0$.

9.3.2 *Lattice action*

Following aspects of the discussion in [Kla07], we first introduce an important set of dimensionless constants by

$$J_{k,l} \equiv \frac{1}{2s+1} \delta_{k,l \in \{k \cup k_{nn}\}} \, , \qquad (9.20)$$

where $\delta_{k,l}$ is a Kronecker delta. This notation means that an equal weight of $1/(2s+1)$ is given to the $2s+1$ points in the set composed of k and its $2s$ nearest neighbors (nn) in the spatial sense only; $J_{k,l} = 0$ for all other points in that spatial slice. [Specifically, we define $J_{k,l} = 1/(2s+1)$ for the points $l = k = (k_0, k_1, k_2, \ldots, k_s)$, $l = (k_0, k_1 \pm 1, k_2, \ldots, k_s)$, $l = (k_0, k_1, k_2 \pm 1, \ldots, k_s), \ldots, l = (k_0, k_1, k_2, \ldots, k_s \pm 1)$.] This definition implies that $\Sigma'_l J_{k,l} = 1$.

We next recall the lattice action for the full theory, including the quartic nonlinear interaction as well as the proposed counterterm, as

$$I(\phi, \hbar, N) = \tfrac{1}{2} \sum_k \sum_{k^*} (\phi_{k^*} - \phi_k)^2 \, a^{n-2} + \tfrac{1}{2} m_0^2 \sum_k \phi_k^2 \, a^n$$
$$+ g_0 \sum_k \phi_k^4 \, a^n + \tfrac{1}{2} \hbar^2 \sum_k \mathcal{F}_k(\phi) \, a^n \, , \qquad (9.21)$$

where k^* denotes the n nearest neighbors to k in the positive sense, i.e., $k^* \in \{ (k_0 + 1, k_1, \ldots, k_s), \ldots, (k_0, k_1, \ldots, k_s + 1) \}$. The last term, which represents the heart of the present procedure, is the suggested counterterm and, once again, is given (with all the following sums over the spatial slice at fixed k_0) by

$$\mathcal{F}_k(\phi) \equiv \tfrac{1}{4} (1 - 2ba^s)^2 a^{-2s} \left(\sum_t' \frac{J_{t,k} \phi_k}{[\Sigma'_m J_{t,m} \phi_m^2]} \right)^2$$
$$- \tfrac{1}{2} (1 - 2ba^s) a^{-2s} \sum_t' \frac{J_{t,k}}{[\Sigma'_m J_{t,m} \phi_m^2]}$$
$$+ (1 - 2ba^s) a^{-2s} \sum_t' \frac{J_{t,k}^2 \phi_k^2}{[\Sigma'_m J_{t,m} \phi_m^2]^2} \, . \qquad (9.22)$$

Observe that we have included the proper dependence on $\hbar$ for the counterterm implying that its contribution disappears in the classical limit in which

$\hbar \to 0$. It may be noticed that each of the separate parts of the counterterm scales as the *inverse square* of the overall field magnitude. One reason has already been given above for choosing this counterterm, and below we show that the counterterm may be considered to arise from a suitable factor ordering ambiguity of the conventional theory.

One feature of the counterterm is the fact that each term involves up to next-nearest-neighbor, spatially separated lattice points. This feature is part of the regularization in the lattice formulation of the model. However, if a second-order phase transition is achieved in the continuum limit, then such a regularization should still lead to a relativistic theory in the limit.

We emphasize once again that as $g_0 \to 0$ and the quartic interaction is turned off, the lattice action does *not* pass to that of the usual free theory but to that of the free theory plus the original counterterm. We have called such a limit a *pseudofree theory* [Kla00], and we shall again show that the interacting theory, with $g_0 > 0$, exhibits a divergence-free perturbation series about the pseudofree theory.

In conventional quantum field theory, counterterms are chosen to deal with the *emergence* of divergences; in the approach adopted in this monograph, the counterterm is chosen to deal with the *cause* of divergences. Relative to conventional treatments, therefore, it is safe to say that using the new counterterm *changes everything* relative to what one normally expects based on the usual free theory. In particular, for local field powers, there is no 'normal ordering'; instead, there is 'multiplicative renormalization'.

9.3.3 *Generating function*

One important ingredient has been left out of the lattice action, and that is the factor Z representing the field-strength renormalization. We introduce this factor most simply by adopting the following expression for the lattice space generating function:

$$S_{latt}(h) \equiv M_0 \int e^{Z^{-1/2}\Sigma_k h_k \phi_k a^n/\hbar \, - \, I(\phi, a, N)/\hbar} \, \Pi_k d\phi_k \, , \quad (9.23)$$

where $\{h_k\}$ determines an appropriate test sequence, the counterterm is implicitly included in I, and the normalization factor M_0 ensures that $S(0) = 1$. By a field rescaling, i.e., $\phi_k \to Z^{1/2}\phi_k$, the factor Z can be removed from the source term and introduced into the lattice action; we shall have occasion to use both forms of this integral. The form of the generating function in terms of physical fields is given later in Eq. (9.79).

The continuum limit will be taken as

$$S(h) \equiv \lim_{a,L} S_{latt}(h) \, , \tag{9.24}$$

as $a \to 0$ and $L \to \infty$ together such that, as discussed above, $C \equiv La$ remains large but finite. The argument of S involves suitable limiting test functions, i.e., smooth functions $h_k \to h(x)$ where $ka \to x \in \mathbb{R}^n$. For sufficiently large C, it may be unnecessary to take the final limit $C \to \infty$.

9.4 Rationale for the Counterterm

From the lattice action it is a simple step to write down the lattice Hamiltonian operator

$$\mathcal{H} \equiv -\tfrac{1}{2}\hbar^2 a^{-s} {\sum_k}' \frac{\partial^2}{\partial \phi_k^2} + \mathcal{V}(\phi)$$

$$\equiv -\tfrac{1}{2}\hbar^2 a^{-s} {\sum_k}' \frac{\partial^2}{\partial \phi_k^2} + \mathcal{V}_0(\phi) + \tfrac{1}{2}\hbar^2 {\sum_k}' \mathcal{F}_k(\phi) \, a^s$$

$$= -\tfrac{1}{2}\hbar^2 a^{-s} {\sum_k}' \frac{\partial^2}{\partial \phi_k^2} + \tfrac{1}{2}{\sum_{k,k^*}}' (\phi_{k^*} - \phi_k)^2 a^{s-2} + \tfrac{1}{2}m_0^2 {\sum_k}' \phi_k^2 \, a^s$$

$$+ \lambda_0 {\sum_k}' \phi_k^4 \, a^s + \tfrac{1}{2}\hbar^2 {\sum_k}' \mathcal{F}_k(\phi) \, a^s - E_0 \, . \tag{9.25}$$

In addition, we introduce the ground state $\Psi_0(\phi)$ for this Hamiltonian operator. The constant E_0 is chosen so that $\Psi_0(\phi)$ satisfies the Schrödinger equation

$$\mathcal{H} \, \Psi_0(\phi) = 0 \, , \tag{9.26}$$

which implies that the Hamiltonian operator can also be written as

$$\mathcal{H} = -\frac{\hbar^2}{2} a^{-s} {\sum_k}' \frac{\partial^2}{\partial \phi_k^2} + \frac{\hbar^2}{2} a^{-s} {\sum_k}' \frac{1}{\Psi_0(\phi)} \frac{\partial^2 \Psi_0(\phi)}{\partial \phi_k^2} \, . \tag{9.27}$$

Since the ground state $\Psi_0(\phi)$ does not vanish, it can be written in the generic form

$$\Psi_0(\phi) = \frac{e^{-U(\phi, a, \hbar)/2}}{D(\phi)} \, , \tag{9.28}$$

and thus (using the abbreviation $X_{,k} \equiv \partial X/\partial \phi_k$ and the summation convention for spatial labels)

$$\mathcal{V}(\phi) = \tfrac{1}{2}\hbar^2 a^{-s} D \, e^{U/2} [D^{-1} e^{-U/2}]_{,kk} \tag{9.29}$$

$$= \tfrac{1}{2}\hbar^2 a^{-s} [\, \tfrac{1}{4}(U_{,k})^2 - \tfrac{1}{2}U_{,kk} + D^{-1} U_{,k} \, D_{,k} + 2D^{-2}(D_{,k})^2 - D^{-1} D_{,kk} \,] \, .$$

We insist that the atypical counterterm $\mathcal{F}_k(\phi)$ is determined by the denominator D alone by requiring that

$$
\begin{aligned}
\tfrac{1}{2}\hbar^2{\sum_k}' \mathcal{F}_k(\phi)\, a^s &\equiv \tfrac{1}{2}\hbar^2\, a^{-s}\, D\, D^{-1}{,}_{kk} \\
&= \tfrac{1}{2}\hbar^2\, a^{-s}[2D^{-2}(D{,}_k)^2 - D^{-1}D{,}_{kk}] \,.
\end{aligned} \tag{9.30}
$$

There are various solutions to this equation which lead to ground-state functions that are locally square integrable near the origin in field space. However, the only solution consistent with a nowhere vanishing ground state is given (up to an overall factor) by

$$
D(\phi) = \Pi_k'\,[{\Sigma_l}' J_{k,l}\,\phi_l^2]^{(1-2ba^s)/4} \,. \tag{9.31}
$$

In point of fact, D was chosen *first*, and the counterterm was then *derived* from D by this very differential equation. Why we have chosen this specific form for D has been discussed earlier in this chapter.

The ground-state wave function $\Psi_0(\phi)$ leads to the ground-state distribution

$$
\Psi_0(\phi)^2 \equiv K\,\frac{e^{-U(\phi,a,\hbar)}}{\Pi_k'\,[{\Sigma_l}' J_{k,l}\,\phi_l^2]^{(1-2ba^s)/2}}\,, \tag{9.32}
$$

where K accounts for normalization of this expression given the additional assumption that $U(0,a,\hbar) = 0$. The normalization integral itself then reads

$$
K\int \frac{e^{-U(\phi,a,\hbar)}}{\Pi_k'\,[{\Sigma_l}' J_{k,l}\,\phi_l^2]^{(1-2ba^s)/2}}\,\Pi_k'\,d\phi_k = 1 \,. \tag{9.33}
$$

Before commenting on this integral further, we discuss several simpler integrals.

9.4.1 *A discussion of many-dimensional integrals*

Consider the family of N'-Gaussian integrals, on a spatial lattice, given (at some fixed k_0) by

$$
I_G(2p) \equiv \int [{\Sigma_k}'\,\phi_k^2]^p\, e^{-A{\Sigma_k}'\,\phi_k^2}\,\Pi_k'\,d\phi_k \,, \tag{9.34}
$$

where $p \in \{0,1,2,3,\ldots\}$, and we assume that A is of 'normal size', e.g., $0.1 < A < 10$. Although these integrals can be evaluated explicitly, we prefer to study the qualitative behavior of such integrals for *large values of* N', i.e., when $N' \gg 1$. For this purpose it is highly instructive to introduce *hyperspherical coordinates* [Kla96] defined by

$$
\begin{aligned}
\phi_k \equiv \kappa\eta_k \,, &\qquad 0 \le \kappa < \infty \,, \qquad -1 \le \eta_k \le 1 \,, \\
\kappa^2 \equiv {\Sigma_k}'\,\phi_k^2 \,, &\qquad 1 \equiv {\Sigma_k}'\,\eta_k^2 \,.
\end{aligned} \tag{9.35}
$$

Here κ acts as a hyperspherical radius variable and the $\{\eta_k\}$ variables constitute an N'-dimensional 'direction field'. Note well that $\kappa \equiv \sqrt{\Sigma'_k \phi_k^2}$ is the 'radius' of *all the field variables in a given spatial slice of the lattice at some fixed k_0*. In terms of these variables, it follows that

$$I_G(2p) = 2 \int [\kappa^2]^p \, e^{-A\kappa^2} \, \kappa^{N'-1} d\kappa \, \delta(1 - \Sigma'_k \eta_k^2) \, \Pi'_k \, d\eta_k \, . \tag{9.36}$$

Observe that the integrand depends on the radius κ, but, other than in the measure, it does not depend on the angular variables $\{\eta_k\}$. For very large N', the integral over κ can be studied by steepest descent methods. To leading order, the integrand is narrowly peaked at values of $\kappa \simeq (N'/2A)^{1/2}$, namely at large values of κ. As a consequence, for each value of A, and as $N' \to \infty$, the integrand is supported on a *disjoint* set of κ. This well-known property [Hid70] leads to divergences in perturbation calculations. For example, let us study

$$I_G^\star(2) = \int [\Sigma'_k \phi_k^2] e^{-A^\star \Sigma'_k \phi_k^2} \, \Pi'_k d\phi_k \, , \tag{9.37}$$

which is the same type of Gaussian integral for a different value of A. For this study, we introduce the perturbation series

$$I_G^\star(2) = I_G(2) - \Delta A \, I_G(4) + \tfrac{1}{2}(\Delta A)^2 \, I_G(6) - \cdots \, , \tag{9.38}$$

where $\Delta A \equiv A^\star - A$. Since $I_G(2p)/I_G(2) = O(N'^{(p-1)})$, this series exhibits *divergences* as $N' \to \infty$. It is not difficult to convince oneself that such divergences are analogous to those that appear in quantum field theory.

If the factor $\kappa^{(N'-1)}$ were replaced by $\kappa^{(R-1)}$, $0 < R < \infty$ being fixed and finite, we would be led to consider

$$I_G'(2p) = 2 \int [\kappa^2]^p \, e^{-A\kappa^2} \, \kappa^{(R-1)} \, d\kappa \, \delta(1 - \Sigma'_k \eta_k^2) \, \Pi'_k \, d\eta_k \, . \tag{9.39}$$

Now, the integrand is broadly supported and no longer favors extremely large κ values when $N' \gg 1$. Consequently, a perturbation series evaluation of

$$I_G'^\star(2) = 2 \int \kappa^2 e^{-A^\star \kappa^2} \, \kappa^{(R-1)} \, d\kappa \, \delta(1 - \Sigma'_k \eta_k^2) \, \Pi'_k \, d\eta_k$$

$$= I_G'(2) - \Delta A \, I_G'(4) + \tfrac{1}{2}(\Delta A)^2 \, I_G'(6) - \cdots \tag{9.40}$$

exhibits no divergences since $I_G'(2p)/I_G'(2) = O(N'^0) = O(1)$.

Unlike the original integrals over κ, integrals over the direction field variables $\{\eta_k\}$ cannot diverge under normal circumstances since each variable satisfies $-1 \le \eta_k \le 1$. We will encounter such variables in the denominator of (9.33); however, the form of that denominator has been specifically chosen to ensure that such integrals converge when any number of η_k variables pass through zero together.

9.4.2 Relevance for the ground-state distribution

The normalization integral for the ground-state distribution, Eq. (9.33), expressed in terms of hyperspherical coordinates and with $R = 2ba^s N'$, becomes

$$2K \int \frac{e^{-U(\kappa\eta, a, \hbar)}}{\Pi'_k [\Sigma'_l J_{k,l}\, \eta_l^2]^{(1-2ba^s)/2}} \, \kappa^{(R-1)} \, d\kappa \, \delta(1 - \Sigma'_k \eta_k^2)\, \Pi'_k \, d\eta_k = 1 \, . \quad (9.41)$$

Note well that the change of the factor $\kappa^{(N'-1)}$ to $\kappa^{(R-1)}$ in this expression is a direct result of the counterterm in the lattice Hamiltonian, which in turn gave rise to the denominator factor D in the ground state $\Psi_0(\phi)$. Just like the elementary examples in which $\kappa^{(N'-1)} \to \kappa^{(R-1)}$, there is no longer any peaking of the integrand in κ. Therefore, for integrals such as

$$K \int [\Sigma'_k \phi_k^2]^p \, \frac{e^{-U(\phi, a, \hbar)}}{\Pi'_k [\Sigma'_l J_{k,l}\phi_l^2]^{(1-2ba^s)/2}} \, \Pi'_k \, d\phi_k \quad (9.42)$$

it is clear that the κ-dependence of the integrand is spread rather broadly; this conclusion would be false if the factor $\kappa^{(N'-1)}$ had not been canceled by part of the term D^2.

9.5 Correlation Functions and their Bounds

In this section, following [Kla07], we wish to show that the full spacetime correlation functions can be controlled by their sharp-time behavior along with a suitable choice of test sequences.

Let the notation

$$\phi_u \equiv \Sigma_k u_k \phi_k \, a^n \quad (9.43)$$

denote the full spacetime summation over all lattice sites where $\{u_k\}$ denotes a suitable test sequence. We also separate out the temporal part of this sum in the manner

$$\phi_u \equiv \Sigma_{k_0} a \, \phi_{u'} \equiv \Sigma_{k_0} a \, \Sigma'_k u_k \phi_k \, a^s \, . \quad (9.44)$$

Observe that the notation $\phi_{u'}$ (with the prime) implies a summation *only* over the spatial lattice points for a fixed (and implicit) value of the temporal lattice value k_0.

Let the notation $\langle\langle \cdot \rangle\rangle$ denote full spacetime averages with respect to the field distribution determined by the lattice action, and then let us consider full spacetime correlation functions such as

$$\langle \phi_{u^{(1)}} \phi_{u^{(2)}} \cdots \phi_{u^{(2q)}} \rangle = \Sigma_{k_0^{(1)}, k_0^{(2)}, \ldots, k_0^{(2q)}} \, a^{2q} \, \langle \phi_{u'^{(1)}} \phi_{u'^{(2)}} \cdots \phi_{u'^{(2q)}} \rangle \, ,$$

$$(9.45)$$

where $q \geq 1$ and the expectation on the right-hand side is over products of fixed-time summed fields, $\phi_{u'}$, for possibly different times, which are then summed over their separate times. All odd correlation functions are assumed to vanish, and furthermore, $\langle 1 \rangle = 1$ in this normalized spacetime lattice field distribution. It is also clear that

$$|\langle \phi_{u^{(1)}} \phi_{u^{(2)}} \cdots \phi_{u^{(2q)}} \rangle| \leq \Sigma_{k_0^{(1)}, k_0^{(2)}, \ldots, k_0^{(2q)}} a^{2q} |\langle \phi_{u'^{(1)}} \phi_{u'^{(2)}} \cdots \phi_{u'^{(2q)}} \rangle| .$$

$$(9.46)$$

At this point we turn our attention toward the spatial sums alone.

We initially appeal to straightforward inequalities of the general form

$$\langle AB \rangle^2 \leq \langle A^2 \rangle \langle B^2 \rangle . \tag{9.47}$$

In particular, it follows that

$$\langle \phi_{u'^{(1)}} \phi_{u'^{(2)}} \phi_{u'^{(3)}} \phi_{u'^{(4)}} \rangle^2 \leq \langle \phi_{u'^{(1)}}^2 \phi_{u'^{(2)}}^2 \rangle \langle \phi_{u'^{(3)}}^2 \phi_{u'^{(4)}}^2 \rangle , \tag{9.48}$$

and, in turn, that

$$\langle \phi_{u'^{(1)}} \phi_{u'^{(2)}} \phi_{u'^{(3)}} \phi_{u'^{(4)}} \rangle^4 \leq \langle \phi_{u'^{(1)}}^2 \phi_{u'^{(2)}}^2 \rangle^2 \langle \phi_{u'^{(3)}}^2 \phi_{u'^{(4)}}^2 \rangle^2$$
$$\leq \langle \phi_{u'^{(1)}}^4 \rangle \langle \phi_{u'^{(2)}}^4 \rangle \langle \phi_{u'^{(3)}}^4 \rangle \langle \phi_{u'^{(4)}}^4 \rangle . \tag{9.49}$$

By a similar argument, it follows that

$$|\langle \phi_{u'^{(1)}} \phi_{u'^{(2)}} \cdots \phi_{u'^{(2q)}} \rangle| \leq \Pi_{j=1}^{2q} [\langle \phi_{u'^{(j)}}^{2q} \rangle]^{1/2q} , \tag{9.50}$$

which has bounded any particular mixture of spatial correlation functions at possibly different times, by a suitable product of higher-power expectations each of which involves field values ranging over a spatial level, all at a single fixed lattice time. By time-translation invariance of the various single-time correlation functions we can assert that $\langle \phi_{u'^{(j)}}^{2r} \rangle$, which is defined at time $k_0^{(j)}$, is actually independent of the time and, therefore, the result could be calculated at any fixed time. In particular, we can express such correlation functions as

$$\langle \phi_{u'}^{2q} \rangle = \int \phi_{u'}^{2q} \, \Psi_0(\phi)^2 \, \Pi_k' \, d\phi_k . \tag{9.51}$$

Thus we see that a bound on full spacetime correlation functions may be given in terms of sharp-time correlation functions in the ground-state distribution.

9.6   The Continuum Limit

Before focusing on the limit $a \to 0$ and $L \to \infty$, let us note some important facts about ground-state averages of the direction-field variables $\{\eta_k\}$. First, we assume that such averages have two important symmetries: (i) averages of an odd number of η_k variables vanish, i.e.,

$$\langle \eta_{k_1} \cdots \eta_{k_{2p+1}} \rangle = 0 \,, \tag{9.52}$$

and (ii) such averages are invariant under any spacetime translation, i.e.,

$$\langle \eta_{k_1} \cdots \eta_{k_{2p}} \rangle = \langle \eta_{k_1+l} \cdots \eta_{k_{2p}+l} \rangle \tag{9.53}$$

for any $l \in \mathbb{Z}^n$ due to a similar translational invariance of the lattice action. Second, we note that for any ground-state distribution, it is necessary that $\langle \eta_k^2 \rangle = 1/N'$ for the simple reason that $\Sigma_k' \eta_k^2 = 1$. Hence, $|\langle \eta_k \eta_l \rangle| \le 1/N'$ as follows from the Schwarz inequality. Since $\langle [\Sigma_k' \eta_k^2]^2 \rangle = 1$, it follows that $\langle \eta_k^2 \eta_l^2 \rangle = O(1/N'^2)$. Indeed, similar arguments show that for any ground-state distribution

$$\langle \eta_{k_1} \cdots \eta_{k_{2p}} \rangle = O(1/N'^p) \,, \tag{9.54}$$

which will be useful in the sequel.

Next, we choose to study the pseudofree model, namely, when the coupling constant $g_0 = 0$. This is as close as we can get to the free model itself. Unfortunately, we cannot solve Eq. (9.29) when $\mathcal{V}$ has the desired form. The best we can do is choose a form for $U(\phi, a, \hbar)$ in (9.28) that leads to an *approximate form* of the pseudofree model. In particular, we choose

$$U(\phi, a, \hbar) = (1/\hbar)\Sigma_{k,l}' \phi_k A_{k-l} \phi_l \, a^{2s} \,. \tag{9.55}$$

This expression is taken to be the form for U for the free model as if there were no counterterm and consequently D was a constant. This ensures us that the potential $\mathcal{V}(\phi)$ that follows from (9.29) agrees with the desired free model to leading order in $\hbar$. Specifically, with the given choice for U and D, it follows that

$$\mathcal{V}(\phi) = \tfrac{1}{2}\Sigma_{k,l,m}' \phi_k A_{k-l} A_{l-m} \phi_m \, a^{3s} - \tfrac{1}{2}\hbar A_0 \, N' a^s + \tfrac{1}{2}\hbar^2 \Sigma_k' \mathcal{F}_k(\phi) \, a^s$$
$$+ \hbar[(1 - 2b a^s)/4]\Sigma_{k,r,m}' J_{r,k} \phi_k A_{k-m} \phi_m /[\Sigma_l' J_{r,l} \phi_l^2] \, a^{2s} \,. \tag{9.56}$$

We choose the matrix A_{k-l} so that the first term in (9.56) yields the desired gradient and a suitable mass term in the Hamiltonian expression (9.25). In particular, we note that to match the quadratic, spatial lattice-derivative terms in the Hamiltonian (typically the most singular of the quadratic terms), we can do so by choosing the elements of the matrix

$A_{k-l} = O(a^{-(s+1)})$ (a value influenced by an intermediate sum). The terms we have chosen are valid for very small-field values (D) and very large-field values (A_{k-l}); only intermediate-field values are misrepresented. Although this leads to an approximate form for the pseudofree model, it is sufficient for our limited purpose at present, namely, to determine the parameters of the model starting with the field-strength renormalization constant Z.

Field-strength renormalization

We now take up the question of the sharp-time averages given by

$$\int Z^{-p} \left[\Sigma'_k h_k \phi_k a^s\right]^{2p} \Psi_{\widetilde{pf}}(\phi)^2 \, \Pi'_k d\phi_k \,, \tag{9.57}$$

where Z denotes the field-strength renormalization factor, $\{h_k\}$ represents a suitable spatial test sequence, and $\Psi_{\widetilde{pf}}(\phi)$ denotes our approximate pseudofree ground state. These are exactly the kinds of expression that should become well-behaved in the continuum limit for a proper choice of Z. Our approximate ground-state representation should be adequate to determine whether an integral that involved the true pseudofree ground-state expression will converge or diverge. Thus, we are led to consider

$$K \int Z^{-p} \left[\Sigma'_k h_k \phi_k a^s\right]^{2p} \frac{e^{-\Sigma'_{k,l} \phi_k A_{k-l} \phi_l a^{2s}/\hbar}}{\Pi'_k \left[\Sigma'_l J_{k,l} \phi_l^2\right]^{(1-2ba^s)/2}} \, \Pi'_k d\phi_k$$

$$= 2K \int Z^{-p} \kappa^{2p} \left[\Sigma'_k h_k \eta_k a^s\right]^{2p} \frac{e^{-\kappa^2 \Sigma'_{k,l} \eta_k A_{k-l} \eta_l a^{2s}/\hbar}}{\Pi'_k \left[\Sigma'_l J_{k,l} \eta_l^2\right]^{(1-2ba^s)/2}}$$

$$\times \kappa^{(R-1)} \, d\kappa \, \delta(1 - \Sigma'_k \eta_k^2) \, \Pi'_k d\eta_k \,. \tag{9.58}$$

Our goal is to use this integral to determine a value for the field strength renormalization constant Z. To estimate this integral we first replace two factors with η variables by their appropriate averages. In particular, the expression in the exponent is estimated by

$$\kappa^2 \Sigma'_{k,l} \eta_k A_{k-l} \eta_l a^{2s} \simeq \kappa^2 \Sigma'_{k,l} N'^{-1} A_{k-l} a^{2s} \propto \kappa^2 N' a^{2s} a^{-(s+1)} \,, \tag{9.59}$$

and the expression in the integrand is estimated by

$$\left[\Sigma'_k h_k \eta_k a^s\right]^{2p} \simeq N'^{-p} \left[\Sigma'_k h_k a^s\right]^{2p} \,. \tag{9.60}$$

The integral over κ is then estimated by first rescaling the variable $\kappa^2 \to \kappa^2/(N' a^{s-1})$, which then leads to an overall integral-estimate proportional to

$$Z^{-p} \left[N' a^{s-1}\right]^{-p} N'^{-p} \left[\Sigma'_k h_k a^s\right]^{2p} \,; \tag{9.61}$$

the remaining integral no longer includes the parameters a and N'. Finally, for this result to be meaningful in the continuum limit, we are led to choose $Z = N'^{-2} a^{-(s-1)}$. However, Z must be dimensionless, so we introduce a fixed positive quantity q with dimensions of an inverse length, which allows us to set

$$Z = N'^{-2}(qa)^{-(s-1)} . \qquad (9.62)$$

This is a fundamental and important relation in our analysis.

Mass renormalization

With a notation where $\langle (\cdot) \rangle$ denotes a full spacetime lattice-space average based on the lattice action for the pseudofree theory, an expansion of the mass term (or part of the mass term) leads to a series of terms of the form

$$\langle [m_0^2 \Sigma_k \phi_k^2 a^n]^p \rangle , \qquad (9.63)$$

which in turn can be expressed as

$$m_0^{2p} \Sigma_{k_0^{(1)}, k_0^{(2)}, \dots, k_0^{(p)}} a^p \langle [\Sigma'_{k^{(1)}} \phi_{k^{(1)}}^2 a^s][\Sigma'_{k^{(2)}} \phi_{k^{(2)}}^2 a^s] \cdots [\Sigma'_{k^{(p)}} \phi_{k^{(p)}}^2 a^s] \rangle . \qquad (9.64)$$

Based on the inequality

$$\langle \Pi_{j=1}^p A_j \rangle \le \Pi_{j=1}^p \langle A_j^p \rangle^{1/p} , \qquad (9.65)$$

valid when $A_j \ge 0$ for all j, it follows that

$$\langle [m_0^2 \Sigma_k \phi_k^2 a^n]^p \rangle \le m_0^{2p} \Sigma_{k_0^{(1)}, k_0^{(2)}, \dots, k_0^{(p)}} a^p \qquad (9.66)$$

$$\times \{ \langle [\Sigma'_{k^{(1)}} \phi_{k^{(1)}}^2 a^s]^p \rangle \langle [\Sigma'_{k^{(2)}} \phi_{k^{(2)}}^2 a^s]^p \rangle \cdots \langle [\Sigma'_{k^{(p)}} \phi_{k^{(p)}}^2 a^s]^p \rangle \}^{1/p} .$$

This leads us to consider

$$\langle [m_0^2 \Sigma'_k \phi_k^2 a^s]^p \rangle = 2 m_0^{2p} a^{sp} K \int \kappa^{2p} \frac{e^{-\kappa^2 \Sigma'_{k,l} \eta_k A_{k-l} \eta_l a^{2s}}}{\Pi_k [\Sigma'_l J_{k,l} \eta_l^2]^{(1-2ba^s)/2}} d\kappa$$

$$\times \kappa^{(R-1)} \delta(1 - \Sigma'_k \eta_k^2) \Pi d\eta_k , \qquad (9.67)$$

which, in the manner used previously, can be estimated as

$$\langle [m_0^2 \Sigma'_k \phi_k^2 a^s]^p \rangle \propto \frac{m_0^{2p} a^{sp}}{[N' a^{(s-1)}]^p} \qquad (9.68)$$

times an integral that no longer depends on a and N'. To make sense in the continuum limit, this leads us to identify

$$m_0^2 = N'(qa)^{-1} m^2 , \qquad (9.69)$$

with m being the physical mass. Moreover, it is noteworthy that

$$Z m_0^2 = [N'^{-2}(qa)^{-(s-1)}][N'(qa)^{-1}] m^2 = [N'(qa)^s]^{-1} m^2 , \qquad (9.70)$$

which for a finite spatial volume $V' = N'a^s$ leads to a finite nonzero result for $Z m_0^2$ in the continuum limit.

Coupling constant renormalization

We repeat the previous calculation for an expansion of the quartic interaction term about the approximate pseudofree theory. This leads us to consider terms of the form

$$\langle [g_0 \Sigma_k \phi_k^4 a^n]^p \rangle , \qquad (9.71)$$

which in turn can be expressed as

$$g_0^p \Sigma_{k_0^{(1)}, k_0^{(2)}, \ldots, k_0^{(p)}} a^p \langle [\Sigma'_{k^{(1)}} \phi_{k^{(1)}}^4 a^s][\Sigma'_{k^{(2)}} \phi_{k^{(2)}}^4 a^s] \cdots [\Sigma'_{k^{(p)}} \phi_{k^{(p)}}^4 a^s] \rangle \quad (9.72)$$

and bounded by

$$\langle [g_0 \Sigma_k \phi_k^4 a^n]^p \rangle \leq g_0^p \Sigma_{k_0^{(1)}, k_0^{(2)}, \ldots, k_0^{(p)}} a^p \qquad (9.73)$$
$$\times \{\langle [\Sigma'_{k^{(1)}} \phi_{k^{(1)}}^4 a^s]^p \rangle \langle [\Sigma'_{k^{(2)}} \phi_{k^{(2)}}^4 a^s]^p \rangle \cdots \langle [\Sigma'_{k^{(p)}} \phi_{k^{(p)}}^4 a^s]^p \rangle \}^{1/p} .$$

This leads us to consider

$$\langle [g_0 \Sigma'_k \phi_k^4 a^s]^p \rangle = 2 g_0^p a^{sp} K \int \kappa^{4p} [\Sigma_k \eta_k^4]^p \frac{e^{-\kappa^2 \Sigma'_{k,l} \eta_k A_{k-l} \eta_l a^{2s}}}{\Pi'_k [\Sigma'_l J_{k,l} \eta_l^2]^{(1-2ba^s)/2}}$$
$$\times \kappa^{(R-1)} d\kappa \, \delta(1 - \Sigma'_k \eta_k^2) \, \Pi d\eta_k , \qquad (9.74)$$

which, in the manner used previously, and up to a suitable integral, can be estimated as

$$\langle [g_0 \Sigma'_k \phi_k^4 a^s]^p \rangle \propto \frac{g_0^p N'^{-p} a^{sp}}{[N' a^{(s-1)}]^{2p}} . \qquad (9.75)$$

To ensure a suitable continuum limit requires us to identify

$$g_0 = N'^3 (qa)^{s-2} g , \qquad (9.76)$$

with g being the physical coupling constant. Moreover, it is noteworthy that

$$Z^2 g_0 = [N'^{-4}(qa)^{-2(s-1)}] [N'^3 (qa)^{s-2}] g = [N'(qa)^s]^{-1} g , \quad (9.77)$$

which for a finite spatial volume $V' = N' a^s$ leads to a finite nonzero result for $Z^2 g_0$.

9.6.1 *Physical version of the generating function*

Based on the previous analysis we are led to reformulate the expression for the lattice-space generating function (9.23). We first make a change of integration variables such that $\phi_k \to Z^{1/2} \phi_k$ leading to the expression

$$S_{latt}(h) \equiv M \int e^{\Sigma_k h_k \phi_k a^n/\hbar - I(Z^{1/2}\phi, a, N)/\hbar} \, \Pi_k d\phi_k , \quad (9.78)$$

where any constant Jacobian factor has been absorbed into a change of the overall normalization factor from M_0 to M. Finally, we introduce the explicit form for the lattice action from (9.21) into (9.78) to yield

$$
\begin{aligned}
S_{latt}(h) = M \int \exp\{ & \Sigma h_k \phi_k \, a^n / \hbar \\
& - \tfrac{1}{2} [N'^2 (qa)^{(s-1)}]^{-1} \textstyle\sum_k \sum_{k^*} (\phi_{k^*} - \phi_k)^2 \, a^{n-2} / \hbar \\
& - \tfrac{1}{2} [N'(qa)^s]^{-1} m^2 \textstyle\sum_k \phi_k^2 \, a^n / \hbar - [N'(qa)^s]^{-1} g \sum_k \phi_k^4 \, a^n / \hbar \\
& - \tfrac{1}{2} \hbar^2 [N'^2 (qa)^{(s-1)}] \textstyle\sum_k \mathcal{F}_k(\phi) \, a^n / \hbar \} \, \Pi_k \, d\phi_k \, .
\end{aligned}
\tag{9.79}
$$

This expression contains a formulation of the lattice-space generating function expressed in terms of physical fields and physical constants.

Comments

In our final expression above there are several noteworthy points to be made. In a finite spatial volume $V' = N' a^s$—which due to our hypercubic assumption for spacetime implies a finite spacetime volume $V = N a^n$—the coefficients of the physical mass m and the physical coupling constant g are both finite and nonzero. We have shown earlier in this section that perturbation in both the quadratic mass term and the quartic nonlinear action leads to a series which is term-by-term finite when perturbed about the pseudofree theory. If the finite spatial volume is taken large enough (e.g., Milky Way sized), then a divergence-free perturbation series is established. In other words, the introduction of the unusual counterterm has resolved any issues with typical ultraviolet divergences.

[Remark: Any theory exhibits infinite-volume divergences for questions of a stationary nature due to time-translation invariance. For example, such divergences even arise already when $n = 1$ and we deal with time alone, as for example with a conventional, stationary Ornstein-Uhlenbeck (O-U) process [Wiki-1] $U(t) \equiv 2^{-1/2} e^{-t} W(e^{2t})$, $-\infty < t < \infty$, where $W(\tau)$, $0 \le \tau < \infty$, denotes a standard, Gaussian, Wiener process for which $W(0) = 0$, $\mathbf{E}(W(\tau)) = 0$, and $\mathbf{E}(W(\tau_1) W(\tau_2)) = \min(\tau_1, \tau_2)$, where $\mathbf{E}$ denotes ensemble average. The O-U paths are concentrated on bounded, continuous paths such that

$$
\mathbf{E}(\textstyle\int_a^{a+1} U(t)^2 \, dt) = C > 0 \, ,
\tag{9.80}
$$

which, thanks to stationarity, is independent of the starting time a. As a consequence, it automatically follows that

$$
\mathbf{E}(\textstyle\int_{-\infty}^{\infty} U(t)^2 \, dt) = \infty
\tag{9.81}
$$

thanks to the stationarity of the process.]

Unlike the quadratic mass and quartic interaction terms, the coefficients of the derivative terms and the inverse-square field counterterm are inverse to one another and do not have finite nonzero limits when $L \to \infty$ and $a \to 0$ such that the spatial volume $V' = (La)^s$ is finite. This aspect is not unexpected since (i) the coefficient of the counterterm is intimately linked to that of the derivative term so that the field power that appears in the denominator factor D^2 in the ground-state distribution is $(1 - 2ba^s)/2$, and (ii) this fact leads to a significant redistribution of probability toward the origin of field space, which then requires an asymptotically small Z factor to reestablish reasonable field averages.

9.7   Additional Discussion and Conclusions

In [Kla00] there is an extensive discussion of two, soluble, scalar nonrenormalizable models that are idealized versions of the relativistic model treated in the present chapter. Such models differ from the relativistic model in that in one case *all* spacetime derivative terms are omitted from the classical action ("Independent-Value Models"), and in the second case *all but one* of the derivative terms are dropped ("Ultralocal Models"), an example also studied in the previous chapter; in this latter case, the remaining term is identified with the eventual time direction in an analog of a Wick rotation. Generally speaking, these models have little to no physics and are principally of academic interest. Nevertheless, from a mathematical viewpoint both of these models lead to trivial, i.e., Gaussian, results if traditional counterterms are introduced, and, conversely, if they are studied perturbatively, they are both nonrenormalizable. Fortunately, both of these models have sufficient symmetry so that they can be rigorously solved on the basis of self-consistency without using any form of perturbation theory or cutoff. One of the results for both models is that as the coupling constant of the nonlinear interaction term is reduced to zero, the theories do *not* return to the appropriate free theory, but, instead, they pass continuously to an appropriate pseudofree theory. Moreover, both theories exhibit meaningful, divergence-free perturbation theories about the pseudofree theory, but definitely not about the customary free theory. These models are explicitly worked out in Chaps. 9 and 10 of Ref. [Kla00], respectively, and the latter model is treated in this monograph in Chap. 8—but there is also a 'natural reason' why such results are plausible.

Nonrenormalizable quantum field theories have exceptionally strong interaction terms. This statement can be quantified as follows: Consider the φ_n^p relativistic scalar theory which has a free (Euclidean) action given by

$$W = \tfrac{1}{2} \int [(\nabla \phi)(x)^2 + m^2 \phi(x)^2] \, d^n x \,, \tag{9.82}$$

where $x \in \mathbb{R}^n$. These theories have the nonlinear interaction term

$$V = \int \phi(x)^p \, d^n x \,, \tag{9.83}$$

where we focus on cases where $p \in \{4, 6, 8, \ldots\}$. Such expressions appear in a formal functional integral such as

$$S_\lambda(h) = \mathcal{N}_\lambda \int e^{\int h \phi \, d^n x - W - \lambda V} \, \mathcal{D}\phi \,. \tag{9.84}$$

If $\lim_{\lambda \to 0} S_\lambda(h) = S_0(h)$, then the interacting theory is continuously connected to the free theory; if, instead, $\lim_{\lambda \to 0} S_\lambda(h) = S_0'(h) \neq S_0(h)$, then the interacting theory is *not* continuously connected to the free theory, but rather it is continuously connected to a pseudofree theory. Under what conditions could this latter situation arise?

Consider the classical Sobolev-like 'inequality' [Kla00] given (for $\phi \not\equiv 0$) by

$$| \textstyle\int \phi(x)^p \, d^n x \,|^{2/p} / \{ \int [(\nabla \phi)(x)^2 + m^2 \phi(x)^2] \, d^n x \} \leq \mathcal{R} \tag{9.85}$$

as a function of the parameters p and n. For $p \leq 2n/(n-2)$, it follows that $\mathcal{R} = (4/3)[m^{n(1-2/p)-2}]$; for $p > 2n/(n-2)$, $\mathcal{R} = \infty$ holds. This dichotomy is *exactly* that between perturbatively renormalizable and perturbatively nonrenormalizable models. Observe that this ratio form of the inequality can also be extended to distributions when they are approached by limits of smooth test functions. But why should this inequality relate to renormalizability?

It is the author's long-held belief that, in the context of a functional integral formulation, the explanation arises from a *hard-core behavior of nonrenormalizable interactions* [Kl73b]. Simply stated, the interaction for such theories is so strong that a set of nonzero measure of the field histories allowed by the free theory alone is projected out when the interaction term is present. For example, $\mathcal{R} = \infty$ means that there are fields for which V is not dominated by W in the same way as when $\mathcal{R} < \infty$. This fact suggests that for positive coupling constant values, some of the field histories, even those that are distributions, are projected out never to return as the coupling constant passes to zero; this result is the effect of a hard core at work. For the first idealized model mentioned above (Independent-Value Models), the analogous ratio is $\{ \int \varphi(x)^4 \, d^n x \}^{1/2} / \{ \int m^2 \varphi(x)^2 \, d^n x \}$, which for any $n \geq 1$ clearly has no finite upper bound, while for the second idealized models (Ultralocal Models), the appropriate ratio reads $\{\int \varphi(x)^4 \, d^n x\}^{1/2} / \{\int [\dot{\varphi}(x)^2 + m^2 \varphi(x)^2] \, d^n x\}$, which, in this case, for any $n \geq 2$ has no finite upper bound. These soluble models—each more singular in principle than the relativistic models—are examples of hard-core interactions that nevertheless have divergence-free perturbations about their own pseudofree model. This set of facts strongly suggests that relativistic scalar fields such as φ_n^4, $n \geq 5$, as we have focused on, also have a corresponding hard-core behavior.

For both idealized models, their unconventional formulation was not assumed, rather, it was *derived*, thanks to a large symmetry for each model. For relativistic models, there is insufficient symmetry to derive the needed counterterm, and thus the counterterm must be postulated, i.e., guessed. There have been several past suggestions by the author that have not lived up to expectations; these include Chap. 11 in [Kla00] and Chap. 11 in [Kl10a] (the latter of which was incomplete since $R = 1$ was the only case studied, while in reality $R \equiv 2ba^s N'$). The present discussion, offered in the current chapter, Chap. 9—a number that, normally, has been the author's 'lucky number'—avoids the previously 'unlucky' number 11. Hopefully, the current presentation offers one more proposal that satisfies all the expected requirements.

Unfortunately, the present model is not (or is most likely not) analytically tractable, probably lacking a technical means to analytically perform perturbation calculations about the pseudofree theory. Nevertheless, it would seem possible that numerical Monte Carlo calculations should be feasible. The first such calculation that should be made is a test for nontriviality that is applied to such theories by testing whether or not the Gaussian property that

$$\langle (\Sigma_k h_k \phi_k \, a^n)^4 \rangle - 3 \langle (\Sigma_k h_k \phi_k \, a^n)^2 \rangle^2 = 0 \tag{9.86}$$

holds true for all choices of $\{h_k\}$ as one approaches the continuum limit; a single violation of this inequality would demonstrate that the continuum limit is not that of a free theory. In view of the connection of such full spacetime correlation functions to those on a single spatial surface, as shown above, it seems unlikely that (9.86) holds true thanks to the chosen form of the counterterm. Indeed, preliminary, but limited, Monte Carlo studies do indeed show a positive, renormalized coupling constant for φ_4^4 models [Sta10]. More studies of that kind surely need to be carried out.

Should the nontriviality test above prove successful for a φ_5^4 nonrenormalizable relativistic model, for example, it would be worthwhile to study various three-dimensional nonrenormalizable models such as φ_3^8, φ_3^{10}, etc. It is noteworthy that if, instead of the interaction $g_0 \, \varphi_n^4$ featured in the present chapter, we had started with $g_0{}_r \, \varphi_n^{2r}$, where $r \in \{3, 4, 5, 6, \ldots\}$, then it follows that such an interaction would also possess a divergence-free perturbation expansion provided we choose

$$g_0{}_r = N'^{(2r-1)} \, a^{(s-1)r - s} \, g_r \, , \qquad (9.87)$$

where g_r is the physical coupling constant. Moreover, it follows that

$$Z^r \, g_0{}_r = [N'(qa)^s]^{-1} \, g_r \, , \qquad (9.88)$$

for all values of r, just as was the case for the mass term in (9.70) and for the quartic coupling constant in (9.77).

9.8 Classical Limit

When dealing with a nonrenormalizable φ_n^4 theory, we have repeatedly argued against choosing those counterterms suggested by a regularized perturbation analysis about the free theory. This was due, in part, to the fact that such quantum theories tend not to have the correct classical limit, namely, the original nonlinear classical theory one started with. This property is clear when perturbative counterterms are considered because, in that case, there is no complete and well-defined quantum theory for which the classical limit can be studied.

One strong test of whether or not the ideas in this chapter have some validity would be to try to take the classical limit of the quantum theory and confirm that the original, motivating nonlinear relativistic theory emerges. The study of this question first requires having some control on the continuum limit, but in support of its possible realization we note that the ultralocal models treated in Chap. 8 of this monograph, namely, the model including just the time derivative of the field, has been shown to have the correct classical limit for the nonlinear ultralocal model in question. These models were also interpreted as proper enhanced quantized systems using affine coherent states. Now, using suitable affine coherent states, and following [Kl10b], we will show that the weak correspondence principle leads to the nonlinearly interacting, motivating classical model in the relativistic case.

Even if our proposal leads to a nontrivial quantum theory, and even if that quantum theory exhibits a correct classical limit, the question may arise whether this proposal for quantization is the "correct" quantization procedure. As in most quantization procedures, in which one starts from a theory with $\hbar = 0$ and constructs a theory with $\hbar > 0$, there is a great deal of latitude in the result. Again we emphasize the role of experiment in choosing the proper form of additional terms for the Hamiltonian. Our procedures do not require any new dimensional parameters, but that is no guarantee of uniqueness. Nevertheless, in the absence of any other satisfactory proposal to deal with nonrenormalizable theories, one might look favorably on a model that offers more than was previously available.

9.9 Affine Coherent States and the Classical/Quantum Connection

9.9.1 _Coherent states for scalar fields_

We generalize the one-dimensional example presented in Chap. 4 regarding affine coherent states, and apply them to the study of covariant scalar fields; in this effort we are partially guided by an analogous story for ultralocal fields that appears in [Kla00] as well as a preliminary study of these questions in [Kl10b]. We introduce an unknown function W that enables us to use the proper lattice-regularized covariant pseudofree theory so we can choose the true ground state for this model as the fiducial vector. Thus we are led to consider (for $\hbar = 1$) the affine coherent states

$$
\langle \phi | p, q \rangle = K \, \Pi'_k |q_k|^{-1/2}
$$
$$
\times \frac{e^{i\Sigma'_k (p_k/2q_k)\phi_k^2 a^s - \frac{1}{2}\Sigma'_{k,l}(\phi_k/|q_k|) A_{k-l}(\phi_l/|q_l|) a^{2s}}}{\Pi'_k [\Sigma'_l J_{k,l}(\phi_l^2/q_l^2)]^{(1-2ba^s)/4}}
$$
$$
\times e^{-\frac{1}{2}W((\phi/|q|)a^{(s-1)/2})} \,. \tag{9.89}
$$

Remark: In this section, we will use ϕ_k as c-number representation variables, which conflicts with our normal choice for coherent state labels of (π, ϕ). Hence, we choose $(p, q) = \{p_k, q_k\}$ as labels of the affine coherent states, and these labels will comprise the variables in the application of the weak correspondence principle. The coherent-state overlap function $\langle p', q' | p, q \rangle$ is given by

$$
\langle p', q' | p, q \rangle = \int \langle p', q' | \phi \rangle \langle \phi | p, q \rangle \, \Pi'_k d\phi_k \,, \tag{9.90}
$$

which is represented by

$$\langle p', q' | p, q \rangle = K^2 \Pi'_k (|q'_k| |q_k|)^{-1/2}$$

$$\times \int \frac{e^{-i\Sigma'_k (p'_k/2q'_k) \phi_k^2 a^s - \frac{1}{2}\Sigma'_{k,l}(\phi_k/|q'_k|) A_{k-l}(\phi_l/|q'_l|) a^{2s}}}{\Pi'_k [\Sigma'_l J_{k,l}(\phi_l^2/q'^2_l)]^{(1-2ba^s)/4}}$$

$$\times e^{-\frac{1}{2} W((\phi/|q'|) a^{(s-1)/2})}$$

$$\times \frac{e^{i\Sigma'_k (p_k/2q_k) \phi_k^2 a^s - \frac{1}{2}\Sigma'_{k,l}(\phi_k/|q_k|) A_{k-l}(\phi_l/|q_l|) a^{2s}}}{\Pi'_k [\Sigma'_l J_{k,l}(\phi_l^2/q_l^2)]^{(1-2ba^s)/4}}$$

$$\times e^{-\frac{1}{2} W((\phi/|q|) a^{(s-1)/2})} \Pi'_k d\phi_k . \tag{9.91}$$

As we approach the continuum limit in this expression, and restricting attention to continuous functions $p_k \to p(x)$ and $q_k \to q(x)$, it follows that

$$\langle p', q' | p, q \rangle = K^2 \Pi'_k (|q'_k| |q_k|)^{-ba^s}$$

$$\times \int \frac{e^{-i\Sigma'_k (p'_k/2q'_k) \phi_k^2 a^s - \frac{1}{2}\Sigma'_{k,l}(\phi_k/|q'_k|) A_{k-l}(\phi_l/|q'_l|) a^{2s}}}{\Pi'_k [\Sigma'_l J_{k,l} \phi_l^2]^{(1-2ba^s)/4}}$$

$$\times e^{-\frac{1}{2} W((\phi/|q'|) a^{(s-1)/2})}$$

$$\times \frac{e^{i\Sigma'_k (p_k/2q_k) \phi_k^2 a^s - \frac{1}{2}\Sigma'_{k,l}(\phi_k/|q_k|) A_{k-l}(\phi_l/|q_l|) a^{2s}}}{\Pi'_k [\Sigma'_l J_{k,l} \phi_l^2]^{(1-2ba^s)/4}}$$

$$\times e^{-\frac{1}{2} W((\phi/|q|) a^{(s-1)/2})} \Pi'_k d\phi_k , \tag{9.92}$$

where we have taken advantage of the fact that for continuous functions we can bring the q' and q factors out of the denominators in the former expression. Although we can not write an analytic expression for the entire continuum limit of this expression, we note that the new prefactor, $\Pi'_k (|q'_k| |q_k|)^{-ba^s}$, has a continuum limit given by

$$\Pi'_k (|q'_k| |q_k|)^{-ba^s} \to e^{-b\int [\ln|q'(x)| + \ln|q(x)|] d^s x} . \tag{9.93}$$

This meaningful partial result for the continuum limit holds only for the affine coherent states; it would decidedly not have led to meaningful results for canonical coherent states [Kl10b]. In other words, measure mashing has had the effect of changing a canonical system into an affine system! This result also favors the choice $R = 2ba^s N'$ as it is compatible with an infinite spatial volume.

The rest of the coherent-state overlap integral in (9.92) is too involved to be simplified, but if we ask only for $\langle p', q | p, q \rangle$ then some progress can be made. In this case we have

$$
\begin{aligned}
\langle p', q | p, q \rangle &= K^2 \Pi'_k |q_k|^{-1} \\
&\times \int \frac{e^{-i\Sigma'_k (p'_k/2q_k)\phi_k^2 a^s - \frac{1}{2}\Sigma'_{k,l}(\phi_k/|q_k|)A_{k-l}(\phi_l/|q_l|)a^{2s}}}{\Pi'_k [\Sigma'_l J_{k,l}(\phi_l^2/q_l^2)]^{(1-2ba^s)/4}} \\
&\quad \times e^{-W((\phi/|q|)a^{(s-1)/2})/2} \\
&\times \frac{e^{i\Sigma'_k (p_k/2q_k)\phi_k^2 a^s - \frac{1}{2}\Sigma'_{k,l}(\phi_k/|q_k|)A_{k-l}(\phi_l/|q_l|)a^{2s}}}{\Pi'_k [\Sigma'_l J_{k,l}(\phi_l^2/q_l^2)]^{(1-2ba^s)/4}} \\
&\quad \times e^{-W((\phi/|q|)a^{(s-1)/2})/2} \, \Pi'_k d\phi_k \\
&= K^2 \int \frac{e^{-i\Sigma'_k ((p'_k - p_k)q_k/2)\phi_k^2 a^s - \Sigma'_{k,l}\phi_k A_{k-l}\phi_l a^{2s}}}{\Pi'_k [\Sigma'_l J_{k,l}\phi_l^2]^{(1-2ba^s)/2}} \\
&\quad \times e^{-W(\phi a^{(s-1)/2})} \, \Pi'_k d\phi_k \ .
\end{aligned}
\tag{9.94}
$$

In this form, we see that $\langle p', q | p, q \rangle = \langle p'q, 1 | pq, 1 \rangle$, from which we learn that the expression $\langle p' | p \rangle \equiv \langle p', 1 | p, 1 \rangle$ contains the same information as contained in $\langle p', q | p, q \rangle$ for all q; we also learn that $\langle p', q | p, q \rangle$ achieves a meaningful continuum limit provided that $\langle p' | p \rangle$ already achieves one, and it is plausible from the form of (9.94) that such a continuum limit holds.

9.9.2 *Classical/quantum connection*

In the same spirit as when dealing with a single affine degree of freedom, we seek to connect the classical action functional with a restricted version of the quantum action functional in the form of an enhanced classical action functional. Thus, still using the special notation introduced in the preceding section, we proceed directly to the expression

$$
A_{Q(R)} = \int [\langle p(t), q(t) | [i\hbar(\partial/\partial t) - \mathcal{H}] | p(t), q(t) \rangle \, dt \ ,
\tag{9.95}
$$

based on the Hamiltonian operator $\mathcal{H}$ for the lattice covariant pseudofree model and the associated coherent states. Following earlier examples dealing with affine coherent states, we are led to the expression (with $\hbar = 1$, $\hat{\phi}_k = \phi_k$, m_0^2 symbolizes the (mass)^2 term, and assuming units are chosen so that $\langle\eta|\phi_k^2|\eta\rangle = \ell^2 = 1$) for (9.95) given by

$$\begin{aligned}
A_{Q(R)} &= \int\{\{\tfrac{1}{2}\Sigma_k'(p_k\dot{q}_k - q_k\dot{p}_k)a^s - \langle\eta|[\tfrac{1}{2}\Sigma_k'(P_k/|q_k| + p_k|q_k|\phi_k/q_k)^2 a^s \\
&\quad -\tfrac{1}{2}\Sigma_{k,k*}'(|q_{k*}|\phi_{k*} - |q_k|\phi_k)^2 a^{s-2} - \tfrac{1}{2}m_0^2\Sigma_k'q_k^2\phi_k^2 a^s \\
&\quad -\tfrac{1}{2}\Sigma_k'\mathcal{F}_k(|q|\phi)a^s]|\eta\rangle + E_{pf}\}\,dt \\
&= \int\{\{\tfrac{1}{2}\Sigma_k'(p_k\dot{q}_k - q_k\dot{p}_k)a^s - \tfrac{1}{2}\Sigma_k'(p_k^2 + \langle P_k^2\rangle q_k^{-2})a^s \\
&\quad -\tfrac{1}{2}\Sigma_k'\Sigma_{k*}'\langle(|q_{k*}|\phi_{k*} - |q_k|\phi_k)^2\rangle a^{s-2} - \tfrac{1}{2}m_0^2\Sigma_k'q_k^2 a^s \\
&\quad -\tfrac{1}{2}\Sigma_k'\langle\mathcal{F}_k(\phi)\rangle q_k^{-2} a^s + E_{pf}\}\,dt\,,
\end{aligned}$$ (9.96)

where in the last line we have made the sum over k^* more explicit. This equation has all the expected ingredients apart from the rather unusual term $\tfrac{1}{2}\Sigma_k'\Sigma_{k*}'\langle(|q_{k*}|\phi_{k*} - |q_k|\phi_k)^2\rangle a^{s-2}$ that we investigate next. We observe that

$$\begin{aligned}
&\langle(|q_{k*}|\phi_{k*} - |q_k|\phi_k)^2\rangle \\
&= q_{k*}^2\langle(\phi_{k*} - \phi_k)^2\rangle + (|q_{k*}| - |q_k|)^2\langle\phi_k^2\rangle \\
&\quad +[(q_{k*}^2 - q_k^2) + (|q_{k*}| - |q_k|)^2]\langle(\phi_{k*} - \phi_k)\phi_k\rangle \\
&\equiv C_1 q_{k*}^2 + (|q_{k*}| - |q_k|)^2 + C_2[(q_{k*}^2 - q_k^2) + (|q_{k*}| - |q_k|)^2]\,,
\end{aligned}$$ (9.97)

where C_j, $j = 1, 2$, are constants due to translation invariance of the ground state. The first term contributes to the mass, the second term and the latter part of the third term contribute to the lattice derivative, $(1+C_2)(q_{k*}-q_k)^2$, since for continuous functions the sign of q_{k*} and q_k are identical except for the possible exception when they are both infinitesimal, in which case they make a negligible contribution to the sum. Moreover, in the continuum limit $C_2 \to 0$, and compared to unity it may be omitted from that factor. The initial part of the last factor sums to zero thanks to the periodic boundary conditions. We observe that $\langle P_k^2\rangle = C_3$ and $\langle\mathcal{F}_k(\phi)\rangle = C_4$ are also constants because of translation invariance. Combining the various separate terms all together leads to a restricted action that has all the expected ingredients. The same remarks about terms of the form $\Sigma_k'\hbar^2 q_k^{-2}$ that held for the one-dimensional affine system also hold in the present case; for this example, it may be suggested to take the limit $\hbar \to 0$ before deriving the classical equations of motion.

The factor E_{pf} that enters the expressions above may be eliminated in a very easy fashion. Since the fiducial vector $|\eta\rangle$ has in this case been chosen as the ground state for the pseudofree model, it follows that

$$\begin{aligned}
\langle p,1|\mathcal{H}_{pf}|p,1\rangle &= \langle[\tfrac{1}{2}\Sigma_k'(P_k + p_k Q_k)^2 a^s + V(Q) - E_{pf}]\rangle \\
&= \tfrac{1}{2}\Sigma_k'p_k^2 a^s + \langle[\tfrac{1}{2}\Sigma_k'P_k^2 a^s + V(Q) - E_{pf}]\rangle \\
&= \tfrac{1}{2}\Sigma_k'p_k^2 a^s\,,
\end{aligned}$$ (9.98)

meaning that the final expression for the pseudofree enhanced classical action functional (with $\hbar = 1$) is given by

$$A_{Q(R)} = \int \{ \tfrac{1}{2} \Sigma'_k (p_k \dot{q}_k - q_k \dot{p}_k) a^s - \tfrac{1}{2} \Sigma'_k [p_k^2 + C_3 (q_k^{-2} - 1)] a^s$$
$$- \tfrac{1}{2} \Sigma'_k \Sigma'_{k*} (q_{k*} - q_k)^2 a^{s-2} - \tfrac{1}{2} m_0^2 \Sigma'_k (q_k^2 - 1) a^s$$
$$- \tfrac{1}{2} \Sigma'_k C_4 (q_k^{-2} - 1) a^s \} \, dt \, . \tag{9.99}$$

If instead of the pseudofree model we dealt with an interacting model, then the ground state would have a different form as would the form of the Hamiltonian operator; the enhanced classical action functional would reflect this difference by adding a nonlinear term to the functional form of the pseudofree enhanced classical action.

Chapter 10

Enhanced Quantization of Gravity

WHAT TO LOOK FOR

When dealing with gravity, a basic physical requirement is the strict positivity of the classical spatial metric 3×3 matrix $\{g_{ab}(x)\}$, $1 \leq a, b \leq 3$, as well as the strict positivity of the quantum spatial metric 3×3 matrix $\{\hat{g}_{ab}(x)\}$ (when smeared with a smooth positive function) composed of the spatial components of the local metric operator. Self-adjoint canonical operators are incompatible with this principle, but they can be replaced by self-adjoint affine operators, leading to an affine quantum gravity. Quantization before reduction means that the initial, kinematical Hilbert space and associated operators lack any sense of correlation between neighboring points leading to an initial ultralocal operator representation. Due to the partially second-class nature of the quantum gravitational constraints, it is advantageous to use the projection operator method for quantum constraints, which treats all constraints on an equal footing. Using this method, enforcement of regularized versions of the gravitational constraint operators is formulated quite naturally by a novel and relatively well-defined functional integral involving only the same set of variables that appears in the usual classical formulation. Although perturbatively nonrenormalizable, gravity may possibly be understood perturbatively taken about an appropriate pseudofree model or nonperturbatively from a hard-core perspective that has proved valuable for other models.

10.1 Introduction and Overview

The quantization of Einstein's theory of gravity is a difficult subject, in part because it is nonrenormalizable when studied by conventional canonical quantization procedures leading to an infinite number of distinct divergent counterterms. To make any progress, therefore, it seems necessary that nonconventional procedures be considered, and that is being done currently with the approach of superstrings and loop quantum gravity; see, e.g., [Wiki-m; Wiki-n]. These distinct approaches have led to many results and each approach makes a claim of being an appropriate quantization scheme. On the other hand, the approach adopted in this chapter offers yet another view on the subject, a view that draws heavily on the new procedures emphasized throughout this monograph. In particular, we regard both classical and quantum gravity as best described by a pseudofree theory, which is not continuously connected to its own 'free theory' [Kla75], similar to what we have encountered many times when dealing with highly singular interactions. The coordinate covariance that permeates the Einstein formulation of classical gravity ensures that there are certain superfluous degrees of freedom that need to be eliminated through the enforcement of several constraints. The relevant constraints are classically 'first class', but they become, partially, 'second class' when quantized; the terms in quotation marks will be explained. Traditional quantization techniques frequently flounder when dealing with second-class constraints, but newer quantization schemes, which will be discussed in a limited fashion, can deal with such constraints. Along with operator techniques, we will also discuss several versions of unusually well-defined functional integrals that offer different perspectives on the subject. Proposals for subspaces that satisfy all gravitational constraints are also suggested. Quantum gravity for the real world refers to 3 space dimensions and 1 time dimension, numbers that will be our sole focus. Certain simplifications occur when dealing with fewer space dimensions, but we will not consider such cases; on the other hand, more space dimensions, while not discussed, may very well be treated as we do for three space dimensions.

10.1.1 *Canonical versus affine variables*

It is important to make clear why affine variables are preferable to canonical variables. Even at a classical level there is a big difference between using canonical variables or affine variables. Of primary importance is the rule of *metric positivity* which means that $u^a g_{ab}(x) u^b > 0$ for all x and all nonvanishing real vectors u^a. Because $\pi^{ab}(x)$ is the generator of translations of $g_{ab}(x)$, this variable may lead to a violation of metric positivity in the classical theory; for example, if $\pi[u] \equiv \int [u_{ab}(x) \pi^{ab}(x)] d^3x$ is used to generate a macroscopic canonical transformation, then

$$e^{\{\cdot, \pi[u]\}} g_{ab}(x) \equiv g_{ab}(x) + \{g_{ab}(x), \pi[u]\} + \tfrac{1}{2}\{\{g_{ab}(x), \pi[u]\}, \pi[u]\} + \cdots$$
$$= g_{ab}(x) + u_{ab}(x) , \tag{10.1}$$

where $\{\cdot, \cdot\}$ denotes the Poisson bracket with the basic relation

$$\{g_{ab}(x), \pi^{cd}(x')\} = \tfrac{1}{2}(\delta_a^c \delta_b^d + \delta_b^c \delta_a^d)\delta(x, x') , \tag{10.2}$$

could violate metric positivity. On the other hand, let us replace the classical momentum tensor $\pi^{ab}(x)$ by the classical *"momentric"* tensor defined by $\pi_c^a(x) \equiv \pi^{ab}(x) g_{bc}(x)$, a variable which is so named to reflect its *momen*tum and me*tric* ingredients. [Remark: The field $\pi_b^a(x)$, called here the momentric, has also been called the dilation field or the scale field by the author.] Unlike the momentum, the momentric rescales the metric tensor in such a way as to preserve metric positivity; specifically, if $\pi(\gamma) \equiv \int [\gamma_b^a(x) \pi_a^b(x)] d^3x$ is now used to generate a macroscopic canonical transformation, then

$$e^{\{\cdot, \pi(\gamma)\}} g_{ab}(x) \equiv g_{ab}(x) + \{g_{ab}(x), \pi(\gamma)\} + \tfrac{1}{2}\{\{g_{ab}(x), \pi(\gamma)\}, \pi(\gamma)\} + \cdots$$
$$= M_a^c(x) g_{cd}(x) M_b^d(x) , \tag{10.3}$$

with $M_a^c(x) \equiv \{\exp[\gamma(x)/2]\}_a^c$ and $\gamma(x) \equiv \{\gamma_a^c(x)\}$. Thus a metric that satisfies metric positivity will still satisfy metric positivity after such a transformation. This satisfactory behavior also holds in the quantum theory of these variables as well.

10.1.2   *Classical action functional for gravity*

Briefly stated, the classical action functional for Einstein gravity [ADM62] is given by [model #(12) in Chap. 1]

$$A_C = \int_0^T \int \{-g_{ab}(x,t) \dot{\pi}^{ab}(x,t) - N^a(x,t) H_a(\pi, g)(x,t)$$
$$- N(x,t) H(\pi, g)(x,t)\} d^3x \, dt , \tag{10.4}$$

where $N^a(x,t)$, and $N(x,t)$ are Lagrange multiplier fields, $H_a(\pi, g)(x,t) = -2\pi_{a\,|\,b}^b(x,t)$ are the classical diffeomorphism constraints, where $|$ denotes a covariant derivative using the spatial metric alone, and $H(\pi, g)(x,t)$ is the classical Hamiltonian constraint, where

$$H(\pi, g)(x,t) = g(x,t)^{-1/2} [\pi_b^a(x,t) \pi_a^b(x,t) - \tfrac{1}{2}\pi_a^a(x,t) \pi_b^b(x,t)]$$
$$+ g(x,t)^{1/2} R(x,t) ; \tag{10.5}$$

here $R(x, t)$ denotes the three-dimensional, spatial scalar curvature.

We begin the present chapter with an overview of affine quantum gravity and how it copes with the various problems this subject faces. It is the author's view that an enhanced affine quantization is surprisingly well suited to tackle the issues encountered in quantizing gravity. Following this initial tour through the 'minefield of quantum gravity', a number of topics are analyzed with additional detail. This chapter is based in part on [Kl99a; WaK00; Kl01a; Kl12b] as well as other related articles; corrections to several previously published equations are incorporated into the present version.

10.2 Basic Principles of Affine Quantum Gravity

The program of affine quantum gravity is founded on *four basic principles* which we briefly review here. *First*, like the corresponding classical variables, the 6 components of the spatial metric field operators $\hat{g}_{ab}(x)\,[=\hat{g}_{ba}(x)]$, $a, b = 1, 2, 3$, form a *positive-definite 3×3 matrix for all x*. *Second*, to ensure self-adjoint kinematical variables when smeared, it is necessary to adopt the *affine commutation relations*. To preserve metric positivity, it follows that the preferred kinematical variables are $\pi_b^a(x)$ and $g_{ab}(x)$, and it is these variables that are promoted to quantum operators in the affine quantum gravity program; specifically (with $\hbar = 1$),

$$[\hat{\pi}_b^a(x), \hat{\pi}_d^c(y)] = i\tfrac{1}{2}[\delta_b^c\,\hat{\pi}_d^a(x) - \delta_d^a\,\hat{\pi}_b^c(x)]\,\delta(x, y)\;,$$
$$[\hat{g}_{ab}(x), \hat{\pi}_d^c(y)] = i\tfrac{1}{2}[\delta_a^c\,\hat{g}_{db}(x) + \delta_b^c\,\hat{g}_{ad}(x)]\,\delta(x, y)\;, \qquad (10.6)$$
$$[\hat{g}_{ab}(x), \hat{g}_{cd}(y)] = 0$$

between the metric and the 9 components of the mixed-index momentum (momentric) field operator $\hat{\pi}_b^a(x)$, the quantum version of the classical variable $\pi_b^a(x) \equiv \pi^{ac}(x)g_{cb}(x)$; these commutation relations are direct transcriptions of conventional Poisson brackets for the classical fields $g_{ab}(x)$ and $\pi_d^c(y)$. The affine commutation relations are like *current commutation relations* and their representations are quite different from those for canonical commutation relations; indeed, the present program is called 'affine quantum gravity' because these commutation relations are analogous to the Lie algebra belonging to the group of affine transformations (A, b), where $\mathsf{x} \to \mathsf{x}' = \mathsf{A}\mathsf{x} + \mathsf{b}$, x, x', and b are real *n*-vectors, and A are real, invertible $n \times n$ matrices. *Third*, the principle of *quantization first and reduction second*, favored by Dirac, suggests that the basic fields $\hat{g}_{ab}$ and $\hat{\pi}_d^c$ are initially realized by *ultralocal representations*, rather like those we encountered in

Chap. 8 and which are briefly explained below. *Fourth*, and last, introduction and enforcement of the gravitational constraints not only leads to the physical Hilbert space but it has the added virtue that all unphysical vestiges of the temporary ultralocal operator representation are replaced by physically acceptable alternatives, as model problems demonstrate. In attacking these basic issues, full use is made of *coherent-state methods* and the *projection-operator method* for the quantization of constrained systems.

10.2.1 *Affine coherent states*

The affine coherent states are defined (for $\hbar = 1$) by

$$|\pi, \gamma\rangle \equiv e^{i\int \pi^{ab}(x)\hat{g}_{ab}(x)\,d^3x}\, e^{-i\int \gamma^a_b(x)\hat{\pi}^b_a(x)\,d^3x}\,|\eta\rangle \quad [\,= |\pi, g\rangle\,] \quad (10.7)$$

for general, smooth, c-number fields $\pi^{ab}(x)\,[= \pi^{ba}(x)]$ and $\gamma^c_d(x)$ of compact support, and the fiducial vector $|\eta\rangle$ is chosen so that the coherent-state overlap functional (derived below) becomes

$$\langle \pi'', g''|\pi', g'\rangle \equiv \exp\left(-2\int b(x)\,d^3x \right. \tag{10.8}$$

$$\times \ln\left\{ \frac{\det\{\frac{1}{2}[g''^{kl}(x) + g'^{kl}(x)] + i\frac{1}{2}b(x)^{-1}[\pi''^{kl}(x) - \pi'^{kl}(x)]\}}{(\det[g''^{kl}(x)])^{1/2}\,(\det[g'^{kl}(x)])^{1/2}} \right\}\right).$$

Observe that the matrices γ'' and γ' do *not* explicitly appear in (10.8) because the choice of $|\eta\rangle$ is such that each $\gamma = \{\gamma^a_b\}$ has been replaced by $g = \{g_{ab}\}$, where

$$g_{ab}(x) \equiv [e^{\gamma(x)/2}]^c_a\,\langle\eta|\hat{g}_{cd}(x)|\eta\rangle\,[e^{\gamma(x)/2}]^d_b\,. \tag{10.9}$$

Note that the functional expression in (10.8) is ultralocal, i.e., specifically of the form

$$\exp\{-\textstyle\int b(x)\,d^3x\,L[\pi''(x), g''(x); \pi'(x), g'(x)]\}\,, \tag{10.10}$$

and thus, *by design*, there are no correlations between spatially separated field values, a neutral position adopted towards spatial correlations before any constraints are introduced. On invariance grounds, (10.8) necessarily involves a *scalar density* $b(x)$, $0 < b(x) < \infty$, for all x; this arbitrary and nondynamical auxiliary function $b(x)$, with dimensions $(\text{length})^{-3}$, should disappear when the gravitational constraints are fully enforced, at which point proper field correlations will arise; see below. In addition, note that the coherent-state overlap functional is *invariant* under general spatial coordinate transformations. Finally, we emphasize that the expression

$\langle \pi'', g'' | \pi', g' \rangle$ is a *continuous functional of positive type* and thus may be used as a *reproducing kernel* to define a *reproducing kernel Hilbert space* (see [Aro43] and later in this chapter) composed of continuous phase-space functionals $\psi(\pi, g)$ on which elements of the initial, ultralocal representation of the affine field operators act in a natural fashion.

10.2.2 Functional integral representation

A functional integral formulation has been developed [Kl99b] that, in effect, *within a single formula* captures the essence of *all four of the basic principles* described above. This "Master Formula" takes the form

$$\langle \pi'', g'' | \mathbb{E} | \pi', g' \rangle$$
$$= \lim_{\nu \to \infty} \mathcal{N}_\nu \int e^{-i \int [g_{ab} \dot{\pi}^{ab} + N^a H_a + N H] d^3x \, dt}$$
$$\times \exp\{-(1/2\nu) \int [b(x)^{-1} g_{ab} g_{cd} \dot{\pi}^{bc} \dot{\pi}^{da} + b(x) g^{ab} g^{cd} \dot{g}_{bc} \dot{g}_{da}] d^3x \, dt\}$$
$$\times [\Pi_{x,t} \, \Pi_{a \le b} \, d\pi^{ab}(x,t) \, dg_{ab}(x,t)] \, \mathcal{D}R(N^a, N) \, . \tag{10.11}$$

Let us explain the meaning of (10.11).

As an initial remark, let us artificially set $H_a = H = 0$, and use the fact that $\int \mathcal{D}R(N^a, N) = 1$. Then the result is that $\mathbb{E} = \mathbb{1}$, and the remaining functional integral yields the coherent-state overlap $\langle \pi'', g'' | \pi', g' \rangle$ as given in (10.8). This is the state of affairs *before* the constraints are imposed, and remarks below regarding the properties of the functional integral on the right-hand side of (10.11) apply in this case as well. We next turn to the full content of (10.11).

The expression $\langle \pi'', g'' | \mathbb{E} | \pi', g' \rangle$ denotes the coherent-state matrix element of a projection operator $\mathbb{E}$ which projects onto a subspace of the original Hilbert space on which the quantum constraints are fulfilled in a regularized fashion. Furthermore, the expression $\langle \pi'', g'' | \mathbb{E} | \pi', g' \rangle$ is another continuous functional of positive type that can be used as a reproducing kernel to generate the reproducing kernel physical Hilbert space on which the quantum constraints are fulfilled in a regularized manner. The right-hand side of equation (10.11) denotes an essentially well-defined ('Feynman-Kac-like') functional integral over fields $\pi^{ab}(x,t)$ and $g_{ab}(x,t)$, $0 < t < T$, designed to calculate this important reproducing kernel for the regularized physical Hilbert space and which entails functional arguments defined by their smooth initial values $\pi^{ab}(x,0) = \pi'^{ab}(x)$ and $g_{ab}(x,0) = g'_{ab}(x)$ as well as their smooth final values $\pi^{ab}(x,T) = \pi''^{ab}(x)$ and $g_{ab}(x,T) = g''_{ab}(x)$, for all x and all a, b. Up to a surface term, the phase factor in the functional integral represents the canonical action for general relativity, and specifically N^a and N denote Lagrange multiplier fields (classically interpreted as the shift and lapse), while H_a and H denote phase-space symbols (since $\hbar > 0$) associated with the quantum diffeomorphism and Hamiltonian constraint field operators, respectively. The ν-dependent factor in the integrand formally tends to unity in the limit $\nu \to \infty$; but prior to that limit, the given expression *regularizes and essentially gives genuine meaning* to the heuristic, formal functional integral that would otherwise arise if such a factor were missing altogether [Kl99b]. The functional form of the given regularizing factor ensures that the metric variables of integration *strictly fulfill* the positive-definite domain requirement. The given form, and in particular the need for the nondynamical, positive definite, arbitrarily chosen scalar density $b(x)$, is very welcome since this form— *and quite possibly only this form*—leads to a reproducing kernel Hilbert space for gravity having the needed infinite dimensionality; a seemingly natural alternative [Kla90] using $\sqrt{\det[g_{ab}(x)]}$ in place of $b(x)$ fails to lead to a reproducing kernel Hilbert space with the required dimensionality [WaK02]. The choice of $b(x)$ determines a specific ultralocal representation for the basic affine field variables, but this unphysical and temporary representation *disappears* after the gravitational constraints are fully enforced (as soluble examples explicitly demonstrate [Kl01b] and which is reviewed later in this chapter). The integration over the Lagrange multiplier fields (N^a and N) involves a *specific measure* $R(N^a, N)$ (generally described in [Kl99b]), which is normalized such that $\int \mathcal{D}R(N^a, N) = 1$. This measure is designed to enforce (a regularized version of) the *quantum constraints*; it is manifestly **not** chosen to enforce the classical constraints, even in a regularized form. The consequences of this choice are *profound* in that no gauge fixing is needed, no ghosts are required, no Dirac brackets are necessary to deal with second-class constraints, etc. In short, *no auxiliary structure of any kind is introduced.* (These facts are general properties of the projection operator method of dealing with constraints [Kla97; Kl01c] and are not limited to gravity. A sketch of this method appears below.)

It is fundamentally important to make clear how Eq. (10.11) was derived and how it is to be used [Kl99b]. The left-hand side of (10.11) is an abstract operator construct in its entirety that came *first* and corresponds to one of the basic expressions one would like to calculate. The functional integral on the right-hand side of (10.11)—which is derived below—came *second* and is a valid representation of the desired expression; its validity derives from the fact that the affine coherent-state representation enjoys a complex polarization that is used to formulate a kind of 'Feynman-Kac realization' of the coherent-state matrix elements of the regularized projection operator [Kl99b]. However, the final goal is to turn that order around and to use the functional integral to *define and evaluate* (at least approximately) the desired operator-defined expression on the left-hand side. In no way should it be thought that the functional integral (10.11) was "simply postulated as a guess as how one might represent the proper expression".

A major goal in the general analysis of (10.11) involves reducing the regularization imposed on the quantum constraints to its appropriate minimum value, and, in particular, for constraint operators that are partially second class, such as those of gravity, the proper minimum of the regularization parameter is *nonzero*; see below. Achieving this minimization involves *fundamental changes* of the representation of the basic kinematical operators, which, as models show (see [Kl01b] and later in this chapter) are so significant that any unphysical aspect of the original, ultralocal representation disappears completely. When the appropriate minimum regularization is achieved, then the quantum constraints are properly satisfied. The result is the reproducing kernel for the physical Hilbert space, which then permits a variety of physical questions to be studied.

We next offer some additional details.

10.3 Quantum Constraints and their Treatment

At a fixed time, $t = 0$, the quantum gravitational constraints, $\mathcal{H}_a(x)$, $a = 1, 2, 3$, and $\mathcal{H}(x)$, formally satisfy (for $\hbar = 1$) the commutation relations

$$[\mathcal{H}_a(x), \mathcal{H}_b(y)] = i\tfrac{1}{2}[\delta_{,a}(x, y)\mathcal{H}_b(y) + \delta_{,b}(x, y)\mathcal{H}_a(x)] ,$$

$$[\mathcal{H}_a(x), \mathcal{H}(y)] = i\delta_{,a}(x, y)\mathcal{H}(y) , \qquad (10.12)$$

$$[\mathcal{H}(x), \mathcal{H}(y)] = i\tfrac{1}{2}\delta_{,a}(x, y)[\hat{g}^{ab}(x)\mathcal{H}_b(x) + \mathcal{H}_b(x)\hat{g}^{ab}(x)$$
$$+ \hat{g}^{ab}(y)\mathcal{H}_b(y) + \mathcal{H}_b(y)\hat{g}^{ab}(y)] .$$

Following Dirac [Dir64], we ask that $\mathcal{H}_a(x)|\psi\rangle_{phys} = 0$ and $\mathcal{H}(x)|\psi\rangle_{phys} = 0$ for all x and a, where $|\psi\rangle_{phys}$ denotes a vector in the proposed physical Hilbert space $\mathfrak{H}_{phys}$. However, these conditions are *incompatible* since, generally, $[\mathcal{H}(x), \mathcal{H}(y)]|\psi\rangle_{phys} \neq 0$ because $[\mathcal{H}_b(x), \hat{g}^{ab}(x)] \neq 0$ and $\hat{g}^{ab}(x)|\psi\rangle_{phys} \notin \mathfrak{H}_{phys}$, even when smeared. This means that the quantum gravity constraints are *partially second class*. While others may resist this conclusion, we accept it for what it is.

One advantage of the projection operator method is that it treats first- and second-class constraints on an *equal footing*; see [Kla97; Kl01c]. The essence of the projection operator method is the following. If $\{\Phi_a\}$ denotes a set of self-adjoint quantum constraint operators, then

$$\mathbb{E} = \mathbb{E}(\Sigma_a \Phi_a^2 \leq \delta(\hbar)^2) = \int \mathbb{T} e^{-i \int \lambda^a(t) \Phi_a \, dt} \, \mathcal{D}R(\lambda) \,, \qquad (10.13)$$

in which $\mathbb{T}$ enforces time ordering and $\mathbb{E}$ denotes a projection operator onto a regularized physical Hilbert space, $\mathfrak{H}_{phys} \equiv \mathbb{E}\mathfrak{H}$, where $\mathfrak{H}$ denotes the original Hilbert space before the constraints are imposed. It is noteworthy that there is a *universal form* for the weak measure R [Kl99b] that depends only on the number of constraints, the time interval involved, and the regularization parameter $\delta(\hbar)^2$; R does *not* depend in any way on the constraint operators themselves! Sometimes, just by reducing the regularization parameter $\delta(\hbar)^2$ to its appropriate size, the proper physical Hilbert space arises. Thus, e.g., if $\Sigma_a \Phi_a^2 = J_1^2 + J_2^2 + J_3^2$, the Casimir operator of $su(2)$, then $0 \leq \delta(\hbar)^2 < 3\hbar^2/4$ works for this first-class example (defined by the fact that the commutator of constraints vanishes when the constraints vanish). If $\Sigma_a \Phi_a^2 = P^2 + Q^2$, where $[Q, P] = i\hbar \mathbb{1}$, then $\hbar \leq \delta(\hbar)^2 < 3\hbar$ covers this second-class example (defined by the fact that the commutator of constraints does not vanish when the constraints vanish). Sometimes, one needs to take the limit when $\delta \to 0$. The example $\Sigma_a \Phi_a^2 = Q^2$ involves a case where $\Sigma_a \Phi_a^2 = 0$ lies in the continuous spectrum. To deal with this case it is appropriate to introduce

$$\langle\langle p'', q'' | p', q' \rangle\rangle \equiv \lim_{\delta \to 0} \langle p'', q'' | \mathbb{E} | p', q' \rangle / \langle \eta | \mathbb{E} | \eta \rangle \,, \qquad (10.14)$$

where $\{|p, q\rangle\}$ are traditional coherent states, as a reproducing kernel for the physical Hilbert space in which "$Q = 0$" holds. It is interesting to observe that the projection operator for *reducible* constraints, e.g., $\mathbb{E}(Q^2 + Q^2 \leq \delta^2)$, or for *irregular* constraints, $\mathbb{E}(Q^{2\Omega} \leq \delta^2)$, $0 < \Omega \neq 1$, leads to the *same reproducing kernel* that arose from the case $\mathbb{E}(Q^2 \leq \delta^2)$. No gauge fixing is ever needed, and thus no global consistency conditions arise that may be violated; see, e.g., [LiK05]. Other cases may be more involved but the principles are similar. The time-ordered integral representation for $\mathbb{E}$ given in (10.13) is useful in path integral representations, and this application explains the origin of $R(N^a, N)$ in (10.11).

10.4   Nonrenormalizability and Symbols

Viewed perturbatively, gravity is nonrenormalizable. However, the (non-perturbative) *hard-core picture of nonrenormalizability* [Kl73b; Kla00] holds that the nonlinearities in such theories are so strong that, from a functional integral point of view, a set of functional histories of nonzero measure that was allowed in the support of the linear theory is now forbidden by the nonlinear interaction. In Chap. 3 we already have seen how very singular potentials make interacting quantum systems pass to a pseudofree model rather than their own free model as the coupling constant is reduced to zero.

Likewise, various, highly specialized, nonrenormalizable quantum field theory models exhibit entirely analogous hard-core behavior, and neverthe-less possess suitable nonperturbative solutions [Kla00]. It is believed that gravity and also φ^4 field theories in high enough spacetime dimensions can be understood in similar terms. As described in Chap. 9, the expectation that nonrenormalizable, self-interacting scalar fields will exhibit hard-core behavior follows from a multiplicative inequality (9.17). A computer study to analyze the φ^4 theory has begun, and there is hope to clarify that partic-ular theory. Any progress in the scalar field case could strengthen a similar argument for the gravitational case as well.

Evidence from soluble examples points to the appearance of a nontradi-tional and nonclassical (proportional to $\hbar^2$) counterterm in the functional integral representing the irremovable effects of the hard core. For the pro-posed quantization of gravity, these counterterms would have an important role to play in conjunction with the symbols representing the diffeomor-phism and Hamiltonian constraints in the functional integral since for them $\hbar > 0$ as well. In brief, the form taken by the symbols H_a and H in (10.11) is closely related to a proper understanding of how to handle the pertur-bative nonrenormalizability and the concomitant hard-core nature of the overall theory. These are clearly difficult issues, but it is equally clear that they may be illuminated by studies of other nonrenormalizable models such as φ^4 in five and more spacetime dimensions.

10.5   Classical Limit

Suppose one starts with a classical theory, quantizes it, and then takes the classical limit. It seems obvious that the classical theory obtained at the end should coincide with the classical theory one started with. However, as noted earlier, there are counterexamples to this simple wisdom! For example, as discussed in the previous chapter, the classical φ^4 theory in five spacetime dimensions has a *nontrivial* classical behavior. But, if one quantizes it as the continuum limit of a natural lattice formulation, allowing for mass, field strength, and coupling constant renormalization, the result is a free (or generalized free) quantum theory the classical limit of which is also free and thus differs from the original theory [Aiz81; Frö82]. This behavior is yet another facet of the nonrenormalizability puzzle. However, as discussed in Chap. 8, the (nonrenormalizable) ultralocal models for which the quantum hard-core behavior has been accounted for do have satisfactory classical limits. The conjectured hard-core nature of φ^4 models is under continuing investigation, and it is expected that a proper classical limit, as discussed in Chap. 9, should arise. It is conjectured that a favorable consequence of clarifying and including the hard-core behavior in gravity will ensure that the resultant quantum theory enjoys the correct classical limit.

An additional remark may be useful. It is a frequent misconception that passage to the classical limit requires that the parameter $\hbar \to 0$. To argue against this view, recall once again that the macroscopic world we know and describe so well by classical mechanics is the same real world in which $\hbar > 0$. In point of fact, classical and quantum formalisms must *coexist*, and this coexistence is very well expressed with the help of coherent states and the program of enhanced quantization. It is characteristic of coherent-state formalisms that classical and quantum "generators", loosely speaking, are related to each other through the *weak correspondence principle* [Kla67]. In the case of the gravitational field, prior to the introduction of constraints, this connection takes the general form

$$\langle \pi, g | \mathcal{W} | \pi, g \rangle = W(\pi, g) \,, \tag{10.15}$$

where $\mathcal{W}$ denotes a quantum generator and $W(\pi, g)$ the corresponding classical generator (which is generally a "symbol" since $\hbar > 0$ still). The simplest examples of this kind are given by $\langle \pi, g | \hat{g}_{ab}(x) | \pi, g \rangle = g_{ab}(x)$ and $\langle \pi, g | \hat{\pi}_a^b(x) | \pi, g \rangle = \pi^{bc}(x) g_{ca}(x) \equiv \pi_a^b(x)$. Moreover, these two examples also establish that the *physical meaning of the c-number labels is that of mean values* of the respective quantum field operators in the affine coherent states.

10.6　Going Beyond the Ultralocal Representation

We started our discussion by choosing an ultralocal representation of the basic affine quantum field operators. Before the constraints were introduced, an ultralocal representation is a very reasonable choice because all the proper spatial correlations are contained in the constraints themselves. Moreover, the chosen ultralocal representation is based on an extremal weight vector of the underlying affine algebra, which has the virtue of leading to affine coherent states that fulfill a complex polarization condition enabling us to obtain a rather well-defined functional integral representation for coherent-state matrix elements of the regularized projection operator. To complete the story, one only needs to eliminate the regularizations! Of course, this is an enormous task. But it should not be regarded as impossible because there are model problems in which just that issue has been successfully dealt with. In [Kl01b] (and below) the quantization of a free field of mass m (among other examples) was discussed starting with a reparametrization-invariant formulation. In particular, by elevating the time variable to a dynamical one, the original dynamics is transformed to the imposition of a constraint. Thus, in the constrained form, the Hamiltonian vanishes, and the choice of the original representation of the field operators is taken as an ultralocal one. Subsequent imposition of the constraint—by the projection operator method—not only eliminated the ultralocal representation but allowed us to focus the final reproducing kernel for the physical Hilbert space on any value of the mass parameter m one desired! It is this kind of procedure used for this relatively simple example of free-field quantization that we have in mind to be used to transform the original ultralocal representation of the quantum gravity story into its final and physically relevant version.

10.7　Derivation of the Affine Coherent-State Overlap

We let the symbols $\hat{\pi}_b^a(x)$ and $\hat{g}_{ab}(x)$ denote local self-adjoint operators, which means that they both become self-adjoint operators after smearing with suitable real test functions. Adopting the algebra of the classical Poisson brackets for these variables, modified by $i\hbar$ (in units where $\hbar = 1$ throughout), the affine commutation relations are given by the formal Lie algebra

$$[\hat{\pi}_k^r(x),\ \hat{\pi}_l^s(y)] = i\,\tfrac{1}{2}\,[\delta_k^s\,\hat{\pi}_l^r(x) - \delta_l^r\,\hat{\pi}_k^s(x)]\,\delta(x,y)\ ,$$
$$[\hat{g}_{kl}(x),\ \hat{\pi}_s^r(y)] = i\,\tfrac{1}{2}\,[\delta_k^r\,\hat{g}_{ls}(x) + \delta_l^r\,\hat{g}_{ks}(x)]\,\delta(x,y)\ , \qquad (10.16)$$
$$[\hat{g}_{kl}(x),\ \hat{g}_{rs}(y)] = 0\ .$$

We are interested in representations of these operators, and since, following Dirac [Dir64], quantization should precede reduction by any constraints, we deliberately choose an ultralocal representation composed solely of independent representations at each spatial point. To regularize such a construction, we again appeal to a spatial lattice regularization in which every site carries an independent representation of the affine algebra. At a generic lattice site the Lie algebra of the affine group has the form [with $\hat{\pi}_b^a \to \tau_b^a\,\Delta^{-1}$, where Δ denotes a uniform cell volume, and $\hat{g}_{ab} \to \sigma_{ab}$] given by

$$[\tau_b^a, \tau_d^c] = i\tfrac{1}{2}(\delta_b^c\tau_d^a - \delta_d^a\tau_b^c)\ ,$$
$$[\sigma_{ab}, \tau_d^c] = i\tfrac{1}{2}(\delta_a^c\sigma_{db} + \delta_b^c\sigma_{ad})\ , \qquad (10.17)$$
$$[\sigma_{ab}, \sigma_{cd}] = 0\ .$$

Following [WaK00] closely, we choose the faithful, irreducible representation for which the operator matrix $\{\sigma_{ab}\}$ is symmetric and positive definite, and which is unique up to unitary equivalence. Furthermore, we choose a representation that diagonalizes $\{\sigma_{ab}\}$ as $k \equiv \{k_{ab}\}$, $k_{ba} = k_{ab} \in \mathbb{R}$, which we refer to as the 'k-representation'.

The positive-definite, symmetric 3×3 matrices $k = \{k_{ab}\}$ are also usefully parameterized in a different manner. Every such matrix can be put in the form $k = Ok O^T$ where k is a positive definite, *diagonal* matrix, $\{\mathsf{k}_A\}_{A=1}^3$, $\mathsf{k}_A > 0$ for all A, and $O \in \mathbf{SO}(3)$, the three-dimensional rotation group. The six parameters of the matrix k are interchangeable with the three parameters of k and the three parameters of $\mathbf{SO}(3)$. The measure $dk \equiv \Pi_{a\leq b}dk_{ab}$ is equivalent, up to a constant scale factor, with the measure $dk \equiv \Pi_A d\mathsf{k}_A \Pi_B dO_B$, where $\{O_B\}$ symbolize the three variables of the three-dimensional rotation group. As an example of this usage, consider the normalization integral given, with $\beta > 0$, by

$$C_0 \int \det[k]^{2\beta-1}\, e^{-2\beta\,\mathrm{Tr}(k)}\, dk = C_0' \int \det[\mathsf{k}]^{2\beta-1}\, e^{-2\beta\,\mathrm{Tr}(\mathsf{k})}\, d\mathsf{k} = 1\ .$$

$$(10.18)$$

Adopting the second form, and with $\int \Pi_B dO_B = \int \sin(\theta)\, d\theta\, d\phi\, d\psi = 8\pi^2$

using Euler angles $(0 \leq \theta \leq \pi;\ 0 \leq \phi, \psi < 2\pi)$, it follows that

$$1 = C_0' \int \det[k]^{2\beta - 1}\, e^{-2\beta\,\mathrm{Tr}(k)}\, dk$$

$$= C_0'\, 8\pi^2 \left\{ \int_0^\infty (k_1)^{2\beta - 1}\, e^{-2\beta\, k_1}\, dk_1 \right\}^3 ,$$

$$= C_0'\, 8\pi^2 \{ (2\beta)^{-2\beta}\, \Gamma(2\beta) \}^3 , \tag{10.19}$$

a relation that determines the value of C_0'. While we generally use the notation k and dk below, such integrals may be easier to evaluate by appealing to the alternative variables $\{k_A, O_B\}$ and dk, which we name the 'k-variables'.

In the associated L^2 representation space, and for arbitrary real matrices $\Pi = \{\Pi^{ab}\}$, $\Pi^{ba} = \Pi^{ab}$, and $\Gamma = \{\Gamma_d^c\}$, it follows that

$$U[\Pi, \Gamma]\psi(k) \equiv e^{i\,\Pi^{ab}\sigma_{ab}}\, e^{-i\,\Gamma_d^c \tau_c^d}\, \psi(k)$$

$$= (\det[S])\, e^{i\,\Pi^{ab} k_{ab}}\, \psi(SkS^T) , \tag{10.20}$$

where $S \equiv e^{-\Gamma/2} = \{S_b^a\}$ and $(SkS^T)_{ab} \equiv S_a^c k_{cd} S_b^d$. The given transformation is unitary within the inner product defined by

$$\int_+ \psi(k)^*\, \psi(k)\, dk , \tag{10.21}$$

where $dk \equiv \Pi_{a \leq b}\, dk_{ab}$, and the "+" sign denotes an integration over only that part of the six-dimensional k-space where the elements form a symmetric, positive-definite matrix, $\{k_{ab}\} > 0$. To define affine coherent states we first choose the fiducial vector as an extremal weight vector [WaK00; Kl01a],

$$\eta(k) \equiv C\, (\det[k])^{\beta - 1/2}\, e^{-\beta\,\mathrm{Tr}[\tilde{G}^{-1}k]} , \tag{10.22}$$

where $\beta > 0$, $\tilde{G} = \{\tilde{G}_{ab}\}$ is a fixed positive-definite matrix, C is determined by normalization, and Tr denotes the trace. This choice of $\eta(k) = \langle k|\eta \rangle$ leads to the expectation values

$$\langle \eta|\sigma_{ab}|\eta \rangle = \int_+ \eta(k)^* k_{ab}\, \eta(k)\, dk = \tilde{G}_{ab} ,$$

$$\langle \eta|\tau_d^c|\eta \rangle = \int_+ \eta(k)^* \tilde{\tau}_d^c\, \eta(k)\, dk = 0 , \tag{10.23}$$

$$\tilde{\tau}_d^c \equiv (-i/2)[(\partial/\partial k_{cb}) k_{db} + k_{db} (\partial/\partial k_{cb})] .$$

In the k-representation, it follows that the affine coherent states are given by

$$\langle k|\Pi, \Gamma \rangle \equiv C\, (\det[S])\, (\det[SkS^T])^{\beta - 1/2}\, e^{i\,\mathrm{Tr}[\Pi k]}\, e^{-\beta\,\mathrm{Tr}[\tilde{G}^{-1} SkS^T]} .$$

$$\tag{10.24}$$

Observe that what really enters the functional argument above is the positive-definite matrix $G^{-1} \equiv S^T \tilde{G}^{-1} S$, where we set $G \equiv \{G_{ab}\}$. Thus, without loss of generality, we can drop the label Γ (or equivalently S) and replace it with G. Hence, the affine coherent states become

$$\langle k | \Pi, G \rangle \equiv C' \left(\det[G^{-1}] \right)^{\beta} \left(\det[k] \right)^{\beta - 1/2} e^{i \operatorname{Tr}[\Pi k]} e^{-\beta \operatorname{Tr}[G^{-1} k]} , \quad (10.25)$$

where C' is a new normalization constant. It is now straightforward to determine the affine coherent-state overlap function

$$\langle \Pi'', G'' | \Pi', G' \rangle = \int_{+} \langle \Pi'', G'' | k \rangle \langle k | \Pi', G' \rangle \, dk$$

$$= \left[\frac{\{\det[G''^{-1}] \det[G'^{-1}]\}^{1/2}}{\det\{\frac{1}{2}((G''^{-1} + G'^{-1}) + i\beta^{-1}(\Pi'' - \Pi'))\}} \right]^{2\beta} . \quad (10.26)$$

In arriving at this result, we have used normalization of the coherent states to eliminate the constant C'.

Suppose we now consider a 3-dimensional spatial lattice of independent sets of matrix degrees of freedom and build the corresponding affine coherent-state overlap as the product of expressions like (10.26). In this section, we adopt the notation where $\mathbf{n}$ labels a lattice site and $\mathbf{n} \in \mathbf{N}$, the set of all spatial lattice sites, which in turn may be considered a finite subset of $\mathbb{Z}^3$; this change of notation conforms with the analysis in [Kl99a] and is needed because k has a different meaning in the present section from its use in Sec. 10.2. With these notational changes, the multi-site, affine coherent-state overlap function is given by

$$\langle \Pi'', G'' | \Pi', G' \rangle_{\mathbf{N}} \quad (10.27)$$

$$= \prod_{\mathbf{n} \in \mathbf{N}} \left[\frac{\{\det[G''^{-1}_{[\mathbf{n}]}] \det[G'^{-1}_{[\mathbf{n}]}]\}^{1/2}}{\det\{\frac{1}{2}((G''^{-1}_{[\mathbf{n}]} + G'^{-1}_{[\mathbf{n}]}) + i\beta^{-1}_{[\mathbf{n}]}(\Pi''_{[\mathbf{n}]} - \Pi'_{[\mathbf{n}]}))\}} \right]^{2\beta_{[\mathbf{n}]}} ;$$

since there is generally no translation symmetry, it is not necessary that the factors $\beta_{[\mathbf{n}]}$ are all the same.

10.7.1 *Affine coherent-state overlap function*

As our next step, we take a limit in which the number of lattice sites with independent matrix degrees of freedom tends to infinity, but also the lattice cell volume tends to zero so that, loosely speaking, the lattice points approach the continuum points of the underlying topological space $\mathcal{S}$. In order for this continuum limit to be meaningful, it is necessary that the exponent $\beta_{[\mathbf{n}]} \to 0$ in a suitable way. Again assuming a uniform cell volume for simplicity, we set

$$\beta_{[\mathbf{n}]} \equiv b_{[\mathbf{n}]} \Delta , \quad (10.28)$$

where the cell volume Δ has the dimensions (length)^3, and thus $b_{[\mathbf{n}]}$ has the dimensions (length)$^{-3}$. In addition, we rename $\Pi_{[\mathbf{n}]}^{ab} \equiv \pi_{[\mathbf{n}]}^{ab}\Delta$, as well as $G_{ab[\mathbf{n}]} \equiv g_{ab[\mathbf{n}]}$, and denote the matrix elements of $G_{[\mathbf{n}]}^{-1}$ by $g_{[\mathbf{n}]}^{ab}$. With these notational changes (10.27) becomes

$$\langle \pi'', g'' | \pi', g' \rangle_{\mathbf{N}} \tag{10.29}$$
$$\equiv \prod_{\mathbf{n} \in \mathbf{N}} \left[\frac{\{\det[g_{[\mathbf{n}]}''^{ab}] \det[g_{[\mathbf{n}]}'^{ab}]\}^{1/2}}{\det\{\frac{1}{2}(g_{[\mathbf{n}]}''^{ab} + g_{[\mathbf{n}]}'^{ab}) + i b_{[\mathbf{n}]}^{-1}(\pi_{[\mathbf{n}]}''^{ab} - \pi_{[\mathbf{n}]}'^{ab})\}} \right]^{2 b_{[\mathbf{n}]} \Delta}.$$

To facilitate the continuum limit, we assume that the various label sets pass to smooth functions, in which case the result, for a compact topological space $\mathcal{S}$, is given by

$$\langle \pi'', g'' | \pi', g' \rangle$$
$$\equiv \exp\left[-2 \int b(x)\, d^3 x \right. \tag{10.30}$$
$$\times \ln \left. \left(\frac{\det\{\frac{1}{2}[g''^{ab}(x) + g'^{ab}(x)] + \frac{1}{2} i b(x)^{-1}[\pi''^{ab}(x) - \pi'^{ab}(x)]\}}{\{\det[g''^{ab}(x)] \det[g'^{ab}(x)]\}^{1/2}} \right) \right].$$

In this way we see how the continuum result may be obtained as a limit starting from a collection of independent affine degrees of freedom. The necessity of ending with an integral over the space $\mathcal{S}$ has directly led to the requirement that we introduce the scalar density function $b(x) > 0$, but, other than smoothness and positivity, this function may be freely chosen. Indeed, the appearance of $b(x)$ in the affine coherent-state overlap function is not unlike the appearance of a Gaussian variance-parameter (often denoted by ω) in the overlap function of canonical coherent states; it is a representation artifact with limited physical significance.

It is important to appreciate that the function $g_{ab}(x)$ that enters the affine coherent-state overlap function is *not* a true metric that has been applied to the topological space $\mathcal{S}$, but it is just a symmetric, two-index covariant tensor field, which we sometime call a metric because of its positivity properties. In like manner, the momentum field $\pi^{ab}(x)$ that enters the affine coherent-state overlap function also carries no special physical meaning and is just a symmetric, two-index contravariant tensor density field—nothing more and nothing less.

The expression (10.30) is not yet in optimal form, especially if one wishes to consider an extension to an infinite spatial volume. In this case, it is useful to incorporate the proper metric and momentum asymptotics for large spatial distances. Assuming that the matrix elements $\gamma_b^a(x)$ and $\pi^{ab}(x)$

are smooth functions with compact support, it follows that for sufficiently large x values, the affine coherent-state matrix elements tend to become $\langle \pi'', g'' | \hat{g}_{ab}(x) | \pi', g' \rangle = \tilde{g}_{ab}(x)$ and $\langle \pi'', g'' | \hat{\pi}_b^a(x) | \pi', g' \rangle = 0$. We can already make use of these facts by freely multiplying the numerator and denominator inside the logarithm in (10.30) by $\det[\tilde{g}_{ab}(x)]$, which leads to the relation

$$\langle \pi'', g'' | \pi', g' \rangle$$

$$= \exp\left[-2 \int b(x)\, d^3x \right. \tag{10.31}$$

$$\left. \times \ln\left(\frac{\det\{\frac{1}{2}[\tilde{g}_b''^a(x) + \tilde{g}_b'^a(x)] + \frac{1}{2} i b(x)^{-1}[\tilde{\pi}_b''^a(x) - \tilde{\pi}_b'^a(x)]\}}{\{\det[\tilde{g}_b''^a(x)]\, \det[\tilde{g}_b'^a(x)]\}^{1/2}} \right) \right].$$

In this equation we have introduced $\tilde{g}_b^a(x) \equiv \tilde{g}_{bc}(x) g^{ac}(x)$ and $\tilde{\pi}_b^a(x) \equiv \tilde{g}_{bc}(x) \pi^{ac}(x)$ for both the $''$ and the $'$ variables. With this formulation, it follows that at large distances $\tilde{g}_b^a(x) \to \delta_b^a$ while $\tilde{\pi}_b^a(x) \to 0$. Moreover, we note that each one of the three determinants in (10.31), e.g., $\det[\tilde{g}_b''^a(x)]$, etc., is *invariant under general coordinate transformations*, and each determinant tends toward unity at large distances. Thanks to the logarithm, this tendency ensures convergence of the contribution from each determinant in an infinite volume provided there is suitable asymptotic behavior of the relevant functions.

Clearly, the case of an infinite spatial volume needs some special care. To focus on the essentials, we now return to our general assumption that we deal with a *finite spatial volume*, unless noted otherwise.

10.8 Measure Mashing in the Affine Coherent-State Overlap

Next, let us derive the coherent-state overlap (10.30) in a different manner. In particular, with $\{C_n''\}$ a presently unimportant set of constants for these integrals, it is clear that

$$\langle \pi'', g'' | \pi', g' \rangle$$

$$= \lim_{\Delta \to 0} \int_+ \Pi_n' \langle \pi_n'', g_n'' | k_n \rangle \langle k_n | \pi_n', g_n' \rangle \, dk_n$$

$$= \lim_{\Delta \to 0} \int_+ \Pi_n' \, C_n'' \, e^{-i\,\mathrm{Tr}[(\pi_n'' - \pi_n') k_n] - \beta_n \, \mathrm{Tr}[(g_n''^{-1} + g_n'^{-1}) k_n]}$$

$$\times \det(k_n)^{2\beta_n - 1} \, dk_n \,, \tag{10.32}$$

which, on transforming to hyperspherical coordinates—note the use of the alternative k-variables—defined here by $k_{\mathbf{n}\,A} \equiv \kappa \eta_{\mathbf{n}\,A}$, for all $\mathbf{n}$, $1 \le A \le 3$, $\kappa^2 \equiv \Sigma'_{\mathbf{n}} \Sigma_A k^2_{\mathbf{n}\,A}$, $1 \equiv \Sigma'_{\mathbf{n}} \Sigma_A \eta^2_{\mathbf{n}\,A}$, with $0 \le \kappa < \infty$ and $-1 \le \eta_{\mathbf{n}\,A} \le 1$, leads to

$$\langle \pi'', g'' | \pi', g' \rangle$$
$$= \lim_{\Delta \to 0} \int_+ \{ \Pi'_{\mathbf{n}} \, C''_{\mathbf{n}} \, e^{-i \operatorname{Tr}[(\pi''_{\mathbf{n}} - \pi'_{\mathbf{n}}) \kappa \eta_{\mathbf{n}}] - \beta_{\mathbf{n}} \operatorname{Tr}[(g''^{-1}_{\mathbf{n}} + g'^{-1}_{\mathbf{n}}) \kappa \eta_{\mathbf{n}}]}$$
$$\times \det(\kappa \eta_{\mathbf{n}})^{2\beta_{\mathbf{n}} - 1} \} \kappa^{(3|\mathbf{N}|-1)} \, d\kappa \, 2\delta(1 - \Sigma'_{\mathbf{n}} \Sigma_A \eta^2_{\mathbf{n}\,A})$$
$$\times \Pi'_{\mathbf{n}\,A} \, d\eta_{\mathbf{n}\,A} \, \Pi'_{\mathbf{n}\,B} \, dO_{\mathbf{n}\,B} \, , \tag{10.33}$$

where $|\mathbf{N}| < \infty$ denotes the total number of lattice sites in this spatial slice. With $\beta_{\mathbf{n}} = b_{\mathbf{n}} \Delta$, it follows that

$$[\Pi'_{\mathbf{n}} \det(\kappa \eta_{\mathbf{n}})^{2 b_{\mathbf{n}} \Delta - 1}] \, \kappa^{(3|\mathbf{N}|-1)} = \Pi'_{\mathbf{n}} \, [\det(\eta_{\mathbf{n}})^{2 b_{\mathbf{n}} \Delta - 1} \, \kappa^{6 b_{\mathbf{n}} \Delta}] \, \kappa^{-1} \, , \tag{10.34}$$

which shows the effects of measure mashing already for which, in the present case, $R \equiv 6\Sigma'_{\mathbf{n}} b_{\mathbf{n}} \Delta < \infty$ provided, as we have assumed, that $\Sigma'_{\mathbf{n}} \Delta = V' < \infty$.

10.8.1   *Significance of already having a mashed measure*

According to the discussion in Chaps. 8 and 9, a mashed measure for the hyperspherical radius variable implies no divergences arise in any perturbation analysis because potentially disjoint measures have been converted into equivalent measures. It also implies that local field powers are defined by means of an operator product expansion and not by normal ordering, as we make clear in the following section. When dealing with quantum field theories, we normally encounter unitarily inequivalent representations of kinematical operators when a parameter is changed in the basic model. For example, in the ultralocal free model discussed in Chap. 8, and for a finite spatial volume as well, representations of the local operators for the momentum $\hat{\pi}(x)$ and the field $\hat{\phi}(x)$ are unitarily inequivalent for different mass values. On the other hand, for a finite spatial volume, the pseudofree model local field operators are unitarily equivalent for different mass values due to measure mashing.

In the present case, we are dealing with the affine gravitational field operators $\hat{\pi}^a_b(x)$ and $\hat{g}_{ab}(x)$, for $a, b = 1, 2, 3$. A change in the scalar density $b(x)$, which enters the representation of the coherent states, could lead to inequivalent representations of the affine kinematical variables, but—*thanks to measure mashing*—that is not the case since all the representations covered by the affine coherent states are unitarily equivalent for different functions $b(x)$. This has the additional consequence that we can construct various operators from the basic kinematical set and these operators will also be unitarily equivalent for different $b(x)$ values as well.

10.9 Operator Realization

In order to realize the metric and momentric fields as quantum operators in a Hilbert space, $\mathfrak{H}$, it is expedient—following the procedures for ultralocal scalar fields in [Kla70; Kla00] and in Chap. 8—to introduce a set of conventional local *annihilation and creation operators*, $A(x, k)$ and $A(x, k)^\dagger$, respectively, with the only nonvanishing commutator given by

$$[A(x, k), A(x', k')^\dagger] = \delta(x, x')\, \delta(k, k')\mathbb{1} \,, \tag{10.35}$$

where $\mathbb{1}$ denotes the unit operator. Here, $x \in \mathbb{R}^3$, while $k \equiv \{k_{rs}\}$ denotes a positive-definite, 3×3 symmetric matrix degree of freedom confined to the domain where $\{k_{rs}\} > 0$; additionally, $\delta(k, k') \equiv \Pi_{a \leq b}\delta(k_{ab} - k'_{ab})$. We introduce a 'no-particle' state $|0\rangle$ such that $A(x, k)\,|0\rangle = 0$ for all arguments. Additional states are determined by suitably smeared linear combinations of

$$A(x_1, k_1)^\dagger\, A(x_2, k_2)^\dagger \cdots A(x_p, k_p)^\dagger\, |0\rangle \tag{10.36}$$

for all $p \geq 1$, and the linear span of all such states is $\mathfrak{H}$ provided, apart from constant multiples, that $|0\rangle$ is the only state annihilated by all the A operators. Thus we are led to a conventional Fock representation for the A and $A^\dagger$ operators. Note that the Fock operators are irreducible, and thus all operators acting in $\mathfrak{H}$ are given as suitable functions of them.

Next, let $c(x, k)$ be a possibly complex, c-number function (defined below) and introduce the translated Fock operators

$$B(x, k) \equiv A(x, k) + c(x, k)\,\mathbb{1} \,,$$
$$B(x, k)^\dagger \equiv A(x.k)^\dagger + c(x, k)^*\,\mathbb{1} \,. \tag{10.37}$$

Evidently, the only nonvanishing commutator of the B and $B^\dagger$ operators is

$$[B(x, k), B(x', k')^\dagger] = \delta(x, x')\, \delta(k, k')\mathbb{1} \,, \tag{10.38}$$

the same as the A and $A^\dagger$ operators. With regard to transformations of the coordinate x, it is clear that $c(x, k)$ (just like the local operators A and B) should transform as a scalar density of weight one-half. Thus we set

$$c(x, k) \equiv b(x)^{1/2}\, d(x, k) \,, \tag{10.39}$$

where $d(x, k)$ transforms as a scalar. The criteria for acceptable $d(x, k)$ are, for each x, that

$$\int_+ |d(x, k)|^2 \, dk = \infty , \tag{10.40}$$

$$\int_+ k_{rs} |d(x, k)|^2 \, dk = \delta_{rs} , \tag{10.41}$$

the latter assuming that $\tilde{g}_{kl}(x) = \langle \eta | \hat{g}_{kl}(x) | \eta \rangle = \delta_{kl}$.

We shall focus only on the case where

$$d(x, k) \equiv \frac{\sqrt{K} \, e^{-\text{Tr}(k)}}{(\det[k])^{1/2}} , \tag{10.42}$$

which is everywhere independent of x; K denotes a positive constant to be fixed by (10.41). The given choice for $d(x, k)$ corresponds to the case where $\tilde{\pi}^{kl}(x) \equiv 0$ and $\tilde{g}_{kl}(x) \equiv \delta_{kl}$. [Remark: For different choices of asymptotic fields it suffices to choose

$$d(x, k) \to \tilde{d}(x, k) \equiv \frac{\sqrt{K} \, e^{-i\tilde{\pi}^{ab}(x) k_{ab}/b(x)} \, e^{-\tilde{g}^{ab}(x) k_{ab}}}{(\det[k])^{1/2}} . \tag{10.43}$$

It would seem that such cases allow one to quantize about a variety of topological backgrounds.]

In terms of the quantities introduced above, the local metric operator is defined by

$$\hat{g}_{ab}(x) \equiv b(x)^{-1} \int_+ B(x, k)^\dagger \, k_{ab} \, B(x, k) \, dk , \tag{10.44}$$

and the local momentric operator is defined by

$$\hat{\pi}_s^r(x) \equiv -i \tfrac{1}{2} \int_+ B(x, k)^\dagger \, (k_{st} \partial^{tr} + \partial^{rt} k_{ts}) \, B(x, k) \, dk . \tag{10.45}$$

Here $\partial^{st} \equiv \partial / \partial k_{st}$, and $\hat{g}_{ab}(x)$ transforms as a tensor while $\hat{\pi}_s^r(x)$ transforms as a tensor density of weight one. It is straightforward to show that these operators satisfy the required affine commutation relations, and moreover that [Kla00; Kla70]

$$\langle 0 | \, e^{i \int \pi^{ab}(x) \hat{g}_{ab}(x) \, d^3x} \, e^{-i \int \gamma_r^s(x) \hat{\pi}_s^r(x) \, d^3x} \, | 0 \rangle$$

$$= \exp\{-K \int b(x) \, d^3x \int [e^{-2\delta^{ab} k_{ab}} - e^{-i\pi^{ab}(x) k_{ab}/b(x)}$$

$$\times e^{-[(\delta^{ab} + g^{ab}(x)) k_{ab}]}] \, dk / (\det[k])\}$$

$$= \exp[-2 \int b(x) \, d^3x \ln([\det(g_{ab}(x))]^{-1/2}$$

$$\times \det\{\tfrac{1}{2}[\delta^{ab} + g^{ab}(x)] - i\tfrac{1}{2} b(x)^{-1} \pi^{ab}(x)\})] , \tag{10.46}$$

where K, as required, has been chosen so that

$$K \int_+ k_{rs} \, e^{-2\,\text{Tr}(k)} \, dk / (\det[k]) = \delta_{rs} . \tag{10.47}$$

An obvious extension of this calculation leads to (10.30).

10.9.1 Local operator products

Basically, local products for the gravitational field operators follow the pattern for other ultralocal quantum field theories in Chap. 8. As motivation, consider the product

$$\hat{g}_{ab}(x)\hat{g}_{cd}(x')$$
$$= [b(x)b(x')]^{-1}\int_+ B(x,k)^\dagger\, k_{ab}\, B(x,k)\, dk \cdot \int_+ B(x',k')^\dagger\, k'_{cd}\, B(x',k')\, dk'$$
$$= [b(x)b(x')]^{-1}\int_+\int_+ B(x,k)^\dagger\, k_{ab}\, [\, B(x,k),\, B(x',k')^\dagger]k'_{cd}\, B(x',k')\, dk\, dk'$$
$$\quad +!\,\hat{g}_{ab}(x)\hat{g}_{cd}(x')\,!$$
$$= b(x)^{-2}\delta(x,x')\int_+ B(x,k)^\dagger k_{ab}\, k_{cd}\, B(x,k)\, dk +!\,\hat{g}_{ab}(x)\hat{g}_{cd}(x')\,!\, , \quad (10.48)$$

where ! ! denotes normal ordering with respect to $B^\dagger$ and B. When $x' \to x$, this relation formally becomes

$$\hat{g}_{ab}(x)\hat{g}_{cd}(x) = b(x)^{-2}\delta(x,x)\int_+ B(x,k)^\dagger k_{ab}\, k_{cd}\, B(x,k)\, dk$$
$$\quad +!\,\hat{g}_{ab}(x)\hat{g}_{cd}(x)\,!\, . \quad (10.49)$$

After formally dividing both sides by the most divergent dimensionless scalar factor, namely, the 'scalar' $b(x)^{-1}\delta(x,x)$, we define the renormalized (subscript R) local product as

$$[\hat{g}_{ab}(x)\hat{g}_{cd}(x)]_R \equiv b(x)^{-1}\int_+ B(x,k)^\dagger\, k_{ab}\, k_{cd}\, B(x,k)\, dk\, ; \quad (10.50)$$

note: this derivation can be made rigorous with test functions and proper limits. Observe that $[\hat{g}_{ab}(x)\hat{g}_{cd}(x)]_R$ is a local operator that becomes a self-adjoint operator when smeared with a suitable real test function just as much as the local operator $\hat{g}_{ab}(x)$ becomes a self-adjoint operator after smearing with a suitable real test function. Higher-order local operator products exist as well, for example,

$$[\hat{g}_{a_1 b_1}(x)\hat{g}^{a_2 b_2}(x)\hat{g}_{a_3 b_3}(x)\cdots\hat{g}_{a_p b_p}(x)]_R$$
$$\equiv b(x)^{-1}\int_+ B(x,k)^\dagger\, (k_{a_1 b_1}\, k^{a_2 b_2}\, k_{a_3 b_3}\cdots k_{a_p b_p})\, B(x,k)\, dk\, , \quad (10.51)$$

which, after contracting on b_1 and b_2, implies that

$$[\hat{g}_{a_1 b}(x)\hat{g}^{a_2 b}(x)\hat{g}_{a_3 b_3}(x)\cdots\hat{g}_{a_p b_p}(x)]_R = \delta_{a_1}^{a_2}\, [\hat{g}_{a_3 b_3}(x)\cdots\hat{g}_{a_p b_p}(x)]_R\, . \quad (10.52)$$

It is in this sense that $[\hat{g}_{ab}(x)\hat{g}^{bc}(x)]_R = \delta_a^c \mathbb{1}$.

Diagonal coherent-state matrix elements are of particular interest. As an example, consider

$$\langle\pi,g|\hat{g}_{ab}(x)|\pi,g\rangle = \langle 0,g|\hat{g}_{ab}(x)|0,g\rangle$$
$$= M(x)_a^c\, \langle k_{cd}\rangle\, M_b^d(x) \equiv g_{ab}(x)\, , \quad (10.53)$$

which determines the meaning of the label function $g_{ab}(x)$, and where we have set

$$\langle\langle\,(\,\cdot\,)\,\rangle\rangle \equiv K \int_+ (\,\cdot\,)\, e^{-2\mathrm{Tr}(k)}\, dk/(\det[k])\;. \qquad (10.54)$$

Likewise,

$$
\begin{aligned}
\langle\pi, g|\,&[\hat{g}_{ab}(x)\,\hat{g}_{ef}(x)]_R\,|\pi, g\rangle \\
&= \langle 0, g|\,[\hat{g}_{ab}(x)\,\hat{g}_{ef}(x)]_R\,|0, g\rangle \\
&= M(x)_a^c\, M_e^g(x)\, \langle k_{cd}\, k_{gh}\rangle\, M_b^d(x)\, M_f^h(x) \\
&= g_{ab}(x)\, g_{ef}(x) \\
&\quad + M(x)_a^c\, M_e^g(x)\, \langle [k_{cd}\, k_{gh} - \langle k_{cd}\rangle\langle k_{gh}\rangle]\rangle\, M_b^d(x)\, M_f^h(x)\;.
\end{aligned}
\qquad (10.55)
$$

The latter expression brings up a question dealing with the classical limit in which $\hbar \to 0$. So far we have kept $\hbar = 1$, and its true role in the story is not evident. The important factor in (10.55) given by

$$\langle [k_{cd}\, k_{gh} - \langle k_{cd}\rangle\langle k_{gh}\rangle]\rangle \equiv \langle k_{cd}\, k_{gh}\rangle^T \qquad (10.56)$$

should be $O(\hbar)$ so that in the classical limit we would find that

$$\lim_{\hbar\to 0} \langle\pi, g|\,[\hat{g}_{ab}(x)\,\hat{g}_{ef}(x)]_R\,|\pi, g\rangle = g_{ab}(x)\, g_{ef}(x)\;. \qquad (10.57)$$

This dependence can be arranged as follows. Restoring $\hbar$ for the moment, we propose that (10.24), the elementary building block of an affine coherent state at a single site in a lattice regularization, is given instead by

$$
\begin{aligned}
\langle k|\Pi, \Gamma\rangle = C\,(\det[S])\,(\det[SkS^T])^{(\tilde{\beta}/\hbar)-1/2} \\
e^{(i/\hbar)\,\mathrm{Tr}(\Pi k)\,-\,(\tilde{\beta}/\hbar)\,\mathrm{Tr}(\tilde{G}^{-1}SkS^T)}\;,
\end{aligned}
\qquad (10.58)
$$

where $\tilde{\beta} \equiv \beta\hbar$ but what is different now is that we treat $\tilde{\beta}$ and $\hbar$ as two *independent* variables and when $\hbar \to 0$, then $\tilde{\beta}$ remains fixed. This has the desired effect of making the truncated expression $\langle k_{cd}\, k_{gh}\rangle^T \propto \hbar \to 0$, as $\hbar \to 0$, leading to (10.57).

A similar story holds for the momentric operator.

10.9.2 *Advantages of the operator product expansion*

The implications of the operator product analysis given above are quite significant in defining the Hamiltonian constraint operator. Ignoring its constraint nature, the classical expression is given (at $t = 0$) by

$$H(x) = g(x)^{-1/2}\,[\pi_b^a(x)\,\pi_a^b(x) - \tfrac{1}{2}\,\pi_a^a(x)\,\pi_b^b(x)] + g(x)^{1/2}\, R(x)\;. \quad (10.59)$$

Quantization of this expression may be interpreted as

$$\mathcal{H}(x) = [\hat{\pi}_b^a(x)\hat{g}(x)^{-1/2}\hat{\pi}_a^b(x)]_R - \tfrac{1}{2}[\hat{\pi}_a^a(x)\hat{g}(x)^{-1/2}\hat{\pi}_b^b(x)]_R$$
$$+[\hat{g}(x)^{1/2}\hat{R}(x)]_R , \qquad (10.60)$$

where each of the terms is a local operator such that when integrated over a finite spatial region each term, as well as the whole expression, leads to a self-adjoint operator. This simple conceptual analysis arises because local operator products are derived from operator product expansions.

Of course, it is not as easy as it might seem to be. The ultralocal representation that applied for the local operators before the constraints are introduced is not suitable when dealing with the full constraints. We need a new representation that allows spatial derivatives. It is quite possible that modifications of the ultralocal representation such as were introduced in Chap. 9 for covariant scalar fields may well be suitable in the case of gravity as well.

10.10 Properties and Virtues of Reproducing Kernel Hilbert Spaces

We have focussed on coherent states and properties of the coherent-state overlap function for a very good reason: such states provide an unusually convenient bridge between the classical and quantum theories. Traditionally, coherent states are associated with a local resolution of unity, but the affine coherent states that are of interest to us do not possess a local resolution of unity; instead, they constitute what we have called a family of *weak coherent states* [KlS85; Kl01d]. To generate a representation of the kinematic Hilbert space of interest without invoking a local integral resolution of unity, we need to appeal to another formulation. The manner in which this is done is quite general and it is convenient to discuss this procedure initially in general terms not limited to the affine coherent states that are the principal focus of our study.

Consider an L-dimensional label space $\mathcal{L}$, composed of points $l \in \mathcal{L}$, that form a topological space; in other words, the space $\mathcal{L}$ admits a notion of continuity, $l' \to l$, in the sense of $\mathcal{L}$. When $L < \infty$, it generally suffices to assume that the label $l = (l_1, l_2, \ldots, l_L)$, with $l_j \in \mathbb{R}$ for all j, and that $\mathcal{L}$ is locally isomorphic to an L-dimensional Euclidean space $\mathbb{R}^L$ so that convergence is simply convergence of each coordinate. Infinite-dimensional label spaces are normally natural generalizations of finite ones.

Next, consider a function $\mathcal{K}(l''; l')$ that is *jointly continuous in both labels*, and which satisfies the condition

$$\Sigma_{i,j=1}^{N,N} \alpha_i^* \alpha_j \mathcal{K}(l_i; l_j) \geq 0 \tag{10.61}$$

for all complex coefficients $\{\alpha_j\}$ and all $N < \infty$. *Such a function is said to be a continuous function of positive type*, and it will be the foundation of a very useful representation space [Aro43]. According to the GNS (Gel'fand, Naimark, Segal) Theorem [Emc72], there is always an abstract, separable Hilbert space with vectors $|l\rangle$ labeled by points $l \in \mathcal{L}$ such that $\mathcal{K}(l''; l') \equiv \langle l''|l'\rangle$. The function $\mathcal{K}(l''; l')$ is called a *reproducing kernel* (for reasons to be made clear below). As the notation suggests, the affine coherent-state overlap function is a continuous function of positive type and therefore qualifies as a reproducing kernel.

To generate an appropriate Hilbert space representation, we envisage two abstract vectors as members of a dense set of vectors given by

$$|\psi\rangle \equiv \sum_{i=1}^{I} \alpha_i |l_{(i)}\rangle \,,$$
$$|\phi\rangle \equiv \sum_{j=1}^{J} \beta_j |l_{[j]}\rangle \,, \tag{10.62}$$

where I and J are both finite. To give a functional representation of these abstract vectors, we appeal to the only set of vectors that we know span the Hilbert space, and that is the set $\{|l\rangle\}$. Thus we introduce

$$\psi(l) \equiv \sum_{i=1}^{I} \alpha_i \langle l|l_{(i)}\rangle = \langle l|\psi\rangle \,,$$
$$\phi(l) \equiv \sum_{j=1}^{J} \beta_j \langle l|l_{[j]}\rangle = \langle l|\phi\rangle \,, \tag{10.63}$$

which yield continuous functional representatives. To define the inner product of two such vectors we appeal to (10.62) and let

$$(\psi, \phi) \equiv \sum_{i,j=1}^{I,J} \alpha_i^* \beta_j \langle l_{(i)}|l_{[j]}\rangle = \langle \psi|\phi\rangle \,. \tag{10.64}$$

The name "reproducing kernel" arises because if $\alpha_i = \delta_{1,i}$, then

$$(\langle \cdot|l'\rangle, \phi) = \sum_{j=1}^{J} \beta_j \langle l'|l_{[j]}\rangle = \phi(l') \,; \tag{10.65}$$

in short, the element $\phi(l)$ is reproduced by this inner product.

Note well that the Hilbert space $L^2(\mathbb{R})$ is *not* a reproducing kernel Hilbert space since the proposed reproducing kernel, $\delta(x - y)$, is neither continuous nor an element of the space.

All that remains to generate a functional Hilbert space is to complete the space by including the limits of all Cauchy sequences. The result is a Hilbert space representation $\mathcal{C}$ composed entirely of continuous functions on $\mathcal{L}$; note there are no sets of measure zero in this approach: every vector is represented by a unique, continuous function. Observe that: *Every aspect of the Hilbert space C is determined by the reproducing kernel!*

Traditional coherent states also generate a reproducing kernel Hilbert space which has the property that it has two independent rules for evaluating inner products: the first by a local integral with a non-negative, absolutely continuous measure, and the second in the manner just described for a reproducing kernel Hilbert space. However, not all reproducing kernel Hilbert spaces involve local integral inner products, and the affine coherent-state overlap function is of that type leading to a continuous function of positive type but not one with a local integral resolution of unity. As already noted, when that happens, we say that we are dealing with weak coherent states. This is already true even for just a single site whenever $0 < \beta \leq 1/2$; to see this, try integrating the absolute square of (10.26) using the left-invariant group measure; see, e.g., [Kl10a].

10.10.1 *Reduction of a reproducing kernel*

It sometimes happens that the space spanned by the set of states $\{|l\rangle\}$ is the same as spanned by a suitable subset of that set, as for example, the set given by $\{|l^*\rangle\}$, where $|l^*\rangle \equiv \int |l\rangle \, \sigma(l) \, dl$, and $\sigma(l) \, dl$ denotes a possibly complex, fixed measure. Stated otherwise, if we denote the linear span of vectors by an overbar, then $\overline{|l^*\rangle} \equiv \overline{\int |l\rangle \, \sigma(l) \, dl} = \overline{|l\rangle}$. A common situation arises when the measure $\sigma \, dl$ is a delta measure fixing one or more of the coordinates. We next discuss a few concrete examples.

To make this discussion more relevant to our general discussion, let us focus on the reproducing kernel given by the affine coherent-state overlap function $\langle \pi'', g'' | \pi', g' \rangle$ as defined in (10.31). Two forms of reduction of this reproducing kernel are worth discussing.

In the first form, we set $\gamma''^a_b = \gamma'^a_b = 0$, which implies that $g''_{ab}(x) = g'_{ab}(x) = \tilde{g}_{ab}(x)$. Thus the first form of reduced reproducing kernel is given by

$$\langle \pi'' | \pi' \rangle \equiv \langle \pi'', \tilde{g} | \pi', \tilde{g} \rangle \qquad (10.66)$$
$$= \exp(-2 \int b(x) \, d^3x \{ \ln \det(\delta^a_b + \tfrac{1}{2} i b(x)^{-1} [\tilde{\pi}''^a_b(x) - \tilde{\pi}'^a_b(x)]) \}) \, .$$

Note carefully: The notation $|\pi\rangle$ as used here does *not* mean a sharp eigenvector for the putative local operator $\hat{\pi}^{ab}(x)$. In the second form, we choose $\pi''^{ab}(x) = \pi'^{ab}(x) = 0$, which leads to the reduced reproducing kernel given by

$$\langle g''|g'\rangle \equiv \langle 0, g''|0, g'\rangle$$
$$= \exp(-2\int b(x)\, d^3x \, \ln\{[\det(\tfrac{1}{2}\tilde{g}_b''^a(x) + \tfrac{1}{2}\tilde{g}_b'^a(x))] \times$$
$$/[\det(\tilde{g}_b''^a(x))\, \det(\tilde{g}_b'^a(x))]^{1/2}\})\,. \qquad (10.67)$$

As before, the notation $|g\rangle$ does *not* refer to an eigenvector of the local operator $\hat{g}_{ab}(x)$.

Both of these expressions serve as reproducing kernels for the same abstract Hilbert space that arose as the abstract Hilbert space associated with the original reproducing kernel. The fact that these reductions still span the original space follows from an examination of the expression

$$\langle k|\Pi, \Gamma\rangle = C\,(\det[S])\,(\det[SkS^T])^{\beta-1/2}\, e^{i\,\mathrm{Tr}(\Pi k)\,-\,\beta\,\mathrm{Tr}(\tilde{G}^{-1}SkS^T)}\,,$$
$$(10.68)$$

which is the elementary building block of an affine coherent state at a single site in a lattice regularization; see (10.24). If the inner product with a general element $\psi(k) = \langle k|\psi\rangle$, given by

$$\langle \psi|\Pi, \Gamma\rangle \equiv C\int_+ \psi(k)^*\, (\det[S])\,(\det[SkS^T])^{\beta-1/2}$$
$$\times e^{i\,\mathrm{Tr}(\Pi k)\,-\,\beta\,\mathrm{Tr}(\tilde{G}^{-1}SkS^T)}\, dk\,, \quad (10.69)$$

vanishes for all matrices Π and $S \equiv \exp(-\Gamma/2)$, then it follows that $\psi(k) = 0$, almost everywhere, thereby determining the span of the affine coherent states as $L_+^2(\mathbb{R}^6, dk)$. Note well that the same result holds if we set either $\Gamma = 0$ or $\Pi = 0$, which establishes that the span of each of the reduced reproducing kernels discussed above leads to the same span as the original reproducing kernel.

10.11   Enforcing the Diffeomorphism Constraints

Let us focus on enforcing the diffeomorphism constraints for gravity given by $\mathcal{H}_a(x) = -2\hat{\pi}^b_{a|b}(x) = 0$ for all a and x, where the notation $_|$ signifies a covariant derivative based on the metric $\hat{g}_{ab}(x)$. Neglecting temporarily the Hamiltonian constraint for gravity, we observe that enforcing the diffeomorphism constraint operators requires that we find states $|\psi\rangle_{phys}$ such that $\mathcal{H}_a(x)|\psi\rangle_{phys} = 0$. Strictly speaking, however, those solutions may not lead to the full solution space since some nonzero eigenvectors may not be normalizable and thus they technically do not lie in the kinematical Hilbert space $\mathfrak{H}$. Unlike some other procedures, the projection-operator method of dealing with quantum constraints [Kla97] can deal with constraint operators that may have their zero in the continuum. Although we have briefly reviewed the projection-operator method above, it suffices, for present purposes, to let $\mathbb{E}$ denote a genuine projection operator ($\mathbb{E}^{\dagger} = \mathbb{E}^2 = \mathbb{E}$) that generates the regularized physical Hilbert space $\mathfrak{H}_{phys} \equiv \mathbb{E}\mathfrak{H}$; here the regularization may be symbolized by a parameter $\delta = \delta(\hbar) > 0$ that enforces the constraints to lie within a (suitably defined) spectral window between $\pm \delta$. This projection operator is characterized by its affine coherent-state matrix elements $\langle \pi'', g' | \mathbb{E} | \pi', g' \rangle$, which is seen to be a continuous function of positive type. If we take the limit $\delta \to 0$ in order to enforce the constraints exactly, the coherent-state matrix elements of $\mathbb{E}$ would possibly vanish. Although different elements may vanish at different rates, the simplest examples may be recovered by a suitable rescaling. To extract the 'germ' carried by such matrix elements, we rescale those matrix elements, e.g., by dividing by $\langle \eta | \mathbb{E} | \eta \rangle$ before taking the limit, which leads us to a nonzero reproducing kernel for a functional Hilbert space in which the diffeomorphism constraints are fully satisfied. In symbols, this construction is given by

$$\langle\!\langle \pi'', g'' | \pi', g' \rangle\!\rangle \equiv \lim_{\delta \to 0} \frac{\langle \pi'', g'' | \mathbb{E} | \pi', g' \rangle}{\langle \eta | \mathbb{E} | \eta \rangle} , \qquad (10.70)$$

yielding a continuous function of positive type, $\langle\!\langle \pi'', g'' | \pi', g' \rangle\!\rangle$, that serves as the reproducing kernel for the Hilbert space in which the diffeomorphism constraints are fully satisfied.

At the present time, we are unable to completely carry out the program outlined above to explicitly determine the entire Hilbert space on which the diffeomorphism constraints are satisfied. However, we can do the next best thing: we can develop functional realizations of numerous subspaces of the diffeomorphism constrained Hilbert space. Let us outline how we intend to generate these subspaces.

10.11.1 *Diffeomorphism invariant reproducing kernels*

For convenience, we temporarily abbreviate $\langle \pi'', g'' | \mathbb{E} | \pi', g' \rangle$ as $\langle '' | \mathbb{E} | ' \rangle$ letting the bra and ket refer to general affine coherent states. As a projection operator, $\mathbb{E}$ admits an expansion—perhaps only partially—as $\Sigma_i |i\rangle\langle i|$, where $\{|i\rangle\}$ denotes a suitable set of orthonormal vectors. Thus, the associated reproducing kernel takes the form $\Sigma_i \langle '' | i \rangle \langle i | ' \rangle$. As a reproducing kernel, the same space of continuous functions arises if we rescale the given vectors $|i\rangle$ leading to a new reproducing kernel $\Sigma_i r_i \langle''|i\rangle\langle i|'\rangle$, where $\{r_i\}$ is a set of positive real numbers; this change amounts to a similarity transformation of the previous Hilbert space representation. Moreover, to create a general reproducing kernel, the states $|i\rangle$ need not be mutually orthogonal, but rather just linearly independent. Thus we are led to consider suitable reproducing kernels that are constructed as

$$\langle\langle''|'\rangle\rangle \equiv \Sigma_i c''^* \langle''|i\rangle\langle i|'\rangle c' , \tag{10.71}$$

where we have generalized the notation so that the states $\{|i\rangle\}$ need only be linearly independent of one another and of arbitrary (nonzero, but finite) norm, as well as adding additional, arbitrary, nonzero coefficients c''^* and c', which can be used—as we do in the following examples—to cancel the normalizing factors in the affine coherent states.

The next question we consider is how are we to chooses vectors $\{|i\rangle\}$ that exhibit diffeomorphism invariance. We first note that the expression (10.31) for the original coherent state overlap $\langle\pi'', g''|\pi', g'\rangle$ is *invariant*, including transforming $b(x)$, under a common coordinate transformation for *both* pairs of labels, i.e., (π'', g'') *and* (π', g'). On the other hand, the idealized diffeomorphism invariant reproducing kernel $\langle\langle\pi'', g''|\pi', g'\rangle\rangle$ should be invariant under *separate* coordinate transformations of each set of the $''$ and $'$ variables; this property embodies the traditional concept that only the "geometry of space" carries the proper physics. How do we capture this aspect? We do so by carefully choosing suitable vectors $\{|i\rangle\}$ that serve the required purpose. Ignoring the coherent-state normalization factor, one conceivable example might be $\langle\pi, g|1\rangle = \int \sqrt{g(x)}\, d^3x$, where as customary we let $g(x) = \det(g_{ab}(x))$. However, this example will not work for the following reason. The affine coherent states $\{|\pi, g\rangle\}$ are, apart from the normalization factor, functions of the *complex* symmetric tensor $g^{(-)ab}(x) \equiv g^{ab}(x) - ib(x)^{-1}\pi^{ab}(x)$. Likewise, the adjoint affine coherent states $\{\langle\pi, g|\}$, again apart from the normalization factor, are functions of the complex symmetric tensor $g^{(+)ab}(x) \equiv g^{ab}(x) + ib(x)^{-1}\pi^{ab}(x)$. Thus, again ignoring the normalization factor in $\langle\pi, g|$, as we shall continue to do, the diffeomorphism invariant state should rather be taken as

$$\langle\pi, g|1\rangle = \int \sqrt{g^{(+)}(x)}\, d^3x . \tag{10.72}$$

[Remark: So long as $\det(g) > 0$ it follows that $\det(g^{(\pm)}) \neq 0$, and therefore $g^{(\pm)}_{ab}(x)$ exists such that $g^{(\pm)}_{ab}(x) g^{(\pm)bc}(x) = \delta^c_a$.] However, (10.72) is not a good state to use for noncompact spatial surfaces. Provided there is suitable asymptotic behavior, we suggest other examples worth considering, such as

$$\langle \pi, g | 2 \rangle = \int R^{(+)}(x) \sqrt{g^{(+)}(x)} \, d^3x \,,$$

$$\langle \pi, g | 3 \rangle = \int R^{(+)\,2}(x) \sqrt{g^{(+)}(x)} \, d^3x \,,$$

$$\langle \pi, g | 4 \rangle = \int R_{ab}^{(+)}(x) \, R^{(+)\,ab}(x) \sqrt{g^{(+)}(x)} \, d^3x \,, \tag{10.73}$$

$$\langle \pi, g | 5 \rangle = \int R_{|ab}^{(+)}(x) \, g^{(+)\,ab}(x) \sqrt{g^{(+)}(x)} \, d^3x \,,$$

$$\langle \pi, g | 6 \rangle = \int R^{(+)\,abcd}(x) \, R_{|abcd}^{(+)}(x) \sqrt{g^{(+)}(x)} \, d^3x \,,$$

etc., where $R^{(+)}(x)$ is the (three-dimensional) scalar curvature constructed from the metric $g_{ab}^{(+)}(x)$, and so on for the Ricci and Riemann tensors and their spatially covariant derivatives denoted by $_|$. Clearly, each of these expressions is invariant under arbitrary coordinate transformations, and they would also admit extension to a noncompact space $\mathcal{S}$ provided the indicated integrals converged, which, for example, would be the case if $\tilde{g}_{ab}(x) = \delta_{ab}$.

The five-dimensional Hilbert space that is generated by the reproducing kernel

$$\langle\!\langle \pi'', g'' | \pi', g' \rangle\!\rangle \equiv \sum_{i=2}^{6} \langle \pi'', g'' | i \rangle \langle i | \pi', g' \rangle \tag{10.74}$$

enjoys complete and independent invariance of both label sets under arbitrary coordinate transformations; it also provides a "toy" example of what is possible to generate by these techniques.

As a further example that incorporates an infinite-dimensional Hilbert space, we offer the reproducing kernel

$$\langle\!\langle \pi'', g'' | \pi', g' \rangle\!\rangle \equiv \exp \Bigg\{ - \int R''^{(+)\,2}(x) \sqrt{g''^{\,(+)}(x)} \, d^3x$$

$$+ \int R''^{(+)}(x) \sqrt{g''^{\,(+)}(x)} \, d^3x \cdot \int R'^{(-)}(y) \sqrt{g'^{(-)}(y)} \, d^3y$$

$$- \int R'^{(-)\,2}(y) \sqrt{g'^{(-)}(y)} \, d^3y \Bigg\} \,. \tag{10.75}$$

Both integrals should involve the same $b(x)$ and be integrated over the same space-like surface. Additional examples appear later.

10.11.2 *Reduction of diffeomorphism invariant reproducing kernels*

Just as we discussed the reduction of reproducing kernels for the original affine coherent state reproducing kernel, we can consider a similar reduction of those reproducing kernels that are designed to satisfy diffeomorphism invariance. As a particularly interesting example, we set $\pi''^{ab}(x) = \pi'^a(x) = 0$ leading, e.g., to $\langle g|1\rangle \equiv \langle 0, g|1\rangle = \int \sqrt{g(x)}\, d^3x$, which involves a real metric. As a further example, we note that

$$\langle g|2\rangle = \int R(x)\sqrt{g(x)}\, d^3x\,, \qquad (10.76)$$

which again involves traditional geometric elements. This same procedure can lead, for example, to

$$\langle\!\langle g''|g'\rangle\!\rangle \equiv \sum_{i=2}^{6} \langle 0, g''|i\rangle\langle i|0, g'\rangle \qquad (10.77)$$

These reproducing kernels do not involve the scalar density $b(x)$ and may therefore be considered to be preferable. On the other hand, analyticity permits an immediate extension of these particular reductions to restore the missing term $\pm i b(x)^{-1}\pi^{ab}(x)$ ensuring that the reduced reproducing kernel spans the same space as does that derived from the original reproducing kernel, $\langle\!\langle \pi'', g''|\pi', g'\rangle\!\rangle$. Of course, other examples of this sort are easy to generate.

10.12   Functional Integral Formulation

The Hamiltonian constraint is related to the embedding of spatial slices nearby one another in the putative time direction, and thus its inclusion within a single spatial slice is not evident. Recall that the Hamiltonian is wholly specified at a single moment of time, but, along with other elements, it can be extended into the time domain by the introduction of a path integral. A similar procedure is available to us.

The analyticity of the arguments of the affine coherent states up to normalization factors, which just was made use of in proposing some diffeomorphism invariant examples of Hilbert spaces, can be put to good use in a rather different manner to build a path-integral representation of the affine coherent-state overlap itself [Kl01d; HaK04]. The success of this procedure stems from the fact that every affine coherent-state representative, $\psi(\pi, g) \equiv \langle \pi, g|\psi\rangle$, satisfies a *complex polarization condition*, namely

$$C_s^r(x)\,\psi(\pi, g) \equiv \left[-ig^{rt}(x)\frac{\delta}{\delta\pi^{ts}(x)} + \delta_s^r \right.$$
$$\left. +b(x)^{-1}g_{st}(x)\frac{\delta}{\delta g_{tr}(x)} \right]\psi(\pi, g) = 0 \qquad (10.78)$$

for all spatial points x and any function $\psi(\pi, g) \in \mathcal{C}$, which is the associated reproducing kernel Hilbert space; we saw a simpler example of this story in

Chap. 3. Equation (10.78) is a first-order functional differential equation because we have chosen the fiducial vector as an extremal weight vector which is determined as the solution of a linear equation in Lie algebra generators. Multiplication by the adjoint of that first-order differential operator, followed by summation over indices and integrated over space, leads to a nonnegative, second-order functional differential operator given by

$$A \equiv \int C_r^s(x)^\dagger \, C_s^r(x) \, b(x) \, d^3x \,, \tag{10.79}$$

with the property that $A \geq 0$, and which annihilates every affine coherent-state function $\psi(\pi, g)$. Thus, with $T > 0$ and as $\nu \to \infty$, it follows that $e^{-\nu TA/2}$ formally becomes a projection operator onto the space $\mathcal{C}$. It is clear that the second-order functional differential operator A is an analog of a Laplacian operator in the presence of a magnetic field, and thus a Feynman-Kac-Stratonovich path (i.e., functional) integral representation may be introduced. In particular, we are led to the formal expression

$$\langle \pi'', g'' | \pi', g' \rangle = \lim_{\nu \to \infty} \mathcal{N} \int \exp[-i \int g_{ab} \dot{\pi}^{ab} \, d^3x \, dt]$$
$$\times \exp\{-(1/2\nu) \int [b(x)^{-1} g_{ab} g_{cd} \dot{\pi}^{bc} \dot{\pi}^{da} + b(x) g^{ab} g^{cd} \dot{g}_{bc} \dot{g}_{da}] \, d^3x \, dt\}$$
$$\times \prod_{x,t} \prod_{a \leq b} d\pi^{ab}(x,t) \, dg_{ab}(x,t) \,. \tag{10.80}$$

It this expression, it is natural to interpret t, $0 \leq t \leq T$, as coordinate "time", and thus on the right-hand side the canonical fields are functions of space and time, that is $g_{ab} = g_{ab}(x,t)$ and $\pi^{ab} = \pi^{ab}(x,t)$, where the overdot ($\dot{}$) denotes a partial derivative with respect to t, and the integration is subject to the boundary conditions that $(\pi(x,0), g(x,0)) = (\pi'(x), g'(x))$ and $(\pi(x,T), g(x,T)) = (\pi''(x), g''(x))$. It is important to note, for any $\nu < \infty$, that underlying the formal expression (10.80) given above, there is a genuine, countably additive measure on (generalized) functions g_{kl} and π^{rs}. Loosely speaking, such functions have Wiener-like behavior with respect to time and δ-correlated, generalized Poisson-like behavior with respect to space.

It is important to understand that although the functional integral (10.80) is over the canonical momentum $\pi^{ab}(x,t)$ and the canonical metric $g_{ab}(x,t)$, this integral has arisen strictly from an affine quantization and *not* from a canonical quantization. It is also noteworthy that the very structure of the ν-dependent regularization factor forces the metric to satisfy metric positivity, i.e., $u^a g_{ab}(x,t) u^b > 0$, for all x and t, provided that $\Sigma_a(u^a)^2 > 0$.

Phase-space path integrals with Wiener-measure regularization were introduced in [DaK85] and are discussed in [Kl10a] as well as in Chap. 3 of this monograph. The regularization involves a phase-space metric, and on examining (10.80) it is clear that the phase-space metric, as given, is almost uniquely determined.

10.13   Imposition of All Constraints

Gravity has four constraints at every point $x \in \mathcal{S}$, and, when expressed in suitable units, they are the familiar spatial (diffeomorphism) and temporal (Hamiltonian) constraints, all densities of weight one, given, at $t = 0$, by [ADM62; MTW71]

$$H_a(x) = -2\pi^b_{a\,|b}(x) \, , \tag{10.81}$$

$$H(x) = g(x)^{-1/2}[\pi^a_b(x)\pi^b_a(x) - \tfrac{1}{2}\pi^c_c(x)\pi^d_d(x)] + g(x)^{1/2}\,R(x) \, ,$$

where, as has been our custom, $R(x)$ is the three-dimensional scalar curvature. While spatial constraints are comparatively easy to incorporate as we have seen before, this is not the case for the temporal constraint. A detailed account of how all the constraints can be accommodated in constructing a regularized projection operator $\mathbb{E}^*$ for them has already been given in [Kl01a] and will be briefly reviewed later. The notation $\mathbb{E}$ and (with a new notation) $\mathbb{E}^*$ is meant here to distinguish the strictly diffeomorphism constraint projection operator $\mathbb{E}$ from the all-constraint projection operator $\mathbb{E}^*$, respectively.

Briefly stated, the introduction of the all-constraint projection operator begins with the continuous-time regularized functional integral representation of the affine coherent-state reproducing kernel $\langle \pi'', g'' | \pi', g' \rangle$ for the kinematical Hilbert space. The reproducing kernel for the regularized physical Hilbert space is given, in turn, by the expression $\langle \pi'', g'' | \mathbb{E}^* | \pi', g' \rangle$, where $\mathbb{E}^*$ refers to a projection operator that includes all constraint operators. In order to give this latter expression a functional integral representation we initially assume that

$$\int [N^a \mathcal{H}_a + N \mathcal{H}]\,d^3x \tag{10.82}$$

is a time-dependent "Hamiltonian" for some fictitious theory, in which case

$$\langle \pi'', g'' | \mathbb{T}\, e^{-i\int [N^a \mathcal{H}_a + N \mathcal{H}]\,d^3x\,dt} | \pi', g' \rangle$$

$$= \lim_{\nu \to \infty} \overline{\mathcal{N}}_\nu \int \exp\{-i\int [g_{ab}\dot{\pi}^{ab} + N^a\,H_a + N\,H]\,d^3x\,dt\}$$

$$\times \exp\{-(1/2\nu)\int [b(x)^{-1}g_{ab}g_{cd}\dot{\pi}^{bc}\dot{\pi}^{da} + b(x)g^{ab}g^{cd}\dot{g}_{bc}\dot{g}_{da}]\,d^3x\,dt\}$$

$$\times \Pi_{x,t}\,\Pi_{a\leq b}\,d\pi^{ab}(x,t)\,dg_{ab}(x,t) \, . \tag{10.83}$$

In this expression, there appear symbols $H_a(\pi, g)$ and $H(\pi, g)$ corresponding to the quantum operators $\mathcal{H}_a$ and $\mathcal{H}$. We do not discuss the details of these symbols, which are unlikely to be, simply, the classical constraint expressions due to the fact that $\hbar > 0$ within this functional integral; indeed, these symbols should follow the pattern of much simpler expressions studied in [Kl10a].

To pass from the intermediate stage of our fictitious theory to the final formulation involving the regularized projection operator, we introduce an integration over the variables N^a and N using a carefully constructed measure $\mathcal{D}R(N^a, N)$, which is formally defined in [Kl01a]. This leads to the expression

$$\langle \pi'', g'' | \mathbb{E}^* | \pi', g' \rangle$$
$$= \lim_{\nu \to \infty} \overline{\mathcal{N}}_\nu \int \exp\{-i \int [g_{ab} \dot{\pi}^{ab} + N^a H_a + N H] \, d^3x \, dt\}$$
$$\times \exp\{-(1/2\nu) \int [b(x)^{-1} g_{ab} g_{cd} \dot{\pi}^{bc} \dot{\pi}^{da} + b(x) g^{ab} g^{cd} \dot{g}_{bc} \dot{g}_{da}] \, d^3x \, dt\}$$
$$\times [\Pi_{x,t} \Pi_{a \le b} \, d\pi^{ab}(x, t) \, dg_{ab}(x, t)] \, \mathcal{D}R(N^a, N) . \qquad (10.84)$$

The result of this functional integral, $\langle \pi'', g'' | \mathbb{E}^* | \pi', g' \rangle$, is a continuous function of positive type that can be used as a reproducing kernel for a Hilbert space in which the full set of gravitational constraints are satisfied in a regularized manner. [Remark: Incidentally, omitting the integral over N and setting $H = 0$ altogether would lead to the result $\langle \pi'', g'' | \mathbb{E} | \pi', g' \rangle$, which, as noted previously, is a continuous function of positive type that would serve as a reproducing kernel for the Hilbert space in which only the diffeomorphism constraints are satisfied in a regularized manner.]

Unfortunately, the evaluation of such functional integrals is beyond present capabilities. However, just as we were able to introduce subspaces that satisfy all the diffeomorphism constraints alone, there should exist similar examples of subspaces where all of the constraints are satisfied, and which should have a form not unlike that for the diffeomorphism constraint situation. In fact, in the next subsection we argue that the examples already developed for the diffeomorphism constraints may also work for the case of all the quantum gravity constraints.

10.13.1 *A proposal for reproducing kernels satisfying all constraints*

The affine coherent-state matrix elements of the two projection operators, $\mathbb{E}$ and $\mathbb{E}^*$, have much in common. Apart from affine coherent-state normalization factors, it follows that both $\langle \pi'', g'' | \mathbb{E} | \pi', g' \rangle$ and $\langle \pi'', g'' | \mathbb{E}^* | \pi', g' \rangle$ are functions of $g^{(\pm)\,ab}(x)$. Both functions have to depend on strictly geometric combinations such as $R^{(\pm)}(x)$, etc., so as to be invariant under any coordinate transformation. Moreover, the original coherent states were defined on a space-like surface in the topological carrier space $\mathcal{S}$, but no particular space-like surface has been specified. In other words, the coherent states and their matrix elements of various operators of interest such as $\mathbb{E}$ and $\mathbb{E}^*$, are invariant under arbitrary changes of the space-like surface on which the coherent states have been defined.

It is noteworthy that fulfillment of the diffeomorphism constraints on all space-like surfaces nearly implies that the Hamiltonian constraint is fulfilled as well, according to the following argument kindly supplied by Karel Kuchař. Let $G_{\mu\nu}(x) = R_{\mu\nu}(x) - \frac{1}{2}g_{\mu\nu}(x)R(x)$—where $x \in \mathbb{R}^4$, $0 \leq \mu, \nu \leq 3$, represent the spacetime indices, and $R(x)$ is the four-dimensional scalar curvature—denote the usual Einstein tensor, and the general fulfillment of the diffeomorphism constraints is given by the relation $n_\perp^\mu(x)\sigma_\perp^\nu(x)G_{\mu\nu}(x) = 0$ for a general time-like vector $n_\perp^\mu(x)$ and a general space-like vector $\sigma_\perp^\nu(x)$, which, however, are required to fulfill $n_\perp^\mu(x)\sigma_\perp^\nu(x)g_{\mu\nu}(x) \equiv 0$ (hence the notation with $\perp$). The perpendicularity relation can be accommodated by a Lagrange multiplier field, leading to the relation $n^\mu(x)\sigma^\nu(x)[G_{\mu\nu}(x) + \lambda(x)g_{\mu\nu}(x)] = 0$, now with no restriction that $n^\mu(x)$ and $\sigma^\nu(x)$ are explicitly perpendicular. Since the difference of two time-like vectors may be space like, and the difference of two space-like vectors may be time like, we conclude that $G_{\mu\nu}(x) + \lambda(x)g_{\mu\nu}(x) = 0$. However, a contracted Bianchi identity implies that $G^\mu_{\nu\,;\mu}(x) = 0 = -[\lambda(x)\delta^\mu_\nu]_{;\mu}$, or finally that $\lambda(x) = \Lambda$, a constant; indeed, if the space is asymptotically flat, then the fact that $\lim_{|x|\to\infty} G_{\mu\nu}(x) = 0$ implies that $\Lambda = 0$. In summary, then, the fulfillment of the diffeomorphism constraints on general space-like surfaces implies that the Hamiltonian constraint is also satisfied, up to a generally unspecified cosmological constant. Consequently, we are led to propose that the various diffeomorphism invariant reproducing kernel examples given previously are valid as well for the full set of quantum constraints.

To give one more example, we offer another infinite-dimensional reproducing kernel given by

$$\langle\!\langle \pi'', g'' | \pi', g' \rangle\!\rangle = \iint \{\exp[R''^{(+)}(x)R'^{(-)}(y)] - 1\}$$
$$\times \sqrt{g''^{(+)}(x)g'^{(-)}(y)}\, d^3x\, d^3y, \qquad (10.85)$$

with, however, a common choice of $b(x)$. Observe, that in this expression the two integrals should be over the same space-like surface, but that surface is an arbitrary space-like surface. This reproducing kernel, as well as the other examples given earlier, may be considered as "toy" examples each one being given, effectively, by its own projection operator $\mathbb{E}_T$ (T for toy), which, in a picturesque sense, satisfies, $\mathbb{E}_T \subset \mathbb{E}$. In the same sense, $\mathbb{E}^* \subset \mathbb{E}$, and it is not out of the question that suitable toy models may be related as $\mathbb{E}_T \subset \mathbb{E}^* \subset \mathbb{E}$, and thus those toy examples would fulfill all of the quantum gravity constraints.

10.14   Comments

The program of affine quantum gravity is very conservative in its approach. It adopts as classical variables the spatial momentric $\pi_b^a(x) \equiv \pi^{ac}(x) g_{bc}(x)$ and the spatial metric $g_{ab}(x)$, and taken together this set of kinematical variables is called the affine variables. [Remark: Note that the traditional momentum may readily be found from the momentric by $\pi^{ab}(x) = \pi_c^a(x) g^{bc}(x)$.] The focus on the affine variables arises from the effort to ensure that the chosen kinematical variables universally respect metric positivity, namely, $u^a g_{ab}(x) u^b > 0$, for all x, provided $\Sigma_a(u^a)^2 > 0$. Upon quantization, as well, the affine variables, which satisfy the affine commutation relations, respect metric positivity, unlike the canonical variables which obey the usual canonical commutation relations. The affine commutation relations are like current commutation relations, and as such they admit representations that are bilinear in conventional creation and annihilation operators, which are quite unlike the representations generated by canonical commutation relations. In particular, the bilinear realizations of the affine variables admit straightforward local products by operator product expansion constructions that do not involve normal ordering of any kind. Consequently, just as the spatial metric operator $\hat{g}_{ab}(x)$ becomes a self-adjoint operator when smeared by a real test function, this same property holds for the local product $[\hat{g}_{ab}(x)\hat{g}_{cd}(x)]_R$, and all other operator-product renormalized local operators. Therefore, it is clear, e.g., that the sum of such expressions integrated over a finite volume generates a self-adjoint operator, which might be part of the Hamiltonian. Stated otherwise, the class of representations consistent with an operator product expansion offers profoundly greater opportunities to construct various local operators of interest that are nonlinear functions of the basic kinematical operators. Going hand-in-hand with the bilinear operator representations, is the realization of such operators as well as their local products within one and the same Hilbert space, meaning that one and only one irreducible representation is involved due to the implicit measure mashing that occurs for such representations. While measure mashing is wholly foreign to operators that obey canonical commutation relations, it is—and at first glance surprisingly so—already built into the representation structure of the affine operators. A theory that involves measure mashing leads to functional integral representations which, under changes of its parameters as part of a perturbation analysis, entail different measures that are *equivalent* (equal support) rather than being (at least partially) *mutually singular* (disjoint support), and for that very reason, do not create term-by-term divergences; to observe measure mashing in action see [Kla11].

An interesting development is the ability to create a vast number of reproducing kernels—and implicitly thereby the Hilbert space representations by continuous functions they generate—that are fully invariant under the action of the diffeomorphism constraints. Moreover, the labels of the various reproducing kernels are the familiar classical variables of the canonical theory, namely, a smooth tensor density of weight one, the spatial momentum $\pi^{ab}(x)$, and a smooth spatial metric tensor $g_{ab}(x)$, restricted as always by metric positivity. One can imagine that these subspaces, which can have infinite dimensionality, may be useful in the study of suitable toy models.

The functional integral representation for: (i) the overlap function of the affine coherent states, (ii) the affine coherent-state matrix elements of the projection operator enforcing just the diffeomorphism constraints, and (iii) the affine coherent-state matrix elements of the projection operator enforcing all of the constraints involves a common language within which these three distinct sets of matrix elements can be considered. In particular, the coherent-state overlap function clearly illustrates the functional dependence of the coherent states on their labels. This functional dependence also holds for the matrix elements of the two separate projection operators of interest, namely, $\mathbb{E}$ and $\mathbb{E}^*$. For the diffeomorphism constraints alone, we were able to make logical choices for the functional dependence of suitable subspaces that remain consistent with the requirements involved in the coherent-state overlap function itself.

It is tempting to believe that there is yet another set of proper arguments that would allow us to set up subspaces of the reproducing kernel in which all of the constraints are fulfilled. We have proposed that the several examples of subspaces which satisfy the diffeomorphism constraints also satisfy all the quantum constraints, but that hypothesis needs to be carefully examined further to see if it carries any truth.

One satisfactory procedure to incorporate all the necessary constraints is as follows. Let $\{h_p(x)\}_{p=1}^{\infty}$ denote a complete, orthonormal set of real functions on $\mathcal{S}$ relative to the weight $b(x)$. In particular, we suppose that

$$\int h_p(x)\, h_n(x)\, b(x)\, d^3x = \delta_{pn} \,, \tag{10.86}$$

$$b(x)\, \Sigma_{p=1}^{\infty}\, h_p(x)\, h_p(y) = \delta(x,y) \,. \tag{10.87}$$

Based on this orthonormal set of functions, we next introduce four infinite sequences of constraints

$$H_{(p)\,a} \equiv \int h_p(x)\, H_a(x)\, d^3x \,, \tag{10.88}$$

$$H_{(p)} \equiv \int h_p(x)\, H(x)\, d^3x \,, \tag{10.89}$$

$1 \le p < \infty$, all of which vanish in the classical theory.

For the quantum theory let us assume, for each p, that $\mathcal{H}_{(p)\,a}$ and $\mathcal{H}_{(p)}$ are self adjoint, and even stronger that

$$X_P^2 \equiv \Sigma_{p=1}^{P}\, 2^{-p}\, [\Sigma_{a=1}^{3}(\mathcal{H}_{(p)\,a})^2 + (\mathcal{H}_{(p)})^2] \tag{10.90}$$

is self adjoint for all $P < \infty$. Note well, as one potential example, the factor 2^{-p} introduced as part of a regulator as $P \to \infty$; we comment on this regulator in the next section. For each $\delta \equiv \delta(\hbar) > 0$, let

$$\mathbb{E}_P \equiv \mathbb{E}(X_P^2 \le \delta^2) \tag{10.91}$$

denote a projection operator depending on X_P^2 and δ as indicated. How such projection operators may be constructed is discussed in [Kl99b]. Let

$$S_P \equiv \limsup_{\pi,\, g}\, \langle \pi, g | \mathbb{E}_P | \pi, g \rangle \,, \tag{10.92}$$

which satisfies $S_P > 0$ since $\mathbb{E}_P \not\equiv 0$ when restricted to sufficiently large δ. Finally, we define

$$\langle\!\langle \pi'', g'' | \pi', g' \rangle\!\rangle \equiv \limsup_{P \to \infty}\, S_P^{-1}\, \langle \pi'', g'' | \mathbb{E}_P | \pi', g' \rangle \tag{10.93}$$

as a reduction of the original reproducing kernel. The result is either trivial, say if δ is too small, or it leads to a continuous functional of positive type on the original phase-space variables. We focus on the latter case.

To obtain the final physical Hilbert space, one must study $\langle\!\langle \pi'', g'' | \pi', g' \rangle\!\rangle$ as a function of the regularization parameter δ. Since gravity has an anomaly, there should be a minimum value of δ, which is still positive, that defines the proper theory. Assuming we can find and then use that value, $\langle\!\langle \pi'', g'' | \pi', g' \rangle\!\rangle$ becomes the reproducing kernel for the physical Hilbert space $\mathfrak{H}_{\text{phys}}$. Attaining this goal would then permit the real work of extracting the physics to begin.

10.14.1 *Pedagogical example*

It is pedagogically useful to outline an analogous story for a simpler and more familiar example. Consider general, locally self-adjoint field and momentum operators, $\hat{\phi}(x)$ and $\hat{\pi}(x)$, $x \in \mathbb{R}^3$, which satisfy the canonical commutation relations (with $\hbar = 1$)

$$[\hat{\phi}(x), \hat{\pi}(y)] = i\delta(x - y) . \tag{10.94}$$

Build a set of coherent states

$$|\pi, \phi\rangle \equiv e^{i[\hat{\phi}(\pi) - \hat{\pi}(\phi)]} |\eta\rangle , \tag{10.95}$$

where $\hat{\phi}(\pi) \equiv \int \hat{\phi}(x)\pi(x) \, d^3x$ and $\hat{\pi}(\phi) \equiv \int \hat{\pi}(x)\phi(x) \, d^3x$, with π and ϕ real test functions, and $|\eta\rangle$ is a normalized but otherwise unspecified fiducial vector. Note well that the choice of $|\eta\rangle$ in effect determines the representation of the canonical field operators. We next present a portion of the story from [Kl01b].

We initially choose $|\eta\rangle$ to correspond to an ultralocal representation such that

$$\langle\pi'', \phi''|\pi', \phi'\rangle = \exp\{\tfrac{1}{2}i \int [\phi''(x)\pi'(x) - \pi''(x)\phi'(x)] \, d^3x\} \tag{10.96}$$
$$\times \exp(-\tfrac{1}{4}\int \{M(x)^{-1}[\pi''(x) - \pi'(x)]^2 + M(x)[\phi''(x) - \phi'(x)]^2\} \, d^3x) .$$

Here $M(x)$, $0 < M(x) < \infty$, is an arbitrary (smooth) function of the ultralocal representation with the dimensions of mass. The given ultralocal field-operator representation is in fact unitarily inequivalent for each distinct function $M(x)$. [Remark: Note that $M(x)$ here plays a role similar to $b(x)$ in our discussion about gravity.]

We wish to apply this formulation to describe the *relativistic free field of mass m* for which the Hamiltonian operator is formally given by

$$\mathcal{H} = \tfrac{1}{2} \int : \{\hat{\pi}(x)^2 + [\nabla\hat{\phi}(x)]^2 + m^2 \hat{\phi}(x)^2\} : \, d^3x . \tag{10.97}$$

If we build this operator out of the field and momentum operators in the ultralocal representation, then no matter what vector is used—still preserving ultralocality—to define : :, $\mathcal{H}$ will have only the zero vector in its domain. We need to change the field-operator representation, which means we have to change the fiducial vector from $|\eta\rangle$ to $|0; m\rangle$, the true ground state of the proposed Hamiltonian operator $\mathcal{H}$, which, incedently, is the proper vector with which the normal ordering in (10.97) is computed.

Let us first regularize the formal Hamiltonian $\mathcal{H}$. To that end, let $\{u_n(x)\}$ denote a complete set of real, orthonormal functions on $\mathbb{R}^3$ and define the sequence of kernels, for all $N \in \{1, 2, 3, \ldots\}$, given by

$$K_N(x, y) \equiv \sum_{n=1}^{N} u_n(x)u_n(y) , \tag{10.98}$$

which converges to $\delta(x - y)$ as a distribution when $N \to \infty$. Let

$$\hat{\phi}_N(x) \equiv \int K_N(x, y) \hat{\phi}(y) \, d^3y \,, \tag{10.99}$$

$$\hat{\pi}_N(x) \equiv \int K_N(x, y) \hat{\pi}(y) \, d^3y \,, \tag{10.100}$$

and with these operators build the sequence of regularized Hamiltonian operators

$$\mathcal{H}_N \equiv \tfrac{1}{2} \int \, : \{ \hat{\pi}_N(x)^2 + [\nabla \hat{\phi}_N(x)]^2 + m^2 \hat{\phi}_N(x)^2 \} : \, d^3x \tag{10.101}$$

for all N, where $:\,:$ denotes normal order with respect to the ground state $|0; m\rangle_N$ of $\mathcal{H}_N$.

We would like to have a constructive way to identify the ground state of $\mathcal{H}_N$. For this purpose consider the set

$$S_N \equiv \left\{ \frac{\sum_{j,k=1}^{J} a_j^* a_k \langle \pi_j, \phi_j | e^{-\mathcal{H}_N^2} | \pi_k, \phi_k \rangle}{\sum_{j,k=1}^{J} a_j^* a_k \langle \pi_j, \phi_j | \pi_k, \phi_k \rangle} : J < \infty \right\} \tag{10.102}$$

for general sets $\{a_j\}$ (not all zero), $\{\pi_j\}$, and $\{\phi_j\}$. (How these expressions may be generated is discussed in [Kl01b].) As N grows, the general element in S_N becomes exponentially small, save for elements that correspond to vectors which well-approximate the ground state $|0; m\rangle_N$. Suitable linear combinations can convert the original reproducing kernel $\langle \pi'', \phi'' | \pi', \phi' \rangle$ to the reproducing kernel $\langle \pi'', \phi''; m | \pi', \phi'; m \rangle_N$ which is based on a fiducial vector that has the form $|0; m\rangle_N$ for the first N degrees of freedom and is unchanged for the remaining degrees of freedom. Finally, we may take the limit $N \to \infty$ which then leads to

$$\langle \pi'', \phi''; m | \pi', \phi'; m \rangle = \exp\{\tfrac{1}{2} i \int [\tilde{\phi}''^*(k) \tilde{\pi}'(k) - \tilde{\pi}''^*(k) \tilde{\phi}'(k)] \, d^3k\} \tag{10.103}$$
$$\times \exp(-\tfrac{1}{4} \int \{\omega(k)^{-1} |\tilde{\pi}''(k) - \tilde{\pi}'(k)|^2 + \omega(k)|\tilde{\phi}''(k) - \tilde{\phi}'(k)|^2\} \, d^3k) \,,$$

where $\omega(k) \equiv \sqrt{k^2 + m^2}$ and $\tilde{\pi}(k) \equiv (2\pi)^{-3/2} \int e^{-ik \cdot x} \pi(x) \, d^3x$, etc. The procedure sketched above is referred to as *recentering the coherent states* or equivalently as *recentering the reproducing kernel*. This form of reproducing kernel is no longer ultralocal and contains no trace of the scalar function $M(x)$, whatever form it may have had. Moreover, and this is an important point, the new representation is fully compatible with the Hamiltonian $\mathcal{H}$ being a nonnegative, self-adjoint operator. Indeed, the expression for the propagator is given by

$$\langle \pi'', \phi''; m | e^{-i\mathcal{H}T} | \pi', \phi'; m \rangle = L'' L' \exp[\int \tilde{\zeta}''^*(k) e^{-i\omega(k)T} \tilde{\zeta}'(k) \, d^3k] \,, \tag{10.104}$$

where

$$\tilde{\zeta}(k) \equiv [\omega(k)^{1/2}\,\tilde{\phi}(k) + i\omega(k)^{-1/2}\,\tilde{\pi}(k)]/\sqrt{2}\,, \qquad (10.105)$$

$$L \equiv \exp[-\tfrac{1}{2}\textstyle\int |\tilde{\zeta}(k)|^2\,d^3k]\,. \qquad (10.106)$$

The definition offered by (10.104) is continuous in T, which is the principal guarantor that the expression

$$\mathcal{H} = \tfrac{1}{2}\textstyle\int \, : \{\hat{\pi}(x)^2 + [\nabla\hat{\phi}(x)]^2 + m^2\,\hat{\phi}(x)^2\} :\, d^3x\,, \qquad (10.107)$$

where : : denotes normal ordering with respect to the ground state $|0;m\rangle$ of the operator $\mathcal{H}$, is a self-adjoint operator as desired.

Let us summarize the basic content of the present pedagogical example. Even though we started with a very general ultralocal representation, as characterized by the general function $M(x)$, we have forced a complete change of representation to one compatible with the Hamiltonian operator for a relativistic free field of arbitrary mass m. In so doing all trace of the initial arbitrary function $M(x)$ has disappeared, and in its place, effectively speaking, has appeared the positive, pseudo-differential operator $\sqrt{-\nabla^2 + m^2}$ having only its dimension (mass) in common with the original positive function $M(x)$. The original ultralocal representation is completely gone! [Remark: A moments reflection should convince the reader that a comparable analysis can be made for either the interacting φ_2^4 or φ_3^4 model as well, both of which satisfy (10.94), showing that the general argument is not limited just to free theories; see [Kl01b].]

10.14.2 *Strong coupling gravity*

The discussion in the present chapter has been predicated on the assumption that we are analyzing the gravitational field and therefore the classical Hamiltonian is that given in (10.5). However, it is pedagogically instructive if we briefly comment on an approximate theory—based on the so-called "strong coupling approximation" (SCA) [Pil82; Pil83; FrP85; Hen79]—where the Hamiltonian constraint (10.5) is replaced by the expression

$$H_{SCA}(x) \equiv g(x)^{-1/2}[\pi_b^a(x)\pi_a^b(x) - \tfrac{1}{2}\pi_a^a(x)\pi_b^b(x)] + 2\Lambda g(x)^{1/2}\,, \quad (10.108)$$

in which the term proportional to the three-dimensional scalar curvature $R(x)$ has been dropped, and where we have also introduced the cosmological constant Λ (with dimensions $(\text{length})^{-2}$) and an associated auxiliary term in the Hamiltonian. The result is a model for which the new Hamiltonian constraint (10.108) is indeed compatible with an appropriate form of an ultralocal representation. The proper form of that ultralocal representation may be determined by a similar procedure, e.g., by studying an analogue of the set S_N, and by recentering the reproducing kernel based on ensuring that the quantum version of $H_{SCA}(x)$ is a local self-adjoint operator. These remarks conclude our discussion of strong coupling gravity.

10.15 Final Comments on Nonrenormalizability

10.15.1 *Nonrenormalizable scalar fields*

Consider the case of perturbatively nonrenormalizable, quartic, covariant, self-interacting scalar fields, i.e., the so-called φ_n^4 theories, where the space-time dimension $n \geq 5$. On the one hand, viewed perturbatively, such theories entail an infinite number of distinct counterterms. On the other hand, the continuum limit of a straightforward Euclidean lattice formulation leads to a quasifree theory—a genuinely *noninteracting* theory—whatever choice is made for the renormalized field strength, mass, and coupling constant, as has been discussed in Chap. 9. In the author's view both of these results are unsatisfactory. Instead, it is possible that an *intermediate behavior* holds true, even though that has not yet been fully established. Let us illustrate an analogous but simpler situation where the conjectured intermediate behavior can be rigorously established.

Consider an ultralocal quartic interacting scalar field, which, viewed classically, is nothing but the relativistic φ_n^4 model with all the spatial gradients in the usual free term dropped. As a mathematical model of quantum field theory, an ultralocal model is readily seen to be perturbatively nonrenormalizable, while the continuum limit of a straightforward lattice formulation becomes quasifree, basically because of the vise grip of the Central Limit Theorem. The pair of (i) perturbative nonrenormalizability and (ii) lattice-limit triviality is similar to the behavior for relativistic φ_n^4 models, but for the simpler ultralocal model, an intermediate approach can be rigorously proven to hold as shown in Chap. 8. Roughly speaking, a characterization of this intermediate behavior is the following: From a functional integral standpoint, and for any positive value of the quartic coupling constant, the quartic interaction acts like a *hard-core* in history space projecting out certain contributions that would otherwise be allowed by the free theory alone. This phenomenon takes the form of a nonstandard, nonclassical counterterm in the Hamiltonian that does *not* vanish as the coupling constant of the quartic interaction vanishes. Specifically, for the model in question, the additional counterterm is *a counterterm for the kinetic energy* and is formally proportional to $\hbar^2/\phi(x)^2$, which, in form, is not unlike the centripetal potential that arises in 'spherical coordinates' in multi-dimensional quantum mechanics. In summary, inclusion of a formal additional interaction proportional to $\hbar^2/\phi(x)^2$ in the Hamiltonian density is sufficient to result in a well-defined and nontrivial (i.e., non-Gaussian) quantum theory for interacting ultralocal scalar models. In addition, it was shown in Chap. 8 that the classical limit of such quantum theories reproduces the classical model with which one started.

The foregoing brief summary holds rigorously for the ultralocal scalar fields, and it has been conjectured that a suitable counterterm would lead to an acceptable intermediate behavior for the relativistic models φ_n^4, $n \geq 5$. What form should the counterterm take in the case of the relativistic φ_n^4 models? We have also made a plausible suggestion guided by the following general principle that holds in the ultralocal case: The counterterm should be an ultralocal (because the model is ultralocal) potential term arising from the kinetic energy. For the relativistic field that argument suggests the counterterm should, in a sense, also be proportional to $\hbar^2/\phi(x)^2$. It is also part of this general conjecture that a similar counterterm is not limited to φ_n^4 models, but should be effective for other nonrenormalizable interactions, e.g., such as φ_n^p, $p \in \{6, 8, 10, \ldots\}$, etc. Our best argument at the present time for a suitable counterterm is the one that appears in Chap. 9.

Note well that the hard-core picture of nonrenormalizable interactions leads to such interactions behaving as *discontinuous perturbations*: Once turned on, such interactions cannot be completely turned off! Stated otherwise, as the nonlinear coupling constant is reduced toward zero, the theory passes continuously to a "pseudofree" theory—different from the "free" theory—*which retains certain effects of the hard core*. The interacting theory is *continuously connected* to the pseudofree theory, and may even possess some form of perturbation theory about the pseudofree theory. Evidently, the presence of the hard-core interaction makes any perturbation theory developed about the original unperturbed theory almost totally meaningless.

10.15.2 *Nonrenormalizable gravity*

Although the differences between gravity and a nonrenormalizable scalar interaction are significant in their details, there are certain similarities we wish to draw on. Most importantly, one can argue [Kla75] that the nonlinear contributions to gravity act as a hard-core interaction in a quantization scheme, and thus the general picture sketched above for nonrenormalizable scalar fields should apply to gravity as well. Assuming that the analogy holds further, there should be a nonstandard, nonclassical counterterm that incorporates the dominant, irremovable effects of the hard-core interaction. Accepting the principle that in such cases perturbation theory offers no clear hint as to what counterterm should be chosen, we appeal to the guide used in the scalar case. Thus, as our proposed counterterm, we look for an ultralocal potential arising from the kinetic energy that appears in the Hamiltonian constraint. In fact, the only ultralocal potential that has the right transformation properties is proportional to $\hbar^2 g(x)^{1/2}$. Thus we are led to conjecture that the "nonstandard counterterm" is none other than a term like the familiar cosmological constant contribution! Unlike the scalar field which required an unusual term proportional to $1/\phi(x)^2$, the gravitational case has resulted in suggesting a term proportional to an "old friend", namely $g(x)^{1/2}$. At first glance, it seems absurd that such a harmless looking term could act to "save" the nonrenormalizability of gravity. In its favor we simply note that the analogy with how other nonrenormalizable theories are "rescued" is too strong to dismiss the present proposal out of hand—and of course one must resist any temptation "to think perturbatively".

As one small aspect of this problem, let us briefly discuss how the factor $\hbar^2$ arises in the gravitational case. Merely from a *dimensional* point of view, we note that (the first term of) the local kinetic energy operator has a formal structure given by

$$-\frac{16\pi G}{c^3}\hbar^2\left(\frac{\delta}{\delta g_{cb}(x)}\, g_{ac}(x)\, g(x)^{-1/2}\, g_{bd}(x)\, \frac{\delta}{\delta g_{da}(x)} - \cdots\right), \quad (10.109)$$

where we have restored the factor $16\pi G/c^3$, where G is Newton's constant and c is the speed of light. Thus the anticipated counterterm is proportional to $(G\hbar^2/c^3)g(x)^{1/2}$. We next cast this term into the usual form for a contribution to the potential part of the Hamiltonian constraint, namely, in a form proportional to $(c^3/G)\Lambda g(x)^{1/2}$. Hence, to recast our anticipated counterterm into this form, we need a factor proportional to

$$\frac{G^2\hbar^2}{c^6} \equiv l_{Planck}^4 \approx (10^{-33}cm)^4. \quad (10.110)$$

In the classical symbol for the Hamiltonian constraint operator, this factor is multiplied by an expectation value with dimensions (length)$^{-6}$ originating from the density nature of the two momentum factors and leading to an overall factor with the dimensions (length)$^{-2}$ that is proportional to $\hbar^2$ as claimed. Let us call the resultant counterterm $\Lambda_C \, g(x)^{1/2}$. Since the sign of the DeWitt 'supermetric' [MTW71], which governs the kinetic energy term, is indefinite, it is not even possible to predict the sign of Λ_C. However, one thing appears certain. While the proposed counterterm $\Lambda_C \, g(x)^{1/2}$ is certainly not cosmological in origin, its influence may well be!

The foregoing scenario has assumed the hard-core model of nonrenormalizable interactions applies to the theory of gravity. However, that may well not be the case, and, instead, some other counterterm(s) may be required to cure the theory of gravity. Note well that the general structure of our approach to quantize gravity is largely insensitive to just what form of regularization and renormalization is required. In particular, the use of the affine field variables, the application of the projection operator method to impose constraints, and the development of the nonstandard phase-space functional integral representation for the reproducing kernel of the regularized physical Hilbert space all have validity quite independently of the form in which the Hamiltonian constraint is ultimately turned into a local self-adjoint operator. Although we have outlined one particular version in which the Hamiltonian constraint may possibly be made into a densely defined operator, we are happy to keep an open mind about the procedure by which this ultimately may take place since many different ways in which this process can occur are fully compatible with the general principles of an enhanced quantization scheme for the gravitational field.

Bibliography

[AKT66] H.D.I. Abarbanel, J.R. Klauder, and J.G. Taylor, "Green's Functions for Rotationally Symmetric Models", Phys. Rev. **152**, 198 (1966).

[AdK14] T. Adorno and J.R. Klauder, "Examples of Enhanced Quantization: Bosons, Fermions, and Anyons", Int. Jour. Math. Phys. A **29**, 14501 (2014); arXiv:1403.1786.

[AhB59] Y. Aharonov and D. Bohm, "Significance of Electromagnetic Potentials in Quantum Theory", Phys. Rev. **115**, 485 (1959). "Further Considerations on Electromagnetic Potentials in the Quantum Theory", Phys. Rev. **123**, 1511 (1961).

[Aiz81] M. Aizenman, "Proof of the Triviality of φ_d^4 Field Theory snd Some Mean-Field Features of Ising Models for $d > 4$", Phys. Rev. Lett. **47**, 1, E-866 (1981).

[ADM62] R. Arnowitt, S. Deser, and C. Misner, "The Dynamics of General Relativity", *Gravitation: An Introduction to Current Research*, Ed. L. Witten, (Wiley & Sons, New York, 1962), p. 227; arXiv:gr-qc/0405109.

[Aro43] N. Aronszajn, "La théorie des noyaux reproduisants et ses applications Premire Partie", Proc. Cambridge Phil. Soc. **39**, 133 (1943). "Note additionnelle l'article La théorie des noyaux reproduisants et ses applications", Proc. Cambridge Phil. Soc. **39**, 205 (1943). *Theory of Reproducing Kernels*, (Transactions of the American Mathematical Society **68**, 337 1950). H. Meschkowski, *Hilbertsche Räume mit Kernfunktion*, (Springer Verlag, Berlin, 1962).

[AsK68] E.W. Aslaksen and J.R. Klauder, "Unitary Representations of the Affine Group", J. Math. Phys. **9**, 206 (1968).

[BeK13] J. Ben Geloun and J.R. Klauder, "Enhanced Quantization on a Circle", Phys. Scr. **87**, 035006 (2013); arXiv:1206.1180.

[BGY13] H. Bergeron, J.-P. Gazeau, and A. Youssef, "Are the Weyl and Coherent-State Descriptions Physically Equivalent?", Phys. Lett. A **377**, 598 (2013); arXiv:1102.3556.

[BIP78] E. Brezin, C. Itzykson, G. Parisi, and J.-B. Zuber, "Planar Diagrams", Commun. Math. Phys. **59**, 35 (1978).

[Cal88] D.J.E. Callaway, "Triviality Pursuit: Can Elementary Scalar Particles Exist?", Physics Reports **167**, 241 (1988).

[Cho91] C. Chou, "Multi-Anyon Quantum Mechanics and Fractional Statistics", Phys. Lett. A **155**, 245 (1991).

[Coh61] P.M. Cohn, *Lie Groups*, (Cambridge University Press, London, 1961).

[Col06] G. Collins, "Computing with Quantum Knots", Sci. Am. (Int. Ed.) **294**, 57 (2006).

[DFN05] S. Das Sarma, M. Freedman and C. Nayak, "Topologically Protected Qubits from a Possible Non-Abelian Fractional Quantum Hall State", Phys. Rev. Lett. **94**, 166802 (2005).

[DaK85] I. Daubechies and J.R. Klauder, "Quantum Mechanical Path Integrals with Wiener Measures for All Polynomial Hamiltonians, II", J. Math. Phys. **26**, 2239 (1985).

[DMO07] R. De Pietri, S. Mori, and E. Onofri, "The Planar Spectrum in U(N)-Invariant Quantum Mechanics by Fock Space Methods: I. The Bosonic Case", JHEP 0701:018 (2007); arXiv:hep-th/0610045.

[Dir58] P.A.M. Dirac, *The Principles of Quantum Mechanics*, (Clarendon Press, Oxford, 1958), Fourth Edition.

[Dir64] P.A.M. Dirac, *Lectures on Quantum Mechanics*, (Belfer Graduate School of Science, Yeshiva University, New York, 1964).

[Emc72] G.G. Emch, *Algebraic Methods in Statistical Mechanics and Quantum Field Theory*, (Wiley-Interscience, New York, 1972).

[FaZ13] M. Fanuel and S. Zonetti, "Affine Quantization and the Initial Cosmological Singularity", EPL **101**, 10001 (2013); arXiv:1203.4936v3.

[FFS92] R. Fernández, J. Fröhlich, and A.D. Sokal, *Random Walks, Critical Phenomena, and Triviality in Quantum Field Theory*, (Springer-Verlag, Berlin, 1992).

[FHL89] A.L. Fetter, C.B. Hanna, and R.B. Laughlin, "Random Phase Approximation in the Fractional Statistics Gas", Phys. Rev. B **39**, 9697 (1989).

[Fey48] R.P. Feynman, "Space-time Approach to Non-relativistic Quantum Mechanics", Rev. Mod. Phys. **20**, 367 (1948).

[Fey52] R.P. Feynman, "An Operator Calculus Having Applications in Quantum Electrodynamics", Phys. Rev. **84**, 108 (1952).

[Fey72] R.P. Feynman, *Statistical Mechanics*, (Benjamin/Cummings: Reading, Massachusetts, 1972).

[FrP85] G. Francisco and M. Pilati, "Strong Coupling Quantum Gravity. 3. Quasiclassical Approximation", Phys. Rev. D **31**, 241 (1985).

[FSW82] B. Freedman, P. Smolensky, and D. Weingarten, "Monte Carlo Evaluation of the Continuum Limit of φ_4^4 and φ_3^4", Phys. Lett. B **113**, 481 (1982).

[Frö82] J. Fröhlich, "On the Triviality of $\lambda\varphi_d^4$ Theories and the Approach to the Critical Point in $d \geq 4$ Dimensions", Nucl. Phys. B **200**, 281 (1982).

[Gho95] S. Ghosh, "Spinning Particles in 2+1 Dimensions", Phys. Lett. B **338**, 235 (1994). "Anyons in an Electromagnetic Field and the Bargmann-Michel-Telegdi Equation", Phys. Rev. D **51**, 5827 (1995).

[GiT90] D.M. Gitman and I.V. Tyutin, *Quantization of Fields with Constraints*, (Springer-Verlag, Berlin, 1990). M. Henneaux and C. Teitelboim, *Quantization of Gauge Systems*, (Princeton University Press, Princeton New Jersey 1994). K. Sundermeyer, *Constrained Dynamics with Applications to Yang-Mills Theory, General Relativity, Classical Spin, Dual String Model*, Lecture Notes in Physics 169, (Springer-Verlag, Berlin, 1982).

[GlJ87] J. Glimm and A. Jaffe, *Quantum Physics*, (Springer Verlag, New York, 1987), Second Edition.

[GoT03] K. Gottfried and T.-M. Yan, *Quantum Mechanics: Fundamentals*, (Springer-Verlag, New York, 2003), Second Edition.

[Hal84] B.I. Halperin, "Statistics of Quasiparticles and the Hierarchy of Fractional Quantized Hall States", Phys. Rev. Lett. **52**, 1583 (1984).

[HaK04] L. Hartmann and J.R. Klauder, "Weak Coherent State Path Integrals", J. Math. Phys. **45**, 87 (2004).

[Hen79] M. Henneaux, "Zero Hamiltonian Signature Spacetimes", Bull. Soc. Math. Bel. **31**, 47 (1979).

[Hid70] T. Hida, *Stationary Stochastic Processes*, (Princeton University Press, Princeton, New Jersey, 1970).

[Hoo74] G. 't Hooft, "A Planar Diagram Theory for Strong Interactions", Nucl. Phys. B **72**, 461 (1974).

[HPV10] P. Horváthy, M. Plyushchay and M. Valenzuela, "Bosons, Fermions and Anyons in the Plane, and Supersymmetry", Ann. Phys. **325**, 1931 (2010); arXiv:1001.0274.

[IeL90] R. Iengo and K. Lechner, "Quantum Mechanics of Anyons on a Torus", Nucl. Phys. B **346**, 551 (1990).

[IeL92] R. Iengo and K. Lechner, "Anyon Quantum Mechanics and Chern-Simons Theory", Phys. Rep. **213**, 179 (1992).

[IK84a] C.J. Isham and A.C. Kakas. "A Group Theoretic Approach to the Canonical Quantization of Gravity. 1. Construction of the Canonical Group", Class. Quant. Grav. **1**, 621 (1984).

[IK84b] C.J. Isham and A.C. Kakas, "A Group Theoretical Approach to the Canonical Quantization of Gravity. 2. Unitary Representations of the

Canonical Group", Class. Quant. Grav. **1**, 633 (1984).

[ItZ80] C. Itzykson and J.-B. Zuber, *Quantum Field Theory*, (McGraw Hill, New York, 1980).

[JaN91] R. Jackiw and V.P. Nair, "Relativistic Wave Equation for Anyons", Phys. Rev D **43**, 1933 (1991).

[Kla62] J.R. Klauder, "Restricted Variations of the Quantum Mechanical Action Functional and Their Relation to Classical Dynamics", Helv. Phys. Acta **35**, 333 (1962).

[Kl63a] J.R. Klauder, "Continuous-Representation Theory I. Postulates of Continuous Representation Theory", J. Math. Phys. **4**, 1055 (1963).

[Kl63b] J.R. Klauder, "Continuous-Representation Theory II. Generalized Relation Between Quantum and Classical Dynamics", J. Math. Phys. **4**, 1058 (1963).

[Kla64] J.R. Klauder, "Continuous-Representation Theory III. On Functional Quantization of Classical Systems", J. Math. Phys. **5**, 177 (1964).

[Kla65] J.R. Klauder, "Rotationally-Symmetric Model Field Theories", J. Math. Phys. **6**, 1666 (1965).

[Kla67] J.R. Klauder, "Weak Correspondence Principle", J. Math. Phys. **8**, 2392 (1967).

[Kla70] J.R. Klauder, "Ultralocal Scalar Field Models", Commun. Math. Phys. **18**, 307 (1970). "Ultralocal Quantum Field Theory", Acta Physica Austriaca, Suppl. VIII, 227 (1971).

[KlA70] J.R. Klauder and E.W. Aslaksen, "Elementary Model for Quantum Gravity", Phys. Rev. D **2**, 272 (1970).

[Kl73a] J.R. Klauder, "Ultralocal Spinor Field Models", Ann. Phys. (N.Y.) **79**, 111 (1973).

[Kl73b] J.R. Klauder, "Field Structure through Model Studies: Aspects of Nonenormalizable Theories", Acta. Phys. Austr. Suppl. XI, 341 (1973).

[Kla75] J.R. Klauder, "On the Meaning of a Nonrenormalizable Theory of Gravitation", GRG **6**, 13 (1975).

[Kla78] J.R. Klauder, "Continuous and Discontinuous Perturbations", Science **199**, 735 (1978).

[KlS85] J.R. Klauder and B.-S. Skagerstam, *Coherent States: Applications to Physics and Mathematical Physics*, (World Scientific, Singapore, 1985).

[Kla90] J.R. Klauder, "Quantization = Geometry + Probability," in *Probabilistic Methods in Quantum Field Theory and Quantum Gravity*, eds. P.H. Damgaard, H. Hüffel, and A. Rosenblum, 73, (North-Holland, Amsterdam, 1990).

[Kla96] J.R. Klauder, "Isolation and Expulsion of Divergences in Quantum Field Theory", Int. J. Mod. Phys. B **10**, 1473 (1996).

[Kla97] J.R. Klauder, "Coherent State Quantization of Constraint Systems", Ann. Phys. (N.Y.) **254**, 419 (1997).

[Kl99a] J.R. Klauder, "Noncanonical Quantization of Gravity. I. Foundations of Affine Quantum Gravity", J. Math. Phys. **40**, 5860 (1999).

[Kl99b] J.R. Klauder, "Universal Procedure for Enforcing Quantum Constraints", Nucl. Phys. B **547**, 397 (1999).

[Kla00] J.R. Klauder, *Beyond Conventional Quantization*, (Cambridge University Press, Cambridge, 2000 & 2005).

[Kl01a] J.R. Klauder, "Noncanonical Quantization of Gravity. II. Constraints and the Physical Hilbert Space", J. Math. Phys. **42**, 4440 (2001).

[Kl01b] J.R. Klauder, "Ultralocal Fields and their Relevance for Reparametrization Invariant Quantum Field Theory", J. Phys. A: Math. Gen. **34**, 3277 (2001).

[Kl01c] J.R. Klauder, "Quantization of Constrained Systems", Lect. Notes Phys. **572**, 143 (2001).

[Kl01d] J.R. Klauder, "Coherent State Path Integrals *Without* Resolutions of Unity", Found. Phys. **31**, 57 (2001); quant-ph/008132.

[Kla06] J.R. Klauder, "Overview of Affine Guantum gravity", Int. J. Geom. Meth. Mod. Phys. **3**, 81 (2006); gr-qc/0507113.

[Kla07] J.R. Klauder, "A New Approach to Nonrenormalizable Models", Ann. Phys. **322**, 2569 (2007).

[Kl10a] J.R. Klauder, *A Modern Approach to Functional Integration*, (Birkhauser, Boston, MA, 2010).

[Kl10b] J.R. Klauder, "New Affine Coherent States Based on Elements of Nonrenormalizable Scalar Field Models", Advances in Mathematical Physics, vol. 2010, 191529 (2010).

[Kla11] J.R. Klauder, "Scalar Field Quantization Without Divergences In All Spacetime Dimensions", J. Phys. A: Math. Theor. **44**, 273001 (2011); arXiv:1101.1706. "Divergences in Scalar Quantum Field Theory: The Cause and the Cure", Mod. Phys. Lett. A **27**, 1250117 (2012); arXiv:1112.0803.

[Kl12a] J.R. Klauder, "The Utility of Affine Variables and Affine Coherent States", J. Phys. A: Math. Theor. **45**, 244001 (2012); arXiv:1108.3380.

[Kl12b] J.R. Klauder, "Recent Results Regarding Affine Quantum Gravity", J. Math. Phys. **53**, 082501 (2012); arXiv:1203.0691.

[Kl12c] J.R. Klauder, "Enhanced Quantization: A Primer", J. Phys. A: Math. Theor. **45**, 285304 (2012).

[Kl12d] J.R. Klauder, "Enhanced Quantum Procedures that Resolve Difficult Problems"; arXiv:1206.4017.

[Kl14a] J.R. Klauder, "Matrix Models and their Large-N Behavior", Int. J. Mod. Phys. A **29**, 1450026 (2014); arXiv:1312.0814v2.

[Kl14b] J.R. Klauder, "Nontrivial Quantization of φ_n^4, $n \geq 2$"; arXiv:1405.0332.

[LSU68] O.A. Ladyzenskaja, V. Solonnikov, and N.N. Ural'ceva, *Linear and Quasi-linear Equations of Parabolic Type*, (Am. Math. Soc., Providence, Vol. 23, 1968).

[LaL77] L. Landau and E.M. Lifshitz, *Quantum Mechanics: Non-relativistic Theory*, (Pergamon Press, Oxford, 1977), Third Edition.

[Lau88] R.B. Laughlin, "Superconducting Ground State of Noninteracting Particles Obeying Fractional Statistics", Phys. Rev. Lett. **60**, 2677 (1988).

[LiK05] J.S. Little and J.R. Klauder, "Elementary Model of Constraint Quantization with an Anomaly", Phys. Rev. D **71**, 085014 (2005); gr-qc/0502045.

[Luk70] E. Lukacs, *Characteristic Functions*, (Hafner Publishing Company, New York, 1970).

[MTW71] C. Misner, K. Thorne, and J.A. Wheeler, *Gravitation*, (W.H. Freeman and Co., San Francisco, 1971).

[MLB91] M.V.N. Murthy, J. Law, M. Brack and R.K. Bhaduri, "Quantum Spectrum of Three Anyons in an Oscillator Potential", Phys. Rev. Lett. **67**, 1817 (1991).

[NSS08] C. Nayak, S. H. Simon, A. Stern, M. Freedman and S. Das Sarma, "Non-Abelian Anyons and Topological Quantum Computation", Rev. Mod. Phys. **80**, 1083 (2008).

[Per86] A. Perelomov, *Generalized Coherent States and Their Applications*, (Springer-Verlag, Berlin, 1986).

[Pil82] M. Pilati, "Strong Coupling Quantum Gravity. 1. Solution In A Particular Gauge", Phys. Rev. D **26**, 2645 (1982).

[Pil83] M. Pilati, "Strong Coupling Quantum Gravity. 2. Solution Without Gauge Fixing", Phys. Rev. D **28**, 729 (1983).

[Ply90] M.S. Plyushchay, "Relativistic Model of the Anyon", Phys. Rev. B **248**, 107 (1990).

[Pol91] A. Polychronakos, "Exact Anyonic States for a General Quadratic Hamiltonian", Phys. Lett. B **264**, 362 (1991).

[Ree76] M. Reed, *Abstract Linear Wave Equations*, (Springer-Verlag, Berlin, 1976).

[Roe96] G. Roepstorff, *Path Integral Approach to Quantum Physics*, (Springer-Verlag, Berlin, 1996).

[Sch38] I.J. Schoenberg, "Metric Spaces and Completely Monotone Functions," Ann. Math. **39**, 811 (1938).

[Sch05] L.S. Schulman, *Techniques and Applications of Path Integration*, (Dover Books on Physics, 2005).

[Sim73] B. Simon, "Quadratic Forms and Klauder's Phenomenon: A Remark on Very Singular Perturbations", J. Funct. Anal. **14**, 295 (1973).

[Sta10] J. Stankowicz, private communication.

[Sty02] D.F. Styer, *et al.*, "Nine Formulations of Quantum Mechanics", Am. J. Phys. **70**, 288 (2002).

[Tho79] C.B. Thorn, "Fock Space Description of the $1/N_c$-Expansion of Quantum Chromodynamics", Phys. Rev. D **20**, 1435 (1979).

[Tru75] T.T. Truong, "Weyl Quantization of Anharmonic Oscillator", J. Math. Phys. **16**, 1034 (1975).

[WaK00] G. Watson and J.R. Klauder, "Generalized Affine Coherent States: A Natural Framework for the Quantization of Metric-Like Variables", J. Math. Phys. **41**, 8072 (2000).

[WaK02] G. Watson and J.R. Klauder, "Metric and Curvature in Gravitational Phase Space", Class. Quant. Grav. **19**, 3617 (2002).

[Wiki-a] http://en.wikipedia.org/wiki/Exterior_calculus.

[Wiki-b] http://en.wikipedia.org/wiki/Exterior_derivative.

[Wiki-c] http://en.wikipedia.org/wiki/Exterior_algebra.

[Wiki-d] https://en.wikipedia.org/wiki/Perturbation_theory_(quantum_mechanics).

[Wiki-e] https://en.wikipedia.org/wiki/Dirichlet_boundary_conditions.

[Wiki-f] See, e.g., http://en.wikipedia.org/wiki/Hamiltonian_mechanics.

[Wiki-g] See, e.g., http://en.wikipedia.org/wiki/Fubini-Study_metric.

[Wiki-h] See, e.g., http://en.wikipedia.org/wiki/Poincaré_half-plane_model.

[Wiki-i] http://en.wikipedia.org/wiki/1/N_expansion.

[Wiki-j] http://en.wikipedia.org/wiki/Central_limit_theorem.

[Wiki-k] http://en.wikipedia.org/wiki/Grassmann_number.

[Wiki-l] http://en.wikipedia.org/wiki/Ornstein-Uhlenbeck_process.

[Wiki-m] http://crosstown.org/f.

[Wiki-n] http://en.wikipedia.org/wiki/Loop_quantum_gravity.

[Wil62] F. Wilczek, "Magnetic Flux, Angular Momentum, and Statistics", Phys. Rev. Lett. **48**, 1144 (1982). "Quantum Mechanics of Fractional-Spin Particles", Phys. Rev. Lett. **49**, 957 (1982).

[Wil90] F. Wilczek, *Fractional Statistics and Anyon Superconductivity*, (World Scientifc, Singapore, 1990).

[WiZ83] F. Wilczek and A. Zee, "Linking Numbers, Spin, and Statistics of Solitons", Phys. Rev. Lett. **51**, 2250 (1983).

[Wu84] Y.-S. Wu, "Multiparticle Quantum Mechanics Obeying Fractional Statistics", Phys. Rev. Lett. **53**, 111 (1984).

[ZhK94] C. Zhu and J.R. Klauder, "The Classical Limit of Ultralocal Quantum Fields", J. Math. Phys. **35**, 3400 (1994).

Index

245